Navigating Complexity: Understanding Human Responses to Multifaceted Disasters

Yibin Ao · Homa Bahmani

Navigating Complexity: Understanding Human Responses to Multifaceted Disasters

Springer

Yibin Ao
Chengdu University of Technology
Chengdu, Sichuan, China

Homa Bahmani
Chengdu University of Technology
Chengdu, Sichuan, China

ISBN 978-981-99-8206-6 ISBN 978-981-99-8207-3 (eBook)
https://doi.org/10.1007/978-981-99-8207-3

This Springer imprint is published by the registered company Springer Nature Singapore Pte Ltd.
The registered company address is: 152 Beach Road, #21-01/04 Gateway East, Singapore 189721, Singapore

Paper in this product is recyclable.

Preface I

In light of global climate change, all kinds of extreme disasters occur more frequently, leading to increased disaster losses. Effective, scientific, and rational strategies are essential in this endeavor, underscoring the need for extensive research and policy development in the field of disaster response. Therefore, the emergency response of people affected by disasters requires more attention and analysis from scholars and policymakers. My commitment is to study the relationship between the built environment and residents' activity behavior, with the ultimate goal of creating harmonious and sustainable living environments that prioritize human well-being. Of particular importance is our dedication to addressing the unique challenges faced by underdeveloped areas, such as rural locations. So far, our research team has systematically explored rural residents' daily travel patterns, resource usage, and disaster response behaviors. This book offers a summary and examination of residents' methods for coping with disasters.

China is one of the countries with the most severe earthquake disasters in the world, characterized by their frequent occurrence, extensive distribution, high magnitudes, and devastating consequences. In Sichuan Province, representative of this issue, there have been 15 earthquakes of magnitude six or above recorded between 2001 and 2021. Among these, the catastrophic Wenchuan County earthquake of 8.0 magnitude at 14:28 on May 12, 2008; the 7.0 magnitude earthquake in Jiuzhaigou County, Aba Prefecture, at 21:19 on August 8, 2017, and a 6.0 magnitude earthquake in Changning County, Yibin City, at 22:55 on June 17, 2019, stand out as particularly devastating events, resulting in significant loss of life and extensive economic damage.

Simultaneously, China is one of the countries with the highest frequency of floods globally, with profound economic damages. The scope of flood hazards in the country is vast, posing a perpetual threat to a substantial portion of its territory. In terms of the frequency of occurrence, the scale of the affected population and territory, and the direct economic losses suffered, floods are the most prevalent natural disasters. In China, Sichuan Province frequently suffers from floods and the number of flood victims due to heavy rainfall has been steadily rising in recent years.

Born and raised in Sichuan Province, I have witnessed the effects of earthquakes and floods, and these experiences have made me deeply aware of our vulnerability in the face of natural disasters. Chengdu University of Technology has a solid research foundation in geological disaster prevention and environmental protection. It also boasts a highly regarded State Key Laboratory. The laboratory's research team focuses on the causes and patterns of geological disasters from a natural sciences perspective while we concentrate our research on the disaster response behaviors of affected individuals (residents) from a social science perspective. We believe that residents' scientific and rational disaster preparation, evacuation, and post-disaster recovery actions are essential in curbing and minimizing disaster losses. From 2020–2023, the COVID-19 pandemic swept across the globe. During this timeframe, the region experienced the Luding Earthquake of 6.8 magnitude. This experience allowed me to contemplate the impacts of multi-hazard environments more profoundly and intensified my commitment to advancing this research.

Consequently, this book covers residents' response behaviors in the context of earthquake disasters, floods, the COVID-19 pandemic, and multi-hazard interactions by delving into pre-disaster preparedness, emergency evacuation during a disaster, and post-disaster reconstruction and recovery behaviors. The technique applied in this book is systematic, incorporating multi-disciplinary knowledge and research methodologies such as management, spatial planning, behavior science, and psychology. As an exploratory research study still in its preliminary stages, there remain many deficiencies, which we hope will be improved and supplemented by other scholars, readers, and future research. We will continue to focus on and conduct experimental research on disaster response behavior, aspiring to contribute to global efforts in natural disaster reduction and prevention.

I would like to extend my heartfelt appreciation to the individuals who have significantly contributed to this book: Hongying Zhang, Linchuan Yang, Yan Wang, Igor Martek, Gang Wang, Liyao Tan, Ling Tan, Jinglin Zhong, Tong Wang, Qiqi Feng, Hongfu Li, Yunfeng Chen, Lei Zhao, Sifan Zhou1, Zijun Zhang, Lili Han, Yunhong Liu, Li Han, Hongtai Yang, Hao Zhu, Fanrong Meng, Gui Ye, Na Dong, Liang Zhao, Li Mingyang, Dujuan Yang, and Dongpo Wang. Your invaluable contributions have greatly enriched this work.

We are immensely grateful for the generous support provided by various organizations that made this research possible. Our sincere appreciation goes to the National Natural Science Foundation of China (72171028), China Postdoctoral Science Foundation (2022T150077, 2022M710496), Soft Science Project of Science and Technology Department of Sichuan Province (2023JDR0249), and Sichuan Provincial Social Science Planning Project (22GL086) for supporting this research, for their financial contributions, which have been instrumental in the successful completion of this project. Additionally, we extend our gratitude to the dedicated individuals who actively participated in the surveys, as their collaborative efforts have significantly assisted us in writing this book.

Chengdu, China
September 2023

Yibin Ao

Preface II

In the face of disasters, humanity has always demonstrated remarkable resilience and adaptability, which have been fundamental to our survival throughout history. The ability to respond effectively to catastrophes is deeply ingrained in our nature. This book embarks on a journey to explore the intricate facets of human responses to disasters, ranging from earthquakes to pandemics, and it draws insights from various phases of disaster management.

This book is the culmination of years of research and study, informed by my experiences growing up in Iran, a country prone to natural and man-made disasters. For over two decades, I resided near Damavand Mountain, a once-active volcano. Many days, I observed visual indicators of volcanic activity on Damavand, in addition to the earthquakes I experienced. During my undergraduate studies in architecture, my interest shifted toward understanding earthquakes' impact on buildings' structural integrity. Subsequently, I embarked on a research journey focused on disaster recovery following natural calamities. In 2018, I decided to pursue my Ph.D. in China, enrolling at Huazhong University of Science and Technology in Wuhan, where I concentrated on disaster recovery projects through the lens of project management. During this period, I examined the recovery processes following the Wenchuan and Bam earthquakes. After relocating to Sichuan Province for my postdoctoral research, I experienced the Luding earthquake during a period of significant restrictions that covered the entire province. While I vividly remember the fear and anxiety of that moment, this experience strongly motivated me to explore the study of human responses to disasters, including those occurring during pandemics. Together, we embark on a journey to uncover the intricate facets of human responses to catastrophes, drawing upon my experiences and research to shed light on this complex and vital field.

We start by investigating the impact of earthquake knowledge and risk perception on the preparedness of rural residents and delve into the role of trust and stakeholders' assistance in households' earthquake preparedness behavior. As we progress, we confront the unique challenges presented by the COVID-19 pandemic.

Amidst seismic events, we scrutinize the intricate decisions made during COVID-19 lockdowns and emergency evacuations, probing the underlying motivations during a pandemic. These chapters yield lessons that extend beyond pandemics, providing valuable insights for future crisis management. Our journey culminates in the realm of post-disaster resettlement and the factors influencing villagers' satisfaction with post-disaster reconstruction, using the Wenchuan earthquake as an example to offer critical insights for long-term recovery efforts. The chapters collectively create a rich tapestry of knowledge, painting a comprehensive picture of human behavior in the face of disasters. This collective effort holds relevance for numerous stakeholders, from emergency management agencies and public health authorities to community leaders, educational institutions, researchers, policymakers, NGOs, and the broader public.

At this point, I would like to extend my heartfelt appreciation to the many individuals and groups who have played pivotal roles in the journey leading up to this book. To my esteemed colleagues and mentors, your guidance and support have been invaluable throughout the research process. I am deeply grateful to my husband, whose unwavering encouragement and understanding have sustained me during long hours of research and writing. To my beloved mother, your boundless love and unwavering belief in my capabilities have been a constant source of strength. I would also like to express my gratitude to the dedicated students from the Engineering Management program at Chengdu University of Technology. Your enthusiastic participation in data collection and questionnaire surveys has been instrumental in ensuring the accuracy and reliability of the research findings presented in this book.

Finally, as we stand at the intersection of research, practice, and policy, we recognize that this book represents a preliminary step in our quest to understand and enhance human responses to disasters. While the contributions within these pages are significant, they only scratch the surface of a vast field ripe for exploration. We invite our readers to join us on this journey, not as passive observers but as active participants in the ongoing quest to build resilience, enhance preparedness, and, ultimately, ensure a safer and more secure future for all.

Chengdu, China
September 2023

Homa Bahmani

Contents

Chapter 1
Current Research Status of Disasters and Human Response

Abstract This chapter provides a comprehensive examination of disasters and human behavior, drawing from previous research that encompasses three phases of the disaster life-cycle: pre-disaster, during-disaster, and post-disaster stages. It serves as an introductory overview of the research context presented in this book. The chapter initiates by clarifying the distinctions between emergencies and disasters, emphasizing the crucial role of proactive human responses in disaster mitigation. Furthermore, it delves into the complex interplay of factors that influence human behavior across these phases. Utilizing extensive empirical research and case studies, this chapter offers valuable insights into the comprehension and improvement of human behavior during each stage. These insights hold significant implications for developing more effective disaster management strategies and cultivating community resilience in the face of adversity.

Keywords Disaster response · Human behavior · Risk perception · Evacuation decision-making · Emergency preparedness

1.1 Introduction

The literature usually uses the words "emergency" and "disaster" interchangeably. Despite their similarity in sudden nature and damage to human society, some scholars believe in the existence of dissimilarities. According to the United Nations Office for Disaster Risk (UNISDR), the term "disaster" refers to "a severe disruption of the functioning of a community or a society involving widespread human, material, economic or environmental losses and impacts, which exceeds the ability of the affected community or society to cope using its resources" (UNISDR, 2017). Disasters result from hazards being exposed to an insufficiently prepared society that cannot manage the hazards' imminent impacts (Gregorio & Soares, 2017). On the other hand, "emergency" is "a state in which normal procedures are suspended and extra-ordinary measures are taken in order to avert a disaster" (Organization, 2002). Based on the systematic review conducted by Al-dahash et al., emergencies have

Y. Ao and H. Bahmani, *Navigating Complexity: Understanding Human Responses to Multifaceted Disasters*, https://doi.org/10.1007/978-981-99-8207-3_1

fixed procedures to deal with, contrary to disasters, which may require rapid public policy change (Al-Dahash et al., 2016). Another difference exists in their magnitude and time frame: an emergency can become a disaster if not adequately addressed, and an emergency may affect smaller groups. In this study, the term "disasters" is preferred to "emergency" because of its broader scope.

Disasters can be divided into natural and man-made (Zibulewsky, 2001). While earthquakes, volcanoes, hurricanes, floods, and fires are natural disasters, man-made disasters account for war, terrorist attacks, pollution, nuclear explosions, fire accidents, hazardous materials exposures, and transportation accidents (Raphael, 1988). As observed in recent times, the heightened intensity of natural hazards poses a significant risk to the well-being of human societies. In the year 2021, the cumulative count of natural disasters and their associated economic ramifications, encompassing phenomena such as droughts, earthquakes, extreme temperatures, floods, landslides, storms, volcanic activities, and wildfires, exceeded the average annual figures recorded between 2001 and 2020 (CRED, 2021). In prior years, the occurrence of global natural hazards from 2008 to 2017 was almost similar to those observed in 2018 (Alexander, 2018). The Asia–Pacific region has borne the brunt of natural hazards, impacting around six million individuals, with the death toll surpassing two million between 1970 and 2014 (Johnson et al., 2016). Figures 1.1 and 1.2 render an overview of global natural disasters and their impacts on societies in the last decades.

The occurrence of disasters and the threats they pose to human lives are inescapable realities that transcend geographical boundaries and societal structures. From natural calamities like earthquakes, floods, and hurricanes to human-induced incidents such as industrial accidents and wars, these crises exact a heavy toll on communities worldwide. We must acknowledge the gravity of these threats and recognize that proactive and well-coordinated human responses are paramount

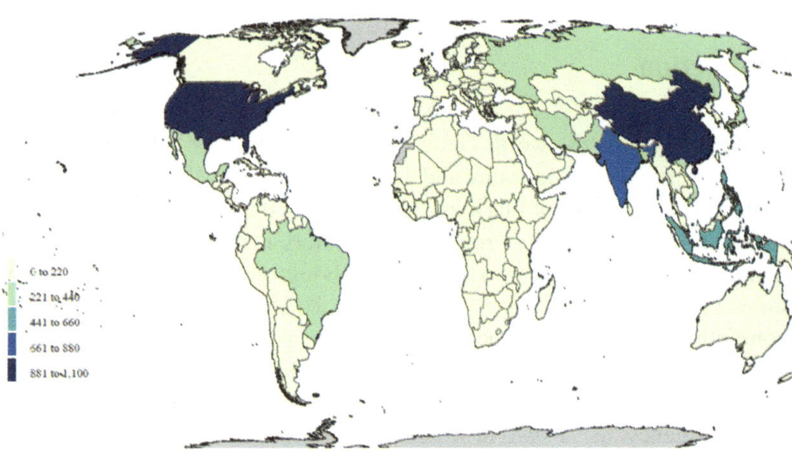

Fig. 1.1 Global occurrence of natural disasters, source: EM-DAT, CRED/UCLouvain, Brussels, Belgium www.emdat.be

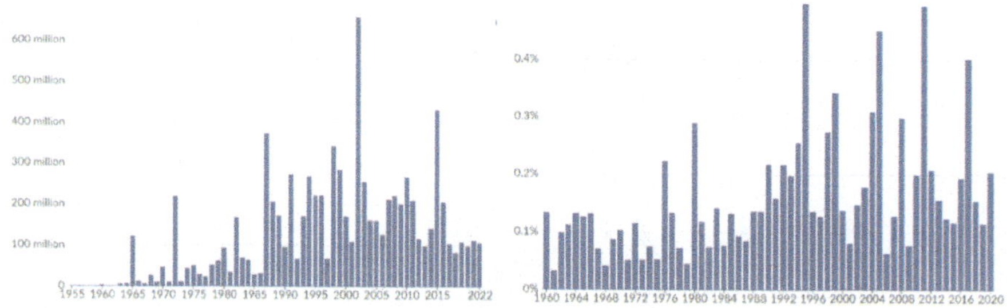

Fig. 1.2 (Left) Total number of people affected by disasters, (Right) Total economic damage from disasters as a share of GPD, source: EM-DAT, CRED/UCLouvain, Brussels, Belgium- www.emdat.be. ***Disasters include earthquakes, volcanic activity, landslides, droughts, wildfires, storms, and flooding

to mitigate their impact. The effectiveness of disaster and emergency management hinges on a multifaceted approach that encompasses preparedness, mitigation, response, and recovery (Akiyama, 2019). Apart from the role of governments and policy-makers in disaster management (He, 2019), the pivotal role of individuals becomes most apparent during the crucial phase of the emergency itself. Individuals play a critical role in disaster preparedness and mitigation, proactively reducing risks and enhancing community resilience before a disaster occurs. From raising awareness about potential hazards (Ismail et al., 2014) to participating in community-based initiatives (Yilmaz et al., 2013), their involvement strengthens the foundation of effective disaster management.

During the emergency phase, the collective behavior of individuals in responding to directives, maintaining calm, and following safety protocols dramatically influences the overall efficacy of emergency response efforts (Raphael, 2005). Public fear or panic can lead to critical challenges in coordination and resource allocation, potentially impeding the smooth flow of emergency operations. The Tohoku earthquake and tsunami that struck Japan in 2011 serve as an example of how public behavior can complicate emergency response efforts, highlighting the need for effective communication and strategies to address psychological aspects of emergency management (Neureiter, 2014).

On the other hand, several studies have explored the proper life-cycle for disaster management. Some studies suggest that this life-cycle focus should be shifted from recovery-oriented to mitigation-oriented policies (Platt, 2018). However, effective pre-disaster planning heavily relies on the cooperation of people and experts from different fields (Edgington, 2011). This study has considered the disaster management life-cycle proposed by Bahmani et al. (Bahmani & Zhang, 2022a) and will delve into providing a comprehensive overview of human response to disasters within three stages: before, during, and after the disaster. Figure 1.3 walks readers through various steps of human disaster response and manifests the factors that affect each step according to the literature.

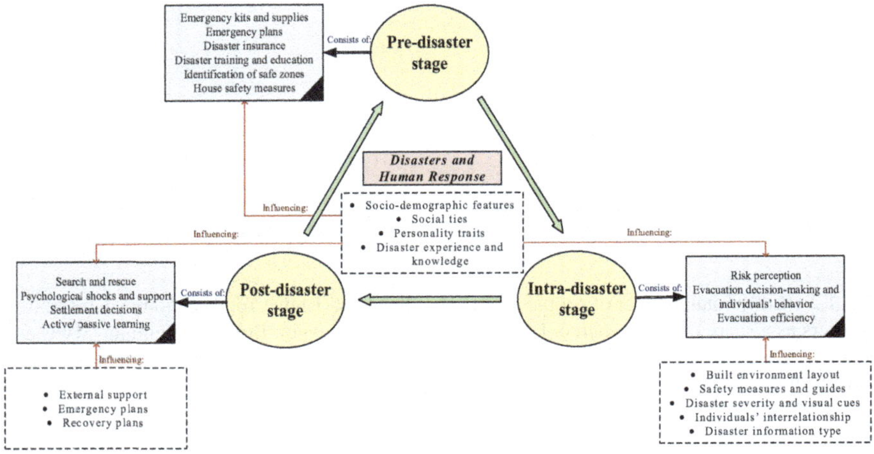

Fig. 1.3 Disasters and human response

1.2 Human Behavior Facing a Disaster

1.2.1 Human Behavior Before a Disaster: Preparedness and Mitigation

The cyclic disaster management framework applied in this chapter reveals the necessity of planning before the disaster. This observation also agrees with the results reported by Bosher et al. (2021), as they have demonstrated the need to pay more attention to disaster mitigation and preparedness rather than emergency management (Bosher et al., 2021). Furthermore, the more resources are being used in the mitigation and preparedness stage, the more sustainable the communities will be. Recent studies divided a pre-disaster stage into "mitigation" and "preparedness" (Altay & Green, 2006), (Xu et al., 2014). Another study split disaster preparedness into three categories: "emergency disaster preparedness", "knowledge and skills preparedness", and "physical disaster prevention preparedness" (Yong et al., 2020).

1.2.1.1 Socio-Demographic Features

The existing literature has extensively explored the link between socio-demographic characteristics and variations in disaster preparedness levels. Researchers have sought to identify the groups of individuals who might be more vulnerable in the face of an impending disaster due to their lack of preparedness. Factors such as age, gender, ethnicity, educational attainment, marital status, income, place of residence, occupation, duration of residency or place of birth, and health status are among the variables that have been previously examined for their impact on disaster preparedness among various groups of individuals. However, many controversial findings surround the relationship between socio-demographic features and disaster preparedness. Studies pointed to the significant impact of age on preparedness level (Ao et al., 2021), (Patel & Etminani-Ghasrodashti, 2021); however, given the specific age-related challenges among older adults, such as limited physical activity, care-giving for the spouse, and loss of driving skills (Tuohy et al., 2015), a study specifically focused on the older adults found no significant association between age and preparedness level (Kim & Zakour, 2017). Income is another contentious topic; while rural families with higher incomes were less prepared against flood, disaster preparedness among individuals with various income ranges did not significantly differ (Kim & Zakour, 2017). On the contrary, being a female gender (Patel & Etminani-Ghasrodashti, 2021), having higher education (Ao et al., 2021), (Ao et al., 2020a, 2020b), and having a professional career (Teo et al., 2018) have been proven to impact the disaster preparedness of people positively. To understand the influence of socio-economic features on population disaster preparedness, (Teo et al., 2018) conducted a comparative study on Australia's low socio-economic (LSE) and non-LSE populations. The results have further illustrated that disaster awareness levels among the former group can be predicted by English proficiency level, job, and familiarity with the local environment, whereas disaster awareness within the non-LSE group is primarily influenced by gender.

Researchers also examined different segments of the population and combinations of socio-demographic characteristics along with other factors to assess their level of disaster preparedness. A survey conducted at the University of Texas Arlington illustrated that apart from the students' socio-demographic features affecting their disaster preparedness, their perspective toward the government and university's responsibility during a disaster could significantly influence their disaster awareness and preparedness (Patel & Etminani-Ghasrodashti, 2021). Within the context of familial relationships, Hung sought to showcase how effective disaster preparedness is among couples by analyzing the key decision-makers at each stage of the preparedness process (Hung, 2017). The findings highlighted a robust and positive influence of shared decision-making at all phases on the disaster preparedness of the couples. Households' disaster preparedness in South Korea was the subject of another study (Kim & Kim, 2022). The authors could successfully model the sample's disaster preparedness by surveying 1234 households and considering socio-demographics, disaster knowledge and experience, and community factors. This study reiterated the greater significance of community resilience compared to the demographic attributes of respondents in influencing household preparedness levels, which aligns with prior findings presented by Tang (Tang & Feng, 2018). In another research, sets of various dissimilar elements influenced the disaster preparedness of households, communities, and workplaces. The residency duration and marital status affected household preparedness, while gender and education, as well as income level, greatly influenced communities' and workplaces' disaster preparedness, respectively (Castañeda et al., 2020).

1.2.1.2 Emergency Plans and Supplies

Developing emergency plans can effectively reduce vulnerability against disasters by influencing evacuation intentions (Lazo et al., 2015) and choosing appropriate proactive countermeasures during a disaster (Cong et al., 2014; Tanim et al., 2022). Based on the guidelines of the Federal Emergency Management Agency (FEMA), household emergency plans should include risk planning (assessing the risks and identifying local government emergency information channels), considering the needs of family members (creating a personal network and supplying emergency supplies), and planning for safe locations (identifying assembly locations after a disaster) (Date & Planning, 2013). Residency status, race, gender, and prior disaster experience can create obstacles when establishing household emergency plans (Rivera, 2020). Previous research has also shown that when family members engage in discussions about their emergency plans, it can enhance effectiveness, particularly in areas with infrequent occurrences of natural hazards (Cong et al., 2021). The inclusion of emergency or disaster insurance within emergency plans is a conceivable aspect, yet its adoption remains inadequate in developing countries (Lu & Xu, 2015). Through the implementation of emergency plans, both households and individuals can greatly improve their capacity to respond to disasters effectively. In addition, physical safety measurements, including building reinforcement and installation of safety items such as handrails and emergency lights, can be counted as household-level disaster mitigation activities (Chai et al., 2021; Yong et al., 2020).

1.2.1.3 Disaster Knowledge and Previous Experience

There exists extensive evidence in the literature on the positive significant impact of disaster knowledge (awareness) on individuals' preparedness levels (Dixit et al., 2018; Okada et al., 2018; Platt, 2018). Disaster knowledge encompasses both explicit and implicit forms. The former can be acquired through structured educational avenues like disaster training and workshops, while the latter, a more intricate kind of disaster knowledge, is typically gained experientially and unconsciously (Nakanishi & Black, 2018). In nations prone to frequent natural disasters, past encounters with such adversities play a pivotal role in determining the level of disaster preparedness among the populace. Based on the observations in two disaster-prone regions—the Sichuan province, China and the Chilean coast, Chile—it becomes evident that prior experience with disaster and increased exposure to natural calamities are linked to better disaster preparedness (Castañeda et al., 2020) (Wei et al., 2021) and enhanced survival aptitude (PRC, 2008). Another recent study further reveals that public recovery information emergency efficiently enhanced in Bam, Iran, among neighborhoods trained before the earthquake (Bahmani & Zhang, 2022b). A recent study evaluated the impact of the nature of disaster exposure on disaster knowledge, examining a sequence of natural disasters occurring in Kobe, Japan, from 1995 to 2011 (Kato, 2021). The findings reveal that individuals with direct disaster experience had better disaster knowledge than those without such direct exposure despite both groups belonging to the same generation. Over time, the disaster knowledge of the subsequent generation, who had not encountered calamities firsthand and relied solely on accounts from the last generation, notably decreased.

In addition, low public awareness (knowledge) of hazard prevention and safe structures can cause further danger (Ao et al., 2021; Platt, 2018; Rehman et al., 2019; UNISDR, 2015). The Bam earthquake in Iran in 2008 offered a valuable chance to improve local people's knowledge about safe rebuilding because the householders were the supervisors and contractors of their housing project. Although this knowledge transmission was informal and no one controlled the validity of this process (Kopaei, 2009), having knowledge of safe construction is better than knowing nothing (Farnaz, 2018). In contrast to Bam's recovery process, the Chinese government primarily spearheaded the reconstruction efforts after the Wenchuan earthquake in China in 2008. This emphasis on reconstruction has not resulted in a substantial enhancement of people's awareness regarding secure reconstruction and posed a threat to their lives as they continue to reside in vulnerable structures, rendering them susceptible to future disasters (Ge et al., 2010). Studies have also discussed the role of disaster knowledge in the effectiveness of evacuation and survival (Feng & Ji, 2014). Therefore, raising public awareness through training, workshops, and meetings should be considered (Tauber, 2015).

The previous studies have assessed not only occupants' disaster preparedness by evaluating the impact of disaster training and evacuation drills (Nakano et al., 2020; Petal et al., 2020; Vásquez et al., 2018) but also literature has taken advantage of new technologies such as Artificial Intelligence and gaming to improve the disaster risk reduction knowledge and participants' satisfaction level (Capuano & King, 2015; Daoudi et al., 2021; Delcea & Cotfas, 2019; Hu et al., 2022). Disaster coping knowledge can also be enhanced by the residents' active participation in the communities' resilience building (Ryan et al., 2020).

1.2.2 Human Behavior During a Disaster: Disaster Response

1.2.2.1 Risk Perception

Risk perception during emergencies encompasses how individuals assess the potential harm and danger associated with a given emergency situation (Slovic, 2016). Robust research within the field underscores the pivotal role played by risk perception in the decision-making process during evacuations (Bowser & Cutter, 2015; Thompson et al., 2017; Villegas et al., 2013). Since an individual's risk assessment is a subjective appraisal of the severity and probability of being impacted by a disaster, it is subject to influence by various factors (Shoji et al., 2020).

The influence of demographic characteristics on individuals' risk perception has been a focal point in the literature. Through extensive investigations spanning age, gender, educational attainment, income, property ownership, disaster experience, family composition, and global case studies involving diverse disasters, scholars have scrutinized how these demographic traits can cause differences in human risk perception. Predominantly, research has converged on the gender-related impact, indicating that women tend to perceive a heightened level of disaster risk compared to men (Koshiba & Nakayama, 2021; Yu et al., 2017). An inquiry into the reasons behind the greater likelihood of evacuation among women unveiled that their higher risk perception and different social responsibilities are associated with higher evacuation likelihood (Bateman & Edwards, 2002). Additionally, studies have affirmed a positive correlation between age and education with individuals' perceived risk (Ao et al., 2020a, 2020b; Yu et al., 2017). Family structure also emerged as a significant factor shaping human risk perception, with individuals having dependents, such as children (Matyas et al., 2011) or pets (Hesterberg et al., 2012), expressing a heightened sense of vulnerability to disaster threats.

The impact of various personality traits and one's experience or knowledge of disasters on risk perception has also garnered significant attention from researchers. While a lack of disaster experience has been shown to negatively affect hurricane risk perception among tourists in Central Florida (Matyas et al., 2011), no significant association was observed between disaster experience and evacuation behavior among students (Zheng et al., 2020). Individuals' risk perception is also subject to be influenced by their surrounding environment. A comprehensive study conducted in the rural area of Sichuan province, where residents contend with numerous earthquakes, unveiled that access to well-prepared emergency routes within residential complexes and the urban area positively influences residents' evacuation choices (Ao et al., 2020a, 2020b). In a survey of individuals affected by Natech accidents in Japan 2011, researchers identified that proximity to the accident site corresponded to a heightened risk perception (Yu et al., 2017). Additionally, studies have demonstrated that the nature of the emergency itself plays a role in shaping public risk perception (Thiede & Brown, 2013; Yu et al., 2017).

1.2.2.2 Evacuation Decision-Making and Individuals' Behavior

In the realm of emergency management, the decision-making process surrounding evacuations has been a thoroughly investigated subject, encompassing a wide array of facets. The Protection Motivation Theory (PMT) has emerged as a prominent framework for assessing human behavior in the context of disasters. According to PMT, when confronting a disaster, individuals engage in a cognitive process involving assessing threats, coping abilities, response efficacy, and, ultimately, the decision to undertake protective actions (Rogers, 1983). To refine PMT, Tang et al. introduced the concept that behavioral intentions can serve as mediators, influencing the impact of PMT components on each other (Tang & Feng, 2018). Another study scrutinized the PMT's ability to predict human evacuation decision-making and revealed that situation awareness errors can disrupt the PMT cycle, affecting the final evacuation decision (Kakimoto & Yoshida, 2022).

Beyond the application of PMT, scholars have delved into the intricate relationship between risk perception and protective decision-making facing a disaster (Stein et al., 2010; Thompson et al., 2017). Previous research efforts have put forth comprehensive models encompassing various factors that influence human evacuation decisions and have rigorously tested the accuracy of these models. For example, Lazo et al. employed the Protection Actions Decision Model and substantiated that factors such as pre-existing evacuation plans, the intention to safeguard family members, and heightened perceptions of vulnerability to property damage enhance evacuation intentions (Lazo et al., 2015). Another study, utilizing a conceptual framework that integrates risk perception, survival skills, emotional attributes, and emergency knowledge, investigated the evacuation behavior of passengers at metro platforms (Zheng et al., 2022). Additionally, another research has highlighted the significant influence of media on public evacuation behavior, with findings indicating that media-induced stress can outweigh the positive effects of raising disaster awareness (Stein et al., 2010). Further analysis has probed disaster information channels and their role in shaping individual evacuation decisions, demonstrating that factors such as evacuation orders (Tanim et al., 2022; Petrolia et al., 2011), official warnings (Bowser & Cutter, 2015), and credible sources of disaster information (Thiede & Brown, 2013; Gómez, 2013) can exert a positive impact on the likelihood of evacuation. Moreover, evidence has shown that having a pre-established evacuation plan and prior disaster experience can assist in making more appropriate evacuation decisions (Tanim et al., 2022; Thompson et al., 2017).

Socio-demographic characteristics have also been identified as factors contributing to individuals' evacuation behavior differences. While some studies have identified a positive association between an increase in age and evacuation likelihood (Durage et al., 2014), others find contradictory findings (Tanim et al., 2022; Mohajeri & Mirbaha, 2021). Additionally, individuals from minority groups or those with non-white racial backgrounds, pet owners, married individuals, and males have shown a lower inclination to respond proactively during emergencies (Tanim et al., 2022; Thiede & Brown, 2013; Mohajeri & Mirbaha, 2021). On the other hand, substantial evidence exists regarding the influence of the built environment and housing type on human behavior in the face of disasters (Bowser & Cutter, 2015; Petrolia et al., 2011). The availability of vehicles at the time of the disaster, the perceived severity of the disaster, and the timing of the disaster's occurrence are additional factors exerting a significant impact on human evacuation decisions (Bowser & Cutter, 2015; Petrolia et al., 2011; Mohajeri & Mirbaha, 2021). Groups and individuals interactions have also substantially shaped route choice behaviour (Ding & Sun, 2020).

1.2.2.3 Evacuation Efficiency

Efficiency in evacuation can be gauged using various metrics, including movement time, response time, casualties, injured people, building damage, and evacuee speed. These parameters can be assessed by observing actual disaster evacuations, conducting organized drills, or simulating evacuations involving a representative sample of the population. Therefore, numerous factors can potentially impact the evacuation process's effectiveness in this context.

The effect of population density on evacuation times reveals that herding behavior[1] occurs selectively, manifesting in specific population densities rather than in highly congested or relatively sparsely populated areas (Liu & Mao, 2022). By referring to this finding, evacuation managers can set timely regulations to increase evacuation efficiency. In a comparative study that integrated population socio-demographic factors into the prediction of evacuation times through Agent-Based Modeling, researchers concluded that a higher level of detail regarding the population leads to more accurate models (Barnes et al., 2021). The literature also explores the impact of the built environment on evacuation efficiency. For instance, in a study simulating school evacuations and considering variables like the number and characteristics of evacuees, exit accessibility, and physical obstacles, findings indicated that changing the school layout is essential to reduce casualties and expedite evacuation (Ionescu et al., 2021). Meanwhile, an assessment of response times to a simulated shooting incident using Virtual Reality experiments revealed that implementing countermeasures within the building led to longer evacuation response times (Zhu et al., 2022). Within the context of educational institutions and safety, scholars have noted that not only do evacuation guidelines significantly reduce evacuation times, but classrooms with two entrances also facilitate faster evacuations (Liu et al., 2016). Another study identified that factors such as staircase incline and road terrain significantly affect evacuee speeds (Chen et al., 2022). Evacuation speed can also be influenced by low visibility conditions (Seike et al., 2017), although experiments involving the evacuation of college students in China did not establish a clear correlation between poor visibility conditions and route choice behaviour (Ding & Sun, 2020). Moreover, a study by Dong et al. delved into assessing the impact of evacuation leaders, their positioning, and their numbers on evacuation times (Dong et al., 2022). This study suggested that the presence of more evacuation guides does not consistently result in shorter evacuation times. The scholars have also proposed a practical evacuation model for urban flood management, incorporating microscopic population characteristics, GIS-based urban data, and flood attributes, and by referencing the Believe-Desire-Intention reasoning system[2] (Jigyasu, 2013).

---

[1] Herding behavior refers to following the direction of most of the crowd, often without critically evaluating or independently assessing the situation [109].

[2] "Belief-Desire-Intention" (BDI) reasoning system is a framework that elucidates how intelligent agents (people) arrive at decisions by incorporating their knowledge (beliefs), objectives (desires), and action plans (intentions).

1.2.3 Human Behavior After a Disaster: Recovery and Improvement

1.2.3.1 Search, Rescue, and Settlement Decision

Emergency search and rescue operations typically commence after the onset of a disaster. In the case of extensive catastrophic events, beyond national entities, international organizations and volunteers, including bodies like the International Federation of Red Cross and Red Crescent, actively engage in delivering humanitarian assistance. Given the paramount significance of this phase in life-saving efforts (Taillandier et al., 2021), not only is fast resource mobilization necessary, but also having well-prepared emergency facilities and proficient rescue teams is vital (Jobe, 2011),.

Identifying a secure and suitable shelter becomes another challenging issue for victims, and its importance has been mentioned in previous studies (Gregorio & Soares, 2017). In the instance of substantial calamities, governments, and international humanitarian aid agencies step in to offer secure accommodations, often in the form of tents or conex containers (Farnaz, 2018; Johnson, 2007). Nevertheless, bestowing attention to providing sanitary facilities and ensuring the overarching safety of these shelters and transitional residences remains of utmost significance (Ghomian et al., 2020). For less severe disasters, affected individuals can be accommodated in secure public spaces like undamaged schools and hotels or be hosted by their extended family (Jhonson, 1999; Wagemann, 2017). However, some residents may persist in returning to the affected regions, especially in low-income and socially vulnerable societies. This behavior can often be attributed to factors such as protecting property rights, place attachment, and inadequate awareness of potential risks (Magni et al., 2019).

1.2.3.2 Psychological Shocks and Support

In the aftermath of a disaster, survivors' psychological well-being can exhibit considerable diversity, contingent upon factors such as the scale and severity of the catastrophe, their levels of resilience, and the emotional assistance they receive (Fukuchi & Koh, 2022). Acute Stress Reaction, Post-Traumatic Stress Disorder, Anxiety and Depression, Survivor's Guilt, Grief, Social and Relationship Challenges, as well as Resilience and Post-traumatic Growth, represent a spectrum of psychological responses that survivors might encounter (Brooks et al., 2020; Davidson & McFarlane, 2006; Fukuchi & Koh, 2022; Fullerton et al., 2004).

1.2.3.3 Active/Passive Learning

Besides the immediate and persistent psychological aftermath of disasters, the experience gained from confronting such catastrophic events equips survivors with invaluable implicit disaster knowledge. The profound imprint left by the devastating Wenchuan earthquake of 2008 shaped the disaster awareness of the survivors, influencing their subsequent disaster responses during the Lushan earthquake of 2013 (Wei et al., 2021). Yet, the lessons learned from the Hurricane Katrina impact along the U.S. Gulf Coast in 2004 had a comparatively lesser impact on the evacuation decision-making of survivors during the Hurricane Ivan occurrence along the same U.S. Gulf Coast in 2005(Murray-Tuite et al., 2012). Consequently, the choice between applying lessons acquired from past disasters (active learning) and disregarding such insights (passive learning) varies based on various factors, including the magnitude of the catastrophe, individual resilience, and perceptions of risk.

1.3 Summary

In the face of disaster, human response is a remarkable testament to our resilience, compassion, and collective strength. Before a disaster, well-prepared individuals/communities can anticipate and respond effectively to potential crises. By investing in planning, training, and infrastructure, vulnerabilities can be reduced, and lives can be saved. When calamity strikes, whether a natural catastrophe or a human-made crisis, human physiological responses kick into high gear. Fight-or-flight responses propel individuals to take action, seeking safety or assisting others in immediate danger. This is when emotions, the built environment, interactions with others, and safety measures can play a crucial role in potentially saving lives by guiding evacuees in appropriately responding to the disaster. In the aftermath of a disaster, resilience becomes a vital force driving recovery and rebuilding efforts. Communities come together, sharing resources and expertise, determined to restore what has been lost. However, the journey towards recovery involves not only rebuilding physical infrastructure but also addressing the emotional scars left in the wake of such events and building a more resilient society by applying the lessons learned. Therefore, human response to disasters is a multifaceted subject encompassing various phases, each demanding distinct considerations and strategies. By recognizing our shared vulnerability and interconnectedness, we can harness the strength of collective response to not only overcome adversity but also to build a more prepared, empathetic, and resilient society.

The research covered in this book encompasses a variety of topics. Specifically, it includes the effects of knowledge, risk perception, and trust on disaster preparedness, the influence of livelihood capital on flood prevention choices, post-disaster resettlement and reconstruction, and the impact of COVID-19 on individuals' evacuation behavior, mental health, and learning intentions. The subsequent book chapters will provide the readers with real cases and demonstrate how the human response to disasters has been affected. The findings of this book have practical and educational implications for disaster preparedness, resilience, and response. Additionally, the results are relevant to pandemic response measures and can be integrated into educational curricula, workshops, and policy development efforts to enhance community resilience and public health preparedness.

While this book comprehensively addresses disaster preparedness, resilience, and response, future research directions might explore the psychological and social aspects of disaster preparedness and response, exploring factors like community cohesion, trust-building measures, and the role of emerging technologies in disaster communication. During disasters, exploring the influence of community engagement and social networks and studying the effectiveness of different communication channels in promoting timely evacuations could offer deeper insights into disaster preparedness and response. Hence, the concluding chapter of this book elucidates its research limitations and provides recommendations for conducting more extensive investigations.

References

Akiyama, H. (2019). *Who rebuilds? Civil society in post-disaster recovery in China and Japan.*

Al-Dahash, H., Thayaparan, M., & Kulatunga, U. (2016). Understanding the terminologies: Disaster, crisis and emergency. In *Proceedings of the 32nd annual ARCOM conference*, ARCOM 2016.

Alexander, D. (2018). *Natural disasters.* Routledge.

Altay, N., & Green, W. G., III. (2006). OR/MS research in disaster operations management. *European Journal of Operational Research, 175*(1), 475–493.

Ao, Y., Huang, K., Wang, Y., Wang, Q., & Martek, I. (2020a). Influence of built environment and risk perception on seismic evacuation behavior: Evidence from rural areas affected by Wenchuan earthquake. *International Journal of Disaster Risk Reduction, 46*, 101504.

Ao, Y., Zhang, H., Yang, L., Wang, Y., Martek, I., & Wang, G. (2021). Impacts of earthquake knowledge and risk perception on earthquake preparedness of rural residents. *Natural Hazards, 107*, 1287–1310. https://doi.org/10.1007/s11069-021-04632-w

Ao, Y., Zhou, X., Ji, F., Wang, Y., Yang, L., Wang, Q., & Martek, I. (2020b). Flood disaster preparedness: Experience and attitude of rural residents in Sichuan, China. *Natural Hazards, 104*, 2591–2618. https://doi.org/10.1007/s11069-020-04286-0

Bahmani, H., & Zhang, W. (2022a). A conceptual framework for integrated management of disasters recovery projects. *Natural Hazards, 113*(2), 859–885. https://doi.org/10.1007/s11069-022-05328-5

Bahmani, H., & Zhang, W. (2022b). Why do communities recover differently after socio-natural disasters? Pathways to comprehensive success of recovery projects based on bam's (Iran) neighborhoods' perspective. *International Journal of Environmental Research and Public Health, 19*(2), 678.

Barnes, B., Dunn, S., Pearson, C., & Wilkinson, S. (2021). Improving human behaviour in macroscale city evacuation agent-based simulation. *International Journal of Disaster Risk Reduction, 60*, 102289.

Bateman, J. M., & Edwards, B. (2002). Gender and evacuation: A closer look at why women are more likely to evacuate for hurricanes. *Natural Hazards Review, 3*(3), 107–117.

Bosher, L., Chmutina, K., & van Niekerk, D. (2021). Stop going around in circles: Towards a reconceptualisation of disaster risk management phases. *Disaster Prevention and Management: An International Journal, 30*(4/5), 525–537. https://doi.org/10.1108/DPM-03-2021-0071

Bowser, G. C., & Cutter, S. L. (2015). Stay or go? Examining decision making and behavior in hurricane evacuations. *Environment: Science and Policy for Sustainable Development, 57*(6), 28–41.

Brooks, S., Amlot, R., Rubin, G., & Greenberg, N. (2020). Psychological resilience and post-traumatic growth in disaster-exposed organisations: Overview of the literature. *BMJ Mil Health, 166*(1), 52–56.

Capuano, N., & King, R. (2015). Adaptive serious games for emergency evacuation training. In *2015 International Conference on intelligent networking and collaborative systems*.

Castañeda, J. V., Bronfman, N. C., Cisternas, P. C., & Repetto, P. B. (2020). Understanding the culture of natural disaster preparedness: Exploring the effect of experience and sociodemographic predictors. *Natural Hazards, 103*, 1881–1904.

Chai, L., Han, Y., Han, Z., Wei, J., & Zhao, Y. (2021). Differences in disaster preparedness between urban and rural communities in China. *International Journal of Disaster Risk Reduction, 53*, 102020.

Chen, C., Mostafizi, A., Wang, H., Cox, D., & Cramer, L. (2022). Evacuation behaviors in Tsunami drills. *Natural Hazards, 112*(1), 845–871.

Cong, Z., Liang, D., & Luo, J. (2014). Family emergency preparedness plans in severe tornadoes. *American Journal of Preventive Medicine, 46*(1), 89–93.

Cong, Z., Liang, D., & Luo, J. (2021). Effectiveness of household emergency plans in violent tornadoes. *Disaster Medicine and Public Health Preparedness, 15*(6), 677–678.

CRED. (2021). Extreme events defining our lives Executive summary. *Cent Res Epidemiol Disasters*. https://doi.org/10.1787/eee82e6e-en

Daoudi, I., Chebil, R., Tranvouez, E., Chaari, W. L., & Espinasse, B. (2021). Improving learners' assessment and evaluation in crisis management serious games: An emotion-based educational data mining approach. *Entertainment Computing, 38*, 100428.

Date, R., & Planning, R. 2013. Have an emergency plan for your family. *FEMA*, 2013. https://www.fema.gov/blog/have-emergency-plan-your-family%0A

Davidson, J., & McFarlane, A. C. (2006). The extent and impact of mental health problems after disaster. *Journal of Clinical Psychiatry, 67*(Suppl 2), 9–14.

Delcea, C., & Cotfas, L.-A. (2019). Increasing awareness in classroom evacuation situations using agent-based modeling. *Physica A: Statistical Mechanics and Its Applications, 523*, 1400–1418.

Ding, N., & Sun, C. (2020). Experimental study of leader-and-follower behaviours during emergency evacuation. *Fire Safety Journal, 117*, 103189.

Dixit, A. M., Shrestha, S. N., Guragain, R., Pandey, B. H., Oli, K. S., Adhikari, S. R., Acharya, S. P., Jimee, G. K., Upadhyaya, B. K., Sangachhe, S. B., et al. (2018). Risk management, response, relief, recovery, reconstruction, and future disaster risk reduction. In *Impacts and insights of the Gorkha earthquake* (pp. 95–134). Elsevier.

Dong, S., Huang, P., & Wang, W. (2022). An optimization method for evacuation guidance under limited visual field. *Physica A: Statistical Mechanics and Its Applications, 607*, 128221.

Durage, S. W., Kattan, L., Wirasinghe, S., & Ruwanpura, J. Y. (2014). Evacuation behaviour of households and drivers during a tornado: Analysis based on a stated preference survey in Calgary, Canada. *Natural Hazards, 71*, 1495–1517.

Edgington, D. W. (2011). Reconstruction after natural disasters: The opportunities and constraints facing our cities. *The Town Planning Review, 82*(6), v–xi. https://doi.org/10.3828/tpr.2011.35

Farnaz, F. A. (2018). Organising post-disaster reconstruction processes process. In *Housing reconstruction after the Bam earthquake*. https://doi.org/10.1007/978-3-319-70911-6

Feng, D., & Ji, L. (2014). Development of a self-administered questionnaire to assess the psychological competencies for surviving a disaster. *Disaster Medicine and Public Health Preparedness, 8*(3), 220–228.

Fukuchi, N., & Koh, E. (2022). Children's survivor guilt after the Great East Japan Earthquake and tsunami: A case report. *Educational Psychology in Practice, 38*(2), 115–124.

Fullerton, C. S., Ursano, R. J., & Wang, L. (2004). Acute stress disorder, posttraumatic stress disorder, and depression in disaster or rescue workers. *American Journal of Psychiatry, 161*(8), 1370–1376.

Ge, Y., Gu, Y., & Deng, W. (2010). Evaluating China's national post-disaster plans: The 2008 Wenchuan earthquake's recovery and reconstruction planning. *International Journal of Disaster Risk Science, 1*, 17–27.

Ghomian, Z., Khodadadizadeh, A., Jahangiri, K., & Yousef Nezhad, S. (2020). Bam, Iran, earthquake experiences, life suddenly collapse in a few seconds: A qualitative study. *Journal of Qualitative Research in Health Sciences, 6*(4), 414–426.

Gómez, O. A. (2013). Lessons from international students' reaction to the 2011 Great East Japan Earthquake: The case of the School of Engineering at Tohoku University. *International Journal of Disaster Risk Science, 4*, 137–149.

Gregorio, L. T. D., & Soares, C. A. P. (2017). Post-disaster housing recovery guidelines for development countries based on experiences in the American continent. *International Journal of Disaster Risk Reduction, 24*.

He, L. (2019). Identifying local needs for post-disaster recovery in Nepal. *World Development, 118*, 52–62. https://doi.org/10.1016/j.worlddev.2019.02.005

Hesterberg, U. W., Huertas, G., & Appleby, M. C. (2012). Perceptions of pet owners in urban Latin America on protection of their animals during disasters. *Disaster Prevention and Management: An International Journal, 21*(1), 37–50.

Hu, H., Lai, X., Li, H., & Nyland, J. (2022). Teaching disaster evacuation management education to nursing students using virtual reality mobile game-based learning. *CIN: Computers, Informatics, Nursing, 40*(10), 705–710.

Hung, L.-S. (2017). Married couples' decision-making about household natural hazard preparedness: A case study of hurricane hazards in Sarasota County, Florida. *Natural Hazards, 87*, 1057–1081.

Ionescu, Ş, Nica, I., & Chiriță, N. (2021). Cybernetics approach using agent-based modeling in the process of evacuating educational institutions in case of disasters. *Sustainability, 13*(18), 10277.

Ismail, D., Majid, T. A., Roosli, R., & Ab Samah, N. (2014). Project management success for post-disaster reconstruction projects: International NGOs perspectives. *Procedia Economics and Finance, 18*, 120–127. https://doi.org/10.1016/s2212-5671(14)00921-6

Jhonson, C. (1999). What's the big deal about temporary housing planning. In *Considerations for temporary accommodation after disasters: Example of the 1999 Turkish earthquakes*

Jigyasu, R. (2013). Long-term cultural impacts of disaster decision-making: The case of post earthquake reconstruction in Marathwada, India. *ArchNet-IJAR: International Journal of Architectural Research, 7*(3), 14.

Jobe, K. (2011). Disaster relief in post-earthquake Haiti: Unintended consequences of humanitarian volunteerism. *Travel Medicine and Infectious Disease, 9*(1), 1–5.

Johnson, C. (2007). *Strategic planning for temporary housing: 1999 earthquakes in Turkey.*

Johnson, K., Depietri, Y., & Breil, M. (2016). Multi-hazard risk assessment of two Hong Kong districts. *International Journal of Disaster Risk Reduction, 19*, 311–323. https://doi.org/10.1016/j.ijdrr.2016.08.023

Kakimoto, R., & Yoshida, M. (2022). Evacuation action during torrential rain considering situation awareness error using protection motivation theory. *International Journal of Disaster Risk Reduction, 82*, 103343.

Kato, T. (2021). Sharing experience by residents of past disaster sites: Effects of time from disasters and types of experience. *International Journal of Disaster Risk Reduction, 61*, 102390.

Kim, H., & Zakour, M. (2017). Disaster preparedness among older adults: Social support, community participation, and demographic characteristics. *Journal of Social Service Research, 43*(4), 498–509. https://doi.org/10.1080/01488376.2017.1321081

Kim, Y., & Kim, M. Y. (2022). Factors affecting household disaster preparedness in South Korea. *PLoS ONE, 17*(10), e0275540.

Kopaei, M. G. (2009). *Knowledge transfer in post-disaster reconstruction: The problem of post-post-disaster reconstruction.*

Koshiba, Y., & Nakayama, J. (2021). Intentions of university students and staff members to re-enter chemical storage buildings immediately after a major earthquake: A case study in Japan. *International Journal of Disaster Risk Reduction, 57*, 102150.

Lazo, J. K., Bostrom, A., Morss, R. E., Demuth, J. L., & Lazrus, H. (2015). Factors affecting hurricane evacuation intentions. *Risk Analysis, 35*(10), 1837–1857.

Liu, R., Jiang, D., & Shi, L. (2016). Agent-based simulation of alternative classroom evacuation scenarios. *Frontiers of Architectural Research, 5*(1), 111–125.

Liu, Y., & Mao, Z. (2022). An experimental study on the critical state of herd behavior in decision-making of the crowd evacuation. *Physica A: Statistical Mechanics and Its Applications, 595*, 127087.

Lu, Y., & Xu, J. (2015). Comparative study on the key issues of postearthquake recovery and reconstruction planning: Lessons from the United States, Japan, Iran, and China. *Natural Hazards Review, 16*(3), 04014033.

Magni, M., Pescaroli, G., & Bartolucci, A. (2019). Factors influencing university students' accommodation choices: Risk perception, safety, and vulnerability. *Journal of Housing and the Built Environment, 34*, 791–805.

Matyas, C., Srinivasan, S., Cahyanto, I., Thapa, B., Pennington-Gray, L., & Villegas, J. (2011). Risk perception and evacuation decisions of Florida tourists under hurricane threats: A stated preference analysis. *Natural Hazards, 59*, 871–890.

Mohajeri, F., & Mirbaha, B. (2021). Studying the role of personality traits on the evacuation choice behavior pattern in urban road network in different severity scales of natural disaster. *Advances in Civil Engineering, 2021*, 1–16.

Murray-Tuite, P., Yin, W., Ukkusuri, S. V., & Gladwin, H. (2012). Changes in evacuation decisions between Hurricanes Ivan and Katrina. *Transportation Research Record, 2312*(1), 98–107.

Nakanishi, H., & Black, J. (2018). Implicit and explicit knowledge in flood evacuations with a case study of Takamatsu, Japan. *International Journal of Disaster Risk Reduction, 28*, 788–797.

Nakano, G., Yamori, K., Miyashita, T., Urra, L., Mas, E., & Koshimura, S. (2020). Combination of school evacuation drill with tsunami inundation simulation: Consensus-making between disaster experts and citizens on an evacuation strategy. *International Journal of Disaster Risk Reduction, 51*, 101803.

Neureiter, N. (2014). *Committee on lessons learned from the Fukushima nuclear accident for improving safety and security of US Nuclear Plants*, National Research Council of the National Academies, Lessons Learned from the Fukushima Nuclear Accident for Improving Safety of US Nuclear Plants. The National Academies Press.

Okada, T., Howitt, R., Haynes, K., Bird, D., & McAneney, J. (2018). Recovering local sociality: Learnings from post-disaster community-scale recoveries. *International Journal of Disaster Risk Reduction, 31*, 1030–1042.

Organization, W. H. (2002). *Disasters and emergencies*. World Health Organization.

Petal, M., Ronan, K., Ovington, G., & Tofa, M. (2020). Child-centred risk reduction and school safety: An evidence-based practice framework and roadmap. *International Journal of Disaster Risk Reduction, 49*, 101633.

Petrolia, D. R., Bhattacharjee, S., & Hanson, T. R. (2011). Heterogeneous evacuation responses to storm forecast attributes. *Natural Hazards Review, 12*(3), 117–124.

Platt, S. (2018). Factors affecting the speed and quality of post-disaster recovery and resilience. In *Earthquake engineering and structural dynamics in memory of ragnar sigbjörnsson: Selected topics* (pp. 369–403). https://doi.org/10.1007/978-3-319-62099-2

PRC. (2008). *International experiences and suggestions on post-disaster rehabilitation and reconstruction.*

Patel, R. K., & Etminani-Ghasrodashti, M. T. R. (2021). Exploring disaster preparedness and awareness due to natural hazards using structural equation modelling (SEM). In *ASCE construction research congress* (pp. 33–35).

Raphael, B. (2005). Crowds and other collectives: Complexities of human behaviors in mass emergencies. *Psychiatry: Interpersonal and Biological Processes, 68*(2), 115–120. https://doi.org/10.1521/psyc.2005.68.2.115

Raphael, W. M. A. B. (1988). *Human response to man-made disaster* (pp. 43–48).

Rehman, J., Sohaib, O., Asif, M., & Pradhan, B. (2019). Applying systems thinking to flood disaster management for a sustainable development. *International Journal of Disaster Risk Reduction, 36*, 101101.

Rivera, J. D. (2020). The likelihood of having a household emergency plan: Understanding factors in the US context. *Natural Hazards, 104*(2), 1331–1343.

Rogers, R. W. (1983). Cognitive and physiological processes in fear appeals and attitude change: A revised theory of protection motivation. *Social psychology: A source book* (pp. 153–176).

Ryan, B., Johnston, K. A., Taylor, M., & McAndrew, R. (2020). Community engagement for disaster preparedness: A systematic literature review. *International Journal of Disaster Risk Reduction, 49*, 101655.

Seike, M., Kawabata, N., & Hasegawa, M. (2017). Evacuation speed in full-scale darkened tunnel filled with smoke. *Fire Safety Journal, 91*, 901–907.

Shoji, M., Takafuji, Y., & Harada, T. (2020). Behavioral impact of disaster education: Evidence from a dance-based program in Indonesia. *International Journal of Disaster Risk Reduction, 45*, 101489.

Slovic, P. (2016). *The perception of risk*. Routledge.

Stein, R. M., Dueñas-Osorio, L., & Subramanian, D. (2010). Who evacuates when hurricanes approach? The role of risk, information, and location. *Social Science Quarterly, 91*(3), 816–834.

Taillandier, F., Di Maiolo, P., Taillandier, P., Jacquenod, C., Rauscher-Lauranceau, L., & Mehdizadeh, R. (2021). An agent-based model to simulate inhabitants' behavior during a flood event. *International Journal of Disaster Risk Reduction, 64*, 102503.

Tang, J.-S., & Feng, J.-Y. (2018). Residents' disaster preparedness after the Meinong Taiwan earthquake: A test of protection motivation theory. *International Journal of Environmental Research and Public Health, 15*(7), 1434.

Tanim, S. H., Wiernik, B. M., Reader, S., & Hu, Y. (2022). Predictors of hurricane evacuation decisions: A meta-analysis. *Journal of Environmental Psychology, 79*, 101742.

Tauber, G. (2015). Architects and rural post-disaster housing: Lessons from South India. *International Journal of Disaster Resilience in the Built Environment, 6*(2), 206–224.

Teo, M., Goonetilleke, A., Ahankoob, A., Deilami, K., & Lawie, M. (2018). Disaster awareness and information seeking behaviour among residents from low socio-economic backgrounds. *International Journal of Disaster Risk Reduction, 31*, 1121–1131.

Thiede, B. C., & Brown, D. L. (2013). Hurricane Katrina: Who stayed and why? *Population Research and Policy Review, 32*, 803–824.

Thompson, R. R., Garfin, D. R., & Silver, R. C. (2017). Evacuation from natural disasters: A systematic review of the literature. *Risk Analysis, 37*(4), 812–839.

Tuohy, R., Stephens, C., & Johnston, D. (2015). Disaster preparedness: Older adults' perspectives. *International Journal of Emergency Management, 11*(1), 46–61. https://doi.org/10.1504/IJEM.2015.069516

UNISDR. (2015). *Considerations on developing a system of indicators based on the Sendai framework for disaster risk reduction 2015–2030: A proposal for monitoring progress* (Vol. 41, pp. 1–88).

UNISDR. (2017). *Economic losses, poverty & disasters 1998–2017*. CRED UNISDR.

Vásquez, A., Marinkovic, K., Bernales, M., León, J., González, J., & Castro, S. (2018). Children's views on evacuation drills and school preparedness: Mapping experiences and unfolding perspectives. *International Journal of Disaster Risk Reduction, 28*, 165–175.

Villegas, J., Matyas, C., Srinivasan, S., Cahyanto, I., Thapa, B., & Pennington-Gray, L. (2013). Cognitive and affective responses of Florida tourists after exposure to hurricane warning messages. *Natural Hazards, 66*, 97–116.

Wagemann, E. (2017). *From shelter to home: Flexibility in post-disaster accommodation.*

Wei, B., Su, G., Liu, F., & Tian, Q. (2021). Public cognition and response to earthquake disaster: From the 2008 Mw7.9 Wenchuan to the 2013 Mw6.6 Lushan earthquakes in Sichuan Province, China. *Natural Hazards, 106*, 2751–2774.

Xu, P., Lu, X., Zuo, K., & Zhang, H. (2014). Post-Wenchuan earthquake reconstruction and development in China. In *Disaster and development: Examining global issues and cases* (pp. 427–445). https://doi.org/10.1007/978-3-319-04468-2_25

Yilmaz, D. G., Von Meding, J., & Erk, G. K. (2013). A theoretical approach to the design of a survey instrument in post-disaster reconstruction: Defining indicators for a human-based study in rural built-environment. *ArchNet-IJAR: International Journal of Architectural Research, 7*(3), 40–56. http://web.b.ebscohost.com.pucdechile.idm.oclc.org/ehost/detail/detail?vid=0&sid=5946c3b9-c417-43fb-b7f6-4e88a838c75a%40pdc-v-sessmgr05&bdata=Jmxhbmc9ZXMmc210ZT1laG9zdC1saXZl#AN=97632884&db=asu

Yong, Z., Zhuang, L., Liu, Y., Deng, X., & Xu, D. (2020). Differences in the disaster-preparedness behaviors of the general public and professionals: Evidence from Sichuan Province, China. *International Journal of Environmental Research and Public Health, 17*(14), 5254.

Yu, J., Cruz, A. M., & Hokugo, A. (2017). Households' risk perception and behavioral responses to natech accidents. *International Journal of Disaster Risk Science, 8*, 1–15.

Zheng, L., Guo, Y., Peeta, S., & Wu, B. (2020). Impacts of information from various sources on the evacuation decision-making process during no-notice evacuations in campus environment. *Journal of Transportation Safety & Security, 12*(7), 892–923.

Zheng, Z., Zhou, Z., Wang, Y., & Su, Y. (2022). The prediction of evacuation efficiency on metro platforms based on passengers' decision-making capability. *Applied Sciences, 12*(18), 8992.

Zhu, R., Lucas, G. M., Becerik-Gerber, B., Southers, E. G., & Landicho, E. (2022). The impact of security countermeasures on human behavior during active shooter incidents. *Scientific Reports, 12*(1), 929.

Zibulewsky, J. (2001). *Defining disaster: The emergency department perspective*. Baylor University Medical Center Proceedings.

Chapter 2
Earthquake Knowledge and Risk Perception Impact on Rural Residents' Preparedness

Abstract This chapter examines 5 rural counties and 10 sample villages that were seriously affected by the 2008 Wenchuan (Sichuan) earthquake. Using an on-site survey of residents, earthquake perceptions and their impact on disaster preparedness behavior were examined empirically. Exploratory factor analysis (EFA) and random-effect logistic regression analysis were used. Results reveal that two factors considerably influence disaster preparedness behavior. First, residents with autonomous earthquake information access tend to be more prepared. Second, residents who are more sensitive about earthquakes (high perceived risk of occurrence) tend to be more prepared for earthquakes. Evidently, knowledge and awareness of earthquakes have a positive impact on the disaster preparedness of residents living in rural earthquake-prone regions. Consequently, government agencies should enhance the earthquake education of local residents as part of the national effort to mitigate the adverse effects of future earthquakes.

Keywords Earthquake preparedness · Earthquake knowledge · Risk perception · Exploratory factor analysis · Random-effect logistic regression · Rural region · China

2.1 Introduction

Earthquakes are one of the most severe and unpredictable forms of natural disasters that often result in tremendous casualties and major economic losses. China experiences a large number of serious earthquakes. Efforts into disaster reduction focus primarily on disaster preparedness. In 2017, the Chinese Seismological Bureau released a special report on disaster preparedness and capacity building for earthquakes. The report stated that substantial efforts have been exerted to improve the basic capacity for disaster preparedness, increase public awareness of disaster preparedness, and establish an earthquake disaster management mechanism based on a coordinated network that comprises government, society, and the public. earthquakes can affect their action concerning earthquake preparedness significantly and

Y. Ao and H. Bahmani, *Navigating Complexity: Understanding Human Responses to Multifaceted Disasters*, https://doi.org/10.1007/978-981-99-8207-3_2

directly (McNeill et al., 2013; Paul & Bhuiyan, 2010); moreover, the negative effects of a disaster can be alleviated by scientific measures (Han et al., 2017). However, previous research has mostly focused on developed countries (Becker et al., 2012; Lindell et al., 2016) rather than developing countries (Han et al., 2017; Wu et al., 2018), with the emphasis being on urban rather than rural regions (Xu et al., 2018). In China, the probability of earthquakes in rural areas is much higher than that in large and medium-sized cities. For example, the 2008 M 8.0 Wenchuan earthquake occurred at 14:28 Beijing time (UTC + 8) on May 12, 2008 and had resulted in nearly 70,000 deaths, spanning over 20 days. The 7.0 magnitude earthquake that struck Jiuzhaigou in Aba County, Sichuan Province at 21:19 Beijing time (UTC + 8) on August 8, 2017 resulted in 20 deaths, with an addition of approximately 500 injuries. In addition, the 6.0 magnitude earthquake that occurred in Changning County, Yibin, Sichuan at 22:55 Beijing time on June 17, 2019, at a depth of 16 km, killed 13 people. Thus, Chinese rural regions suffer disproportionately with residents comprising the largest number of victims.

The rural areas affected by the earthquake, such as Sichuan, share a priority need in terms of initiating earthquake prevention and risk reduction programs. Indeed, the potential threat of an earthquake of a magnitude 7 or above remains strong for Sichuan, China. Compared with urban areas, rural areas in China exhibit weak earthquake resilience capacity in terms of their built environment. Thus, residents of these regions sorely need to improve disaster preparedness and implement effective steps to reduce injury, damage, and economic loss.

Accordingly, this study investigates the worst-hit rural areas in the 2008 Wenchuan earthquake to ascertain how residents' knowledge and perception about earthquakes influence disaster preparedness behavior, at this time—circa 10 years after the quake. On the basis of the findings, implications for policies and measures aimed at improving the disaster preparedness ability of rural residents are proposed. The results also offer important theoretical and practical insights into governments for improving disaster resilience.

The remainder of this paper is organized as follows. Section 2.2 reviews the literature pertinent to this study. Section 2.3 presents the research method, including model description, questionnaire design, sample selection, data collection, and variable description. Section 2.4 provides the results and discussion. Finally, Sect. 2.5 draws the conclusions.

2.2 Literature Review

In the aftermath of the Wenchuan earthquake, scholars have considered various perspectives in analyzing this disaster. Some scholars have investigated the building damage that resulted from high-strength earthquakes and offered a theoretical framework regarding building design in disaster-prone areas. Most researchers have studied secondary disasters caused by earthquakes (Huang et al., 2012) and their corresponding traits, such as the damage to and impact on ecosystems in stricken areas

(Cui et al., 2012). Other researchers have reviewed post-earthquake rescue lessons learned from disasters (Zhang et al., 2012). Several studies also consider the influence of earthquakes on social development, such as post-disaster reconstruction (Ge et al., 2010) and disaster vulnerability assessment (Du et al., 2015).

Few researchers, however, have explored people's cognition regarding earthquake risks and related influences (Lo & Cheung, 2015) or the mental health conditions of disaster victims. Although scholars have studied problems that arose after the 2008 Wenchuan earthquake, minimal attention has been paid to the disaster preparedness capability of local rural residents. The limited local and international literature indicates that sociodemographic factors, earthquake knowledge and experience, and perception can influence the disaster preparedness actions of local residents directly (Becker et al., 2012; Paul & Bhuiyan, 2010). Even so, interactions between these factors with respect to the 2008 Wenchuan earthquake have not been explored, to the best of our knowledge. As formerly mentioned, understanding the disaster preparedness behavior of local residents is of great theoretical and practical merit as a prerequisite to improving that preparedness and mitigating future earthquake losses.

2.2.1 Influence of Sociodemographic Variables on Earthquake Preparedness

Many empirical studies have shown that the disaster preparedness of residents is closely related to sociodemographic factors, including but not limited to age, gender, education, and earthquake experience. Differences in disaster prevention preparedness among various age groups and genders have been revealed in earlier studies (Heller et al., 2005; Wu et al., 2018). Older people are less likely to be prepared. Moreover, women perform more household activities than men, and thus, they have a higher tendency to prepare; at least, this observation is the case in China. Several studies found that education level exerts a positive influence on preparedness (Mabuku et al., 2018; Mishra & Suar, 2007), whereas others suggest that they are not correlated (Lindell & Hwang, 2008; Miceli et al., 2008). Experience influences disaster preparation significantly (Baker, 2011; Lindell & Hwang, 2008). The study reported that people with disaster experience were more prepared for future disasters than those without such experience (Mishra & Suar, 2007). People who suffered considerable injury or loss in previous earthquakes are likely to be more prepared (Heller et al., 2005).

2.2.2 Influence of Earthquake Knowledge on Earthquake Preparedness

Existing research indicates that knowledge on earthquakes exerts considerable influences on the disaster preparedness and risk perception (Becker et al., 2012) of residents (Becker et al., 2012; Shaw et al., 2004). Moreover, different degrees of knowledge of disasters give rise to varying levels of disaster preparedness (Soffer et al., 2011). Extensive earthquake knowledge influences preparedness positively (Tekeli-Yeşil et al., 2010). Consistent with this finding, earthquake education also correlates with disaster preparedness (Azim & Islam, 2016). For example, a combination of earthquake awareness education in schools and families and even in communities can develop a disaster preparedness culture that can motivate local residents to make appropriate preparations (Shaw et al., 2004). Therefore, knowledge on earthquakes gained through public education contributes significantly.

However, rural residents in China generally lack sufficient understanding regarding earthquakes and have limited awareness toward appropriate actions that should be taken to safeguard against the negative impacts brought on by earthquakes. Current knowledge on earthquake prevention and disaster reduction is largely inaccurate. Consequently, appropriate earthquake preparation should be communicated, and earthquake mitigation awareness should be raised among local residents to enable them to adopt effective earthquake mitigation measures. Preparedness can also be enhanced through capacity-building training and improved construction (Koshiba et al., 2012; Shah et al., 2019) conducted emergency drills and smoke evacuation experiments among students at the Yokahama National University. Furthermore, (Levac et al., 2012) found that extensive disaster preparedness knowledge and sufficient disaster preparedness motivation and resources exert a positive impact on effective behavior. Overall, people's preparedness can be significantly affected by their access to knowledge and information regarding earthquakes along with their ability to learn.

2.2.3 Influence of Risk Perception on Earthquake Preparedness

Risk perception refers to the public's subjective judgment about a disaster's characteristics and the severity of its potential consequences (Slovic, 1999). Personal earthquake risk perception is commonly based on previous experiences, memory, and awareness. By examining individuals' perceptions, determining how people perceive the threat of extreme events is possible. Perception condition, attitude, and emotions, in turn, affect choices (preparedness) considered in dealing with disasters (Slovic, 1999). Extant studies on earthquake risk perception focus on three aspects: perception of the probability of earthquake occurrence, post-earthquake consequences, and sensitivity to earthquakes (Becker et al., 2012; Paul & Bhuiyan, 2010).

Scholars have studied the impact of risk perception on disaster preparedness. Some believe that risk perception has a marginal influence on disaster preparedness (Lindell & Whitney, 2000; Siegrist & Gutscher, 2006). These researchers believe that people are only willing to prepare for hazards when facing an imminent, direct threat (Brenkert–Smith et al., 2006). However, others believe that risk perception augments prevention preparedness (Martins et al., 2019). Factors that influence preparedness include the intensity of earthquakes that was previously perceived by residents, the degree of threat to family properties (Calvello et al., 2016; Lindell & Hwang, 2008), and personal safety. Given that residents' predictions of a future earthquake can be influenced by an earlier earthquake, several scholars believe that no evident relationship exists between expectations of a possible earthquake occurrence and disaster preparedness (Mabuku et al., 2018). However, other researchers believe that people's degree of expectation regarding earthquake occurrence exerts a positive impact on disaster preparedness (Mulilis & Lippa, 1990). Meanwhile, residents' perception in terms of the severity of the consequences of a future earthquake and their adoption of additional disaster preparedness measures exhibit a significantly positive correlation (Han et al., 2017). Additionally, the residents' level of anxiety regarding earthquakes correlates with daily disaster preparedness (Dooley et al., 1992; Miceli et al., 2008).

Overall, studies on risk perception have focused on several aspects, such as the sense of unease (e.g., anxiety and worry), perceived earthquake consequences (e.g., risk, threat, and influence), and expectations of future likelihood of earthquake occurrence (e.g., risk judgment and probability). However, few studies have been conducted on how risk perception affects preparedness behavior in rural areas that are most affected by the 2008 M 8.0 Wenchuan earthquake. To explore the awareness and ideas of residents in disaster-stricken areas, this study is essential.

2.2.4 Previous Research Methods

The present study mainly conducts statistical analysis through questionnaire survey or database data. Through independent and dependent variables, the influence relationship or influence path is processed. Commonly used methods include binary regression analysis, multiple regression analysis, and structural equation model. The most basic statistical tools beyond univariate frequencies include bivariate cross-tabular analysis using the Pearson chi square, a non-parametric test that compares cross-tabulated data on the observed frequency of values with the expected frequency of values. Tests of significance based on Pearson Chi-square were used to test the null hypothesis to prove that the independent and dependent variables have no relationship (Goltz & Bourque, 2017). The risk information seeking and processing model (RISP) elicits great attention because of the proposed predictors of individual-level information behavior and the universal applicability of risk situations (Griffin et al., 1999). The RISP model and its subsequent iterations have been applied to environmental and health risk issues related to the environment. Furthermore, to date, hardly any studies have tested the RISP model in the seismological fields. Only (Li et al.,

2017) investigated the information-seeking behavior of individuals within the setting of earthquake risk in Chinese society, which successfully expanded on the application of the RISP model and related work in earthquake science (Li et al., 2017).

2.3 Research Method

2.3.1 *Model Specification*

2.3.1.1 Exploratory Factor Analysis

Exploratory factor analysis (EFA) is a statistical method used to simplify and analyze high-dimensional data. The principle is to reduce dimensions, namely, decreasing a large number of complex variables into fewer numbers of factors. The method has the least information loss and can ensure the integrity of the original information to the greatest extent. The general form of EFA is as follows:

$$x_i = \alpha_{i\,1}F_1 + \alpha_{i\,2}F_2 + \ldots + \alpha_{i\,n}F_n + \varepsilon_i(i = 1, 2, \ldots n) \tag{2.1}$$

As a factor load, the essence of αi j is the correlation coefficient between the common factor Fj and variable xi. εi is a special factor that represents influencing factors other than the common factors.

We first collected sample data, extracted five factors (F1 ~ F5) by using EFA and then used random-effect logistic regression to study the impact of five factors on the earthquake preparedness behavior of rural residents.

2.3.1.2 Random-Effect Logistic Regression Model

Let yi (I = 1, 2… N) be an independent binary random variable. yi is 1 when the i-th resident takes an action; otherwise, it is 0. Let pi be the probability of the i-th resident that takes action, then yi obeys a binomial distribution with probability pi.

$$y_i|p_i \sim B(1, p_i). \tag{2.2}$$

The probability of a resident to take action is:

$$P(y_i|p_i) = p_i\, y_i(1 - p_i)1 - y_i. \tag{2.3}$$

Our data are obtained from 10 villages, and respondents living in the same village are likely to have unobservable common characteristics. This intra-village correlation, also known as inter-village heterogeneity, is unsuitable for the independent co-distribution hypothesis of traditional logistic regression models. A regression model logistic random effects would be more ideal. A log-likelihood function of random effects logistic regression can be written as:

$$\ln P(y_1, y_2, \ldots, y_i) = \sigma_0 + \sigma_1 \ln F_1 + \sigma_2 \ln F_2 + \sigma_3 \ln F_3 + \sigma_4 \ln F_4 + \sigma_5 \ln F_5 + \sum_m r_m X_{im} + \varepsilon_i \quad (2.4)$$

α_0 is a constant term, and α_1~α_5 are the regression coefficients of the five corresponding principal factors. X_{im} represents the m-th control variables of the i-th residents, such as sex, age, and marital status. ε_i is the random error items.

2.3.2 *Questionnaire Design*

The questionnaire survey is the most common research method used on natural disaster risk perception and behavioral response. The questionnaire of this study has three parts. The first part mainly comprises basic information, such as gender, age, place of birth, the intensity of earthquake experienced, and the type of residential structure. The second part includes the main body of the questionnaire, namely, two modules of earthquake knowledge, risk perception, and 20 indicators, which are mainly summarized from a previous literature. The indicators are designed using a 5-point Likert scale. Five grades, namely, "totally yes," "yes," "neutral," "no," and "totally no," match different points that range from 5 to 1, respectively. The respondents made judgments according to the degree of consistency between the questions in the questionnaire and their own perceptions. Finally, the core question of this questionnaire is "do you usually take earthquake prevention measures," which can be answered by No (0) and Yes (1). Before the formal survey, we also conducted a preliminary survey on the recent rural areas in December 2018. A total of 40 questionnaires were collected, and the preliminary analysis showed that the internal consistency of the questionnaire data was remarkable. According to the actual situation of the preliminary investigation, combined with the characteristics of rural residents in Sichuan after the earthquake, three questions, including X17 (You will always stay alert after an earthquake), X18 (You sense aftershocks after an earthquake), and X24 (You are confused during an earthquake), are added to complete the final formal survey questionnaire. Please refer to the appendix for the detailed content of the questionnaire.

2.3.3 Sample Selection and Data Collection

2.3.3.1 Sample Selection

Post-disaster reconstruction has witnessed remarkable progress in the past 10 years following the 2008 Wenchuan earthquake. Accordingly, the present work sets out to determine residents' preparedness. This approach aims to investigate five of the worst-hit counties (or county-level city) impacted by the Wenchuan earthquake; namely: Qingchuna County (Guangyuan City), Mianyang City (Mianyang City), Mianzhu County (Deyang City), Dujiangyan City (Chengdu City), and Wenchuan County (the Aba Tibetan and Qiang Autonomous Prefecture) (Liang et al., 2019). Across these representative locales, the impact of earthquake knowledge and perceptions of local residents on daily disaster preparedness behavior is investigated.

Historical seismic data statistics show that more than 90% of earthquakes in China occur in rural areas, with 85% of the economic damage suffered as a result of earthquakes that occur in these regions. Therefore, the scope of this research is restricted to the worst-hit rural areas impacted by the 2008 Wenchuan earthquake. Two villages are randomly selected from each county for subsequent analysis. Overall, 10 sample villages are investigated: Dongfang and Yinping Villages in Qinghuan County, Guangyuan City; Laochang and Qinglin Villages in Beichuan County, Mianyang City; Jixian Community and Hongming Village in Mianzhu County, Deyang City; Heming and Hejia Villages in Dujiangyan City, Chengdu City; and Guojiaba and Lianshanpo Villages in Wenchuan County, Aba Tibetan, and Qiang Autonomous Prefecture. All the 10 villages are situated along the Longmenshan earthquake belt. Their locations and related cities and prefectures are shown in Fig. 2.1.

2.3.3.2 Data Collection

A total of 32 volunteers from the countryside of Sichuan Province were recruited and trained for the data collection task. All the volunteers were familiar with the rural environment and can communicate with local residents effectively. Before the formal fieldwork was undertaken, a pilot study was conducted in a village close to Chengdu, and the results showed that the internal consistency of the questionnaire survey design was acceptable.

The research group entered the villages to conduct the formal face-to-face survey, which was carried out over a 9-day period (from December 28, 2018, to January 5, 2019). Residents in each sample village were selected randomly. When respondents did not wish to participate, the next household was solicited using an ad-hoc approach. The research subjects include adult rural residents, with ages ranging from 18 to 70 years old. Overall, 503 questionnaires were collected during the survey, with an actual recovery of 500 copies. Out of the 500 copies, 476 were valid questionnaires forming the basis of the analysis. The effective questionnaire return ratio was 94.6%.

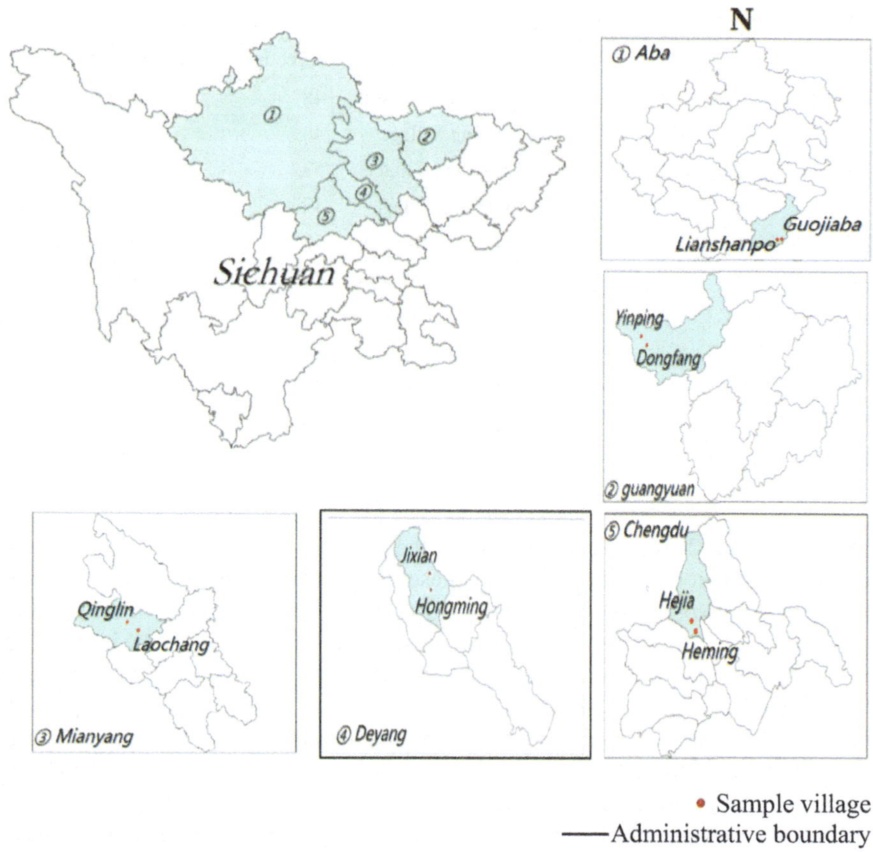

Fig. 2.1 The locations of sample villages

Table 2.1 shows the sample size from each village, revealing no statistical variability between the sample villages.

2.3.4 Specification of Variables

2.3.4.1 Dependent Variable

In accordance with previous research, disaster preparedness is a combination of several aspects: long-term preparedness (e.g., purchasing insurance), emergency protection measures (e.g., mobile cars and safety equipment), material preparation (e.g., emergency kits), and capacity building activities (e.g., acquiring knowledge and participating in exercises) (Lindell & Perry, 2012). Studies have been conducted to rank preparedness levels and to conduct surveys and analyses (Ablah et al., 2009;

Table 2.1 Effective recovery distribution of questionnaires

City and county	Sample village	Effective number	Percentage (%)
Qingchuan County, Guangyuan City	Dongfang village	48	10.09
	Yinping Village	48	10.09
Beichuan County, Mianyang City	Laochang Village	53	11.13
	Qingling Village	42	8.82
Mianzhu County, Deyang City	Ji Xian Community	58	12.18
	Hongming Village	48	10.09
Dujiangyan City, Chengdu City	Heming Village	49	10.29
	Hejia Village	42	8.82
Wenchuan County, Aba Tibetan and Qiang Autonomous Prefecture	Guojiaba Village	40	8.40
	Lianshanpo Village	48	10.09
Total		476	100.00

Baker & Baker, 2010). On the one hand, these studies are based on rich practical research foundation. On the other hand, these studies are closely related to the people with extensive experience or certain cognition under the environment of the research sites. However, China's disaster preparedness level research is still in the initial stage of exploration, and the current residents are also ignorant of the unknown. Such detailed analysis and study of disaster preparedness in terms of specific, degree, or other measures are not applicable to current field studies or current people. Therefore, to further study step-by-step, only a simple random-effect problem is set up in this paper, that is, are earthquake preparedness measures usually taken.

From the survey, only 191 residents, which accounts to 40%, prepare for hazards in their daily routine. Meanwhile, 285 respondents, constituting the other 60%, do not prepare at all. Table 2.2 presents the results.

2.3.4.2 Control Variables

Following previous studies (Goltz & Bourque, 2017; Wei et al., 2019), the socioeconomic factors that are used as control variables include gender, age, education, marriage, and birthplace. Moreover, people's preparedness for a disaster is not only affected by objective factors, such as experienced earthquake intensity, but also by disaster experience, that is, whether a person has experienced an earthquake and suffered from economic losses or injury as a result (Goltz & Bourque, 2017). On the basis of the characteristics of groups from within disaster areas, this study considers past experiences, including whether their house is located near a hill, earthquake intensity, property losses, and casualties. The type of housing structure also influences disaster preparedness (Spittal et al., 2008). The detailed control variables are listed in Table 2.3.

Table 2.2 Proportion of respondents preparing for earthquakes

City and county	Sample village	Do you usually prepare for earthquakes?		Preparedness proportion (%)
		No	Yes	
Qingchuan County, Guangyuan City	Dongfang Village	29	19	39.58
	Yinping Village	21	27	56.25
Beichuan County, Mianyang City	Laochang Village	41	12	22.64
	Qingling Village	23	19	45.24
Mianzhu County, Deyang City	Ji Xian Community	36	22	37.93
	Hongming Village	36	12	25.00
Dujiangyan City, Chengdu City	Heming Village	28	21	42.86
	Hejia Village	20	22	52.38
Wenchuan County, Aba Tibetan and Qiang Autonomous Prefecture	Guojiaba Village	29	11	27.50
	Lianshanpo Village	22	26	54.17
Total			285	191

Table 2.3 shows that women, accounting for 55%, respondents slightly outnumber men in the survey. The distribution across different age groups is consistent with reality in rural regions in Sichuan. That is, the number of teenagers aged 18–20 years is low because most of them study or work outside. Most of the respondents are over 40 years old. The reason is that labor force is mostly engaged outside rural regions or in urban areas. Thus, rural areas are overly represented by a relatively aging population. Moreover, 4.8% and 4.6% of the respondents belong to the two extremes of the educational spectrum (i.e., uneducated and educated at or above the college level, respectively). Rural residents generally have a low educational level because of a serious imbalance in educational resources between urban and rural areas combined with the economic backwardness of rural regions. People living in rural areas also do not recognize the importance of education. Approximately 73% of respondents are permanent residents, whereas the remaining 27% are migrants. Nearly 40% of the population lives near the hill, which is related to the local terrain. With regard to the intensity of earthquake experience, those with a serious encounter account for 87%. This finding shows that apart from the actual rated intensity level of the earthquake, residents generally believe that the earthquake they experienced in 2008 was intense. This factor is also related to personal risk and psychological perceptions (Table 2.4).

Regarding physical harm, 49% suffered a family casualty in all the experienced earthquakes. All the respondents generally believe that the earthquake brought economic damage, with only 8.6% not losing the property. Out of all the respondents, 42% suffered from extremely serious losses, thereby corroborating the findings of other researchers, such as (Wang et al., 2017), who believed that the earthquake

Table 2.3 Statistical table of control variables (n = 476)

Basic information		Percentage (%)	Number	basic information		Percentage (%)	Number
Q1 Gender	Male	44.7	213	Q7 Experienced earthquake intensity	No	0.4	2
	Female	55.3	263		Light	3.2	15
Q2 Age	< 20 years	2.1	10		Secondary	9.4	45
	21 ~ 30 years	11.1	53		Moderate	35.5	169
	31 ~ 40 years	11.3	54		Very strong	51.5	245
	41 ~ 50 years	22.7	108	Q8 Casualties	None	51.4	245
	51 ~ 60 years	23.9	114		Low	16.9	80
	> 61 years	28.8	137		Secondary	8.2	39
Q3 Education	None	4.8	23		Moderate	11.6	55
	Primary	36.8	175		Very high	12	57
	Junior high school	30.9	147	Q9 Property loss	None	8.6	41
	Senior high school	22.9	109		Low	7.8	37
	University +	4.6	22		Secondary	12.2	58
Q4 Marriage	Unmarried	13.4	64		Moderate	29.6	141
	Married	77.1	367		Very high	41.8	199
	Divorced or widowed	9.5	45	Q10 Structure type	Stone–wood	2.4	11
Q5 Born in village	No	26.9	128		Brick–wood	7.3	35
	Yes	73.1	348		Brick–concrete	62.7	298
Q6 Near the hill	No	60.3	287		Steel–concrete	27.5	131
	Yes	39.7	189				

brought considerable damage to the local economy. Housing construction is mostly brick–concrete and steel–concrete, thereby providing houses with strong earthquake resistance and structural stability against shock. In general, the reconstruction of the earthquake-stricken countryside in Wenchuan is much progressed.

Table 2.4 KMO and Bartlett's test results

KMO and Bartlett's test		Earthquake knowledge	Risk perception
KMO		0.873	0.868
Bartlett's test	Chi-square	3276.900	4349.226
	Degrees of freedom	45	91
	p-value	< 0.001	< 0.001

2.3.4.3 Explanatory Variables

The study sets earthquake knowledge and risk perception as independent variables. In the past, information collection, coping cognition, the degree of knowledge understanding, and the scope of earthquake knowledge were used as earthquake knowledge (Becker et al., 2012; Li et al., 2017; Soffer et al., 2011). Earthquake knowledge not only includes the formation of earthquake disasters in the narrow sense, the impact, and self-help methods at a deeper level but also includes the understanding of the earthquake in the surrounding environment and the willingness and ability to learn independently. Therefore, the earthquake knowledge in this study covers the relevant content of existing studies, and a new earthquake knowledge item is added on the basis of the results of interviews with rural residents living in areas significantly affected by the 2008 Wenchuan earthquake. Finally, ten items are included. The details are presented in Table 2.5.

(Slovic, 1993) indicated that the perception of risk is composed of unknown risks and feared emotions. To conform with established precedent, previous research indicators are included in this study. The sources of perceived risk cognition indicators include (Goltz & Bourque, 2017; Li et al., 2017; Miceli et al., 2008; Paul & Bhuiyan, 2010). A total of 11 indicators from previous studies and 3 indicators from the pilot survey are included. Overall, 14 indicators are considered. Table 2.6 presents the selection and sources of indicators.

On the basis of the preliminary breakdown, the average value of the earthquake knowledge scale is 3.14, and the standard deviation (SD) is 0.98, indicating that most people have adequate knowledge about earthquakes. All the collected data were analyzed for reliability using SPSS 24.0 software. The Cronbach's α coefficient of the 10 indicators is 0.908, indicating good internal consistency (Xu et al., 2018). The Cronbach's α coefficient measures the internal consistency between 0 and 1; if the value is less than 0.60, then the result is unsatisfactory; if it is more than 0.70, then the result is rather satisfactory. The average value of the risk perception scale is 3.21, and the SD is 1.07. Most of the respondents exhibit a negative attitude. The Cronbach's α coefficient of the 14 indicators is 0.903, and that of the total amount is 0.880, which satisfies reliability requirements.

Table 2.5 EFA results of earthquake knowledge: factor component matrix

Item		Component	
	Source	Autonomous learning and acquisition of earthquake knowledge (K1)	Understanding of earthquake knowledge and information (K2)
X9 You will actively learn how to reduce disaster losses	(Siriwardena et al., 2013)	0.908	0.192
X8 You will learn to improve your earthquake self-rescue skills actively	Authors	0.895	0.225
X10 You will maintain a life-long learning attitude toward earthquakes preparedness	(Siriwardena et al., 2013)	0.861	0.238
X7 You can seek earthquake information to deal with earthquakes	(Li et al., 2017)	0.733	0.310
X6 You know how to obtain the earthquake information	(Li et al., 2017)	0.649	0.472
X3 You know the possible effects of earthquakes	(Li et al., 2017)	0.146	0.807
X4 You know the number and magnitude of earthquakes in your village	(Li et al., 2017)	0.140	0.788
X2 You are knowledgeable about methods of earthquake rescue	(Li et al., 2017)	0.344	0.764
X1 You know the cause of an earthquake	(Li et al., 2017)	0.436	0.652
X5 You know the local evacuation route and shelter location	(Li et al., 2017)	0.254	0.610
Characteristic value		3.737	3.117

(continued)

Table 2.5 (continued)

Item		Component	
	Source	Autonomous learning and acquisition of earthquake knowledge (K1)	Understanding of earthquake knowledge and information (K2)
Variance percentage (%)		37.368	31.172
Percentage cumulative variance (%)		37.368	68.540

Extraction method: PCA. Rotation method: Caesar normalized maximum variance method. The rotation has converged after four iterations

2.4 Results and Discussion

2.4.1 Exploratory Factor Analysis

In this study, the Kaiser–Meyer–Olkin (KMO) and Bartlett's sphericity tests were used to analyze and test the applicability of factor analysis to earthquake knowledge and risk perception, respectively. In general, when the KMO value is ≥ 0.70 and the statistical significance of Bartlett's sphericity test is < 0.05, the variables are closely correlated and suitable for factor analysis. In this study, the KMO values of earthquake knowledge and risk perception are 0.873 and 0.868, respectively, thereby meeting the correlation requirements. Indeed, the p-value is < 0.001, thereby confirming that the variables are closely correlated and that EFA is suitable.

Subsequently, principal component analysis (PCA) and the maximum variance rotation method were used to extract common factors with eigenvalues greater than 1. In general, when the contribution rate of the accumulated variance of each factor reaches more than 60%, the scale exhibits good validity. The rotated factor load matrices are provided in Tables 2.5 and 2.6, from which the distribution of the indicators in factors are obtained.

In Table 2.5, the earthquake knowledge factor analysis extracted two common factors, namely, autonomous learning and acquisition of earthquake knowledge (K1) and understanding of earthquake knowledge and information (K2). The amount of information interpretation reached 68%. The number of corresponding variables decreased to 20%; that is, the number of variables was reduced by 80%, resulting in only a 31% information loss, which exhibits a good effect.

In Table 2.6, three factors were extracted from the analysis of risk perception factors, namely, earthquake psychological (P1), earthquake result perception (P2), and earthquake occurrence probability perception (P3). X11 (You are sensitive to shaking) points into the earthquake result perception factor (P2), the reason may be that the sensitivity of the residents to shaking is close to the sensitivity to the

Table 2.6 EFA results of risk perception: factor component matrix

Item		Component		
	Source	Earthquake psychological perception (P1)	Earthquake result perception (P2)	Earthquake occurrence probability perception (P3)
X22 You feel alarmed during an earthquake	(Goltz & Bourque, 2017)	**0.861**	0.247	0.217
X24 You feel confused during an earthquake	Authors	**0.855**	0.226	-0.053
X20 You feel anxious during an earthquake	(Li et al., 2017)	**0.845**	0.238	-0.046
X21 You feel worried during an earthquake	(Li et al., 2017)	**0.812**	0.274	0.177
X23 You feel afraid during an earthquake	(Goltz & Bourque, 2017)	**0.808**	0.236	0.2
X19 You become nervous when it comes to earthquakes	(Li et al., 2017)	**0.525**	0.029	0.167
X16 Do you think the earthquake was disastrous?	(Li et al., 2017)	0.091	**0.797**	0.002
X17 You will always stay alert after an earthquake	Authors	0.149	**0.746**	0.092
X14 Do you think you will be directly affected by an earthquake?	(Li et al., 2017; Paul & Bhuiyan, 2010)	0.168	**0.660**	0.345
X18 You sense aftershocks after an earthquake	Authors	0.257	**0.637**	0.128
X15 Do you think extreme earthquakes will have long-term negative effects?	(Li et al., 2017)	0.224	**0.613**	0.147
X11 You are sensitive to shaking	(Li et al., 2017)	0.242	**0.461**	0.293

(continued)

Table 2.6 (continued)

Item		Component		
	Source	Earthquake psychological perception (P1)	Earthquake result perception (P2)	Earthquake occurrence probability perception (P3)
X13 Do you think the risk of a severe earthquake in the future is high?	(Li et al., 2017; Paul & Bhuiyan, 2010)	0.131	0.135	**0.853**
X12 You think the area is prone to earthquakes	(Li et al., 2017)	0.135	0.28	**0.814**
Characteristic value		4.503	2.963	1.905
Variance percentage (%)		32.162	21.162	13.605
Percentage cumulative variance (%)		32.162	53.324	66.929

Extraction method: PCA. Rotation method: Caesar normalized maximum variance method. The rotation has converged after four iterations

consequences of the earthquake. Thus, they think that it is close to X17 (You will always stay alert after an earthquake) or X18 (You sense aftershocks after an earthquake), and the residents' choice of these questions is highly consistent when filling out the questionnaire. Lastly, the EFA turns them all into indicators of earthquake result perception factor (P2). The amount of information interpretation reached 67%. The number of corresponding variables was reduced to 22%, that is, the number of variables was reduced by 78%, thereby resulting in only a 33% information loss, which exhibits a good effect.

The five common factors extracted via factor analysis are used as explanatory variables in the random-effect logistic regression model. F1 ~ F5 in formula (2.4) corresponds to K1, K2, P1, P2, and P3, respectively.

2.4.2 Random-Effect Logistic Regression

The dependent variable in this study is whether residents in earthquake-stricken areas generally undertake disaster preparedness measures for earthquakes. This option is known as random-effect, namely, "no" (0) and "yes" (1). The independent variables correspond to all the aforementioned variables. Stata version 16.0 was used to estimate the random-effect logistic regression model. The Husman results show that

Table 2.7 Model estimation results

Variables	Coefficient	Standard Error	Z	P > |Z|	95% Confidence Intervals	
					Lower	Upper
Q1 Gender	0.030	0.262	0.11	0.910	− 0.483	0.543
Q2 Age	0.397	0.122	3.26	0.001***	0.158	0.637
Q3 Education	0.647	0.109	5.92	0.000***	0.431	0.858
Q4 Marriage	0.761	0.295	2.58	0.010***	0.183	1.339
Q5Born in current village	− 1.094	0.304	− 3.60	0.000***	− 1.691	− 0.498
Q6 Residence near the hill	2.568	0.285	9.01	0.000***	2.009	3.126
Q7 Experienced earthquake intensity	− 0.747	0.164	− 4.55	0.000***	− 1.069	− 0.425
Q8 Casualties	0.150	0.084	1.79	0.073*	− 0.014	0.313
Q9 Property loss	− 0.223	0.194	− 1.15	0.251	− 0.603	0.158
Q10 Structure type	− 0.595	0.141	− 4.22	0.000***	− 0.871	− 0.319
LnK1	0.332	0.130	2.55	0.011**	0.077	0.587
LnK2	0.643	0.114	5.64	0.000***	0.419	0.866
LnP1	0.471	0.134	3.50	0.000***	0.207	0.734
LnP2	0.288	0.130	2.22	0.027***	0.034	0.543
lnP3	0.293	0.150	1.95	0.050**	0.002	0.587
Constant	− 9.362	1.744	− 5.37	0.000***	− 12.780	− 5.946

Note *** *p*-value < 0.01. ** *p*-value < 0.05. * *p*-value < 0.1

Chi-square statistics is 5.77, and the adjoint probability is 0.4496, indicating that the model fits the data. Table 2.7 presents the estimates of the model.

The random-effect logistic regression results (Table 2.7) show that the variables of earthquake knowledge and risk perception and personal sociodemographic characteristics exert significant effects on the earthquake preparedness of rural residents in the study area. Meanwhile, variables, such as sex and property loss information, have no significant impact.

2.4.2.1 Influence of Control Variables on Earthquake Preparedness

Table 2.7 shows that 8 out of 10 variables perform well in terms of basic information. Specifically, age has a positive effect on earthquake preparedness (p-value = 0.001). More adult or more mature people will be more prepared for disaster preparedness. The preparation of older people, such as those over 60 years, should be self-comfort in the heart, regardless of the situation. Education has a positive impact on the disaster preparedness of residents (p-value = 0.000); that is, the higher the education level is, the higher the probability he/she will take earthquake preparedness actions. The question of whether considerable damage can be alleviated and losses reduced even when only minimal preparation has been undertaken before an earthquake needs to be addressed. Consequently, people with a high education level tend to practice disaster preparedness. This finding supports the conclusions of Mishra and Suar (2007) and Mabuku et al. (2018). Marriage also has a positive effect (p-value = 0.010). People with complex kinship relationships are more likely to be prepared for disaster preparedness. This finding may be related to their more emotional or richer experiences. Casualties have a more significant positive impact on residents' preparedness (p-value = 0.073) than property losses (p-value = 0.251). This finding also shows that residents are more concerned about their lives, and when relatives die because of earthquake, they will be more prepared for the earthquake.

Location has an important influence on disaster prevention (Xu et al., 2018). This study shows that whether the birthplace is the current village has an important influence on the disaster preparedness behavior of local residents (p-value = 0.000). Residents who are not born locally and who do not know the surroundings as well as natives do are more prone to conduct disaster preparations. As mentioned earlier, people who have recently moved into a region are more willing to prepare for disasters than long-time locals are (Paul & Bhuiyan, 2010). Whether they reside near mountains also has a particularly positive impact on residents' preparedness (p-value = 0.000). Imagine that your house is located at the foot of the mountain and has experienced a big earthquake in 2008 first hand. Thus, you will be more likely to be prepared for disasters.

Earthquake intensity experience has a very important negative influence on people's preparedness for disasters (p-value = 0.000). This result shows that the higher the intensity of the earthquake that the people have previously experienced, the more they know about the dangers of earthquakes. However, such residents believe that nothing can withstand earthquakes; thus, they are less likely to prepare. Previous disaster experience has been studied greatly and proven to exert mixed influences. In studies conducted in the USA (Goltz et al., 1992), New Zealand, and Japan (Lindell et al., 2016), earthquake experience has no significant impact on preparedness decisions and actions. Some researchers believe that the intensity of earthquakes experienced by individuals is unrelated to earthquake preparedness (Rüstemli & Karanci, 1999), whereas others conclude that people who prepare for disasters are those who have already experienced disasters and suffered serious damage (Heller et al., 2005; Mishra & Suar, 2007). As an empirical study on rural areas affected by the 2008 Wenchuan earthquake, this study concludes that the higher the intensity of an earthquake experienced by the respondents is, the less inclined they are to implement disaster preparedness measures in their daily lives, thereby confirming a negative correlation between earthquake experience and disaster preparedness, which is sometimes called fatalism.

2.4.2.2 Influence of Explanatory Variables on Earthquake Preparedness

Table 2.7 shows that earthquake knowledge and risk perception influence daily disaster preparedness. The principal component K1 is composed of five indicators: knowing how to acquire earthquake information, being able to collect earthquake information to deal with disasters, learning to improve his/her own survival and self-help skills during an earthquake actively, learning relevant knowledge regarding disaster management actively, and keeping a life-long learning attitude. K1 has an evident positive influence on the disaster preparedness behavior of local residents (p-value = 0.011); the more active a person is in improving his/her own self-help skills and learning relevant knowledge, the more likely it is for him/her to fully prepare for an earthquake. Moreover, such disposed people can acquire knowledge on how to deal with earthquakes on their own initiative and are willing to prepare for disasters. The disaster prevention behavior of residents mostly comes from self-learning (Xu et al., 2018), and life-long learning of disaster management is fairly necessary (Siriwardena et al., 2013). Consequently, people should be more active in learning and acquiring earthquake information. Similarly, governments should emphasize the importance of personal self-study.

K2 is composed of five indicators: possession of earthquake knowledge, self-help theory and methods in an earthquake, possible impacts of an earthquake, damage to the present location and evacuation routes, and location of shelters. K2 exerts a significantly positive influence on earthquake preparedness (p-value = 0.000); that is, the more familiar local residents are with earthquake knowledge and local conditions, the more likely they are to take preparedness measures. This observation agrees with the conclusion of (Tekeli-Yeşil et al., 2010), who stated that a high level of possession of earthquake knowledge could be helpful in disaster preparedness. The average value of the earthquake knowledge scale is 3.14, and the standard deviation (SD) is 0.98, indicating that most rural people have medium knowledge about earthquakes. This result is mainly caused by our government. In the 2008 Wenchuan earthquake, a large number of earthquake science work have been conducted. At present, the sound of drills is heard occasionally. Our study shows that rural residents have obtained limited knowledge and information. They believe that human beings are powerless in the face of natural disasters, which is in line with the actual condition of residents' disaster preparedness. Residents with an inclination to self-study can possibly make full preparations for an earthquake, thereby showing that education about earthquakes would be effective if delivered to the public. Earthquake education can promote understanding among residents and improve their self-help skills during an earthquake. This phenomenon would increase the likelihood of them making preparations (Goltz & Bourque, 2017).

P1 constitutes six indicators: feeling nervous when the earthquake is discussed; feeling anxious, worried, panic-stricken, scared, and confused during an earthquake. It exerts a remarkable positive influence on disaster preparedness (p-value = 0.000). That is, the more a person is scared, confused, and overwhelmed (sensitive to an earthquake) during an earthquake, the more possible for him/her to exhibit preparedness behavior. These negative emotions (e.g., being nervous, scared, and anxious) have a positive impact on earthquake preparedness. Existing studies have concluded that emotions, such as being worried and scared, are significantly related to preparedness (Miceli et al., 2008). The more worried people are, the more motivated they are to take measures. This finding is in accordance with the conclusions of other researchers (Aspinwall et al., 2005; Han et al., 2017).

P2 has six indicators: sensitivity to shaking, being directly influenced by an earthquake, long-term negative influences of an earthquake, how disastrous an earthquake is, and always being alert and sensing possible aftershocks after an earthquake. These indicators have a positive effect on disaster preparedness (p-value = 0.027). People who are sensitive to shaking and who believe that an earthquake is a disaster are likely to take measures and may even remain perpetually vigilant once having experienced an earthquake. These residents believe that preparing is necessary because of the perceived high risk of getting hurt in the event of a strong earthquake, which is consistent with the conclusions of existing literature; that is, the results of perception are correlated to disaster preparedness (Han et al., 2017; McNeill et al., 2013).

P3 is composed of two indicators: people's belief in a high probability and risk of possible future earthquakes, which satisfy the significance level test of 1% to show that they have a positive influence on disaster preparedness (p-value = 0.050). That is, people who believe that an increasingly dangerous earthquake may occur are likely to take measures, showing that the probability and risk of earthquakes are closely related to disaster preparedness. This result is consistent with the conclusion of Mulilis and Lippa, who stated that the judgment of an earthquake occurrence rate has a significant positive impact on earthquake preparation behavior (Mulilis & Lippa, 1990).

In summary, the earthquake knowledge, risk perception, and individual sociodemographic variables of the respondents have evident influences on disaster preparedness. The results show that the disaster preparedness of local residents in quake-stricken areas is influenced by their own individual characteristics but not by their families.

2.5 Conclusions and Recommendations

The 2008 Wenchuan earthquake of May 12th in Sichuan Province, which struck with a magnitude of 8.0, resulted in the death of 70,000 people. China is a country that is particularly susceptible to earthquakes, with 90% of them occurring in rural regions, accounting for as much as 85% of all Chinese earthquake-related damage and economic loss.

Earthquake preparedness, however, requires human engagement. Research in this realm is less developed, but the extant literature points to knowledge on earthquakes and cognitive perception of earthquakes as drivers associated with preparedness. However, no work of this kind has been undertaken in the context of China, especially rural China.

Thus, this study examines conditions correlated with human preparedness to identify which drivers do or do not motivate action in this regard. In undertaking this study, a significant database was generated. A total of 10 years on from the devastating earthquake, local citizens from the 10 worst-hit village localities near and around the 2008 Wenchuan earthquake epicenter were investigated. Interview surveys were carried out on location, with a significant 476 valid questionnaires garnered from an initial total of 503. EFA and random-effect logistic regression analysis were used to draw conclusions and recommendations.

Some of the results confirm existing knowledge. By contrast, others disavow earlier research. Specifically, the demographic characteristics of residents have a significant impact on earthquake preparedness. In particular, the more educated the residents are, the more likely they are to take preparatory actions. Unexpectedly, non-local residents (those not born in the locale) are more likely to be prepared for disasters than locals. Simultaneously, the less the intensity of the earthquake previously experienced by residents is, the more likely it is they will prepare. That is, those with a more traumatic history of earthquakes were less likely to consider preparations. Moreover, age, marriage, nearness of their house from the hill, seriousness of the casualties caused by the earthquake, and residential structure type have impacts on residents' disaster preparedness.

The primary finding is that those native to the location and those with previous experience of severe earthquakes do not engage in earthquake preparedness. This phenomenon may be due to trauma, fatalism, or maybe entrenched cultural apathy. At present, understanding the underlying psychology of such people within the earthquake zones becomes a ripe topic for further research.

On the positive side, the findings of this study reveal that knowledge is pivotal to changing behavior. Individuals with informed awareness of earthquakes are more likely to pursue earthquake preparedness. In this regard, educated people are more likely to know more about earthquakes and how to self-inform themselves about such disasters. Furthermore, those who are not born in villages—having come in from outside—may have a better earthquake education than the insulated native population. Again, this observation presents a further question that needs to be pursued in subsequent research.

Overall, the key message is that although improved engineering, infrastructure, and building practices undoubtedly create greater resilience against the next earthquake, which will sooner or later come, human preparedness remains paramount. Governments and local communities must remember that earthquake education is the key to motivating individuals toward preparedness. Governments must educate the local population directly through campaigns and by making appropriate materials available online for self-study. By these means, people will be prepared, and the next earthquake's potentially devastating outcomes may be mitigated.

2.6 Limitations

This paper still has some shortcomings and deficiencies because of the limitations of various conditions and the author's level. First, the village households in Wenchuan earthquake area is the only data source. This location is representative but not universal for the whole of China. Therefore, the next step is to study other regions to verify the accuracy of the conclusion. Second, this paper only uses simple statistical and analysis methods and does not explore the influence path behind earthquake disaster preparedness for village households deeply. Third, the setting and processing of simple random-effect dependent variable are understandable. However, they should be concretized, expanded, and supplemented in the future. The aforementioned problems, which is also the focus of our current work, need to be discussed further in our future research.

Acknowledgements The authors appreciate financial support from Sichuan Province Social Science Planning project (SC19TJ030), Sichuan science and technology department soft science major work support research project (20RKX0412/ 2020JDR0177), Sichuan Rural Community Governance Research Center funding (SQZL2019C01), Ministry of education industry-school cooperative education program (201901098002), Funding for key teachers of Chengdu University of Technology (10912-KYGG2019-02305), the Natural Science Key Project from the Sichuan Provincial Department of Education (18ZA0048), and the Development Research Center of Oil and Gas, Sichuan (CYQK-SKB17-04), the Research Center for Systems Science & Enterprise Development, Key Research of Social Sciences Base of Sichuan Province (Grant No. Xq17B05), the Electronic Commerce and Modern Logistics Research Center Program, Key Research Base of Humanities and Social Science, Sichuan Provincial Education Department (Grant No. DSWL17-13).

Author Contributions All authors contributed to the research presented in this paper. Yibin Ao, Hongying Zhang had the original idea for the study; the co-authors conceived and designed the questionnaire. Linchuan Yang. and Yan Wang. organized the investigation; Hongying Zhang. analyzed the data and drafted the manuscript, which was revised by Igor Martek and Linchuan Yang. All authors have read and approved the final manuscript.

Conflicts of Interest The authors declare that they have no conflict of interest.

Ethical Approval The Office of Humanities of Social Sciences, which is in charge of ethical review, of the Chengdu University of Technology approved this study.

Appendix

Questionnaire survey on disaster preparedness behavior of residents in Wenchuan earthquake-stricken areas

Dear fellow townspeople, we are from the College of Environment and Civil Engineering, Chengdu University of Technology. To understand the residents' level of disaster preparedness, promote the formulation and implementation of local disaster preparedness policies, and improve the residents' awareness toward disaster preparedness, we specially come to ask for your opinions on earthquake disaster preparedness behavior. Please fill it out in a place that suits your own situation or a volunteer will instruct you to fill it out. Thank you for your support and cooperation!

I. Basic information:

1. Gender: ○ Male/ ○Female
2. Age:
3. Education: ○None / ○ Primary school / ○ Junior high school / ○ Senior high school / ○ University+
4. Marital status: ○ Unmarried / ○ Married / ○ Divorced or Widowed
5. Born in the current village: ○ No / ○ Yes
6. Whether the residence is residing near a mountain: ○ No / ○ Yes
7. The strongest earthquake intensity you have experienced: ○No /○ Light /○ Secondary / ○ Moderate / ○ Very strong
8. Casualties in the last earthquake: ○No / ○ Low / ○ Secondary / ○ Moderate / ○ Very high
9. Property loss due to the last earthquake: ○No / ○ Low / ○ Secondary / ○ Moderate / ○ Very high
10. Residential structure type: ○ Stone-wood / ○ Brick-wood / ○ Brick-concrete / ○ Steel -concrete

II. Explanatory variables:

	Item	Options				
		Totally No	No	Neutral	Yes	Totally Yes
X1	You know the cause of an earthquake					
X2	You are knowledgeable about methods of earthquake rescue					
X3	You know the possible effects of earthquakes					
X4	You know the number and magnitude of earthquakes in your village					
X5	You know the local evacuation route and shelter location					
X6	You know how to obtain the earthquake information					
X7	You can seek earthquake information to deal with earthquakes					
X8	You will learn to improve your earthquake self-rescue skills actively					
X9	You will actively learn how to reduce disaster losses					
X10	You will maintain a life-long learning attitude toward earthquake preparedness					
X11	You are sensitive to shaking					

(continued)

(continued)

	Item	Options				
		Totally No	No	Neutral	Yes	Totally Yes
X12	You think the area is prone to earthquakes					
X13	Do you think the risk of a severe earthquake in the future is high?					
X14	Do you think you will be directly affected by an earthquake?					
X15	Do you think extreme earthquakes will have long-term negative effects?					
X16	Do you think the earthquake was disastrous?					
X17	You always stay alert after an earthquake					
X18	You sense aftershocks after an earthquake					
X19	You become nervous when it comes to earthquakes					
X20	You feel anxious during an earthquake					
X21	You feel worried during an earthquake					
X22	You feel alarmed during an earthquake					
X23	You feel afraid during an earthquake					
X24	You feel confused during an earthquake					

III. Independent variable

Do you usually take earthquake preparedness measures: ○ No / ○ Yes.

References

Ablah, E., Konda, K., & Kelley, C. L. (2009). Factors predicting individual emergency preparedness: A multi-state analysis of 2006 BRFSS data. *Biosecurity and Bioterrorism: Biodefense Strategy, Practice, and Science, 7*(3), 317–330.

Aspinwall, L. G., Sechrist, G. B., & Jones, P. R. (2005). Expect the best and prepare for the worst: Anticipatory coping and preparations for Y2K. *Motivation and Emotion, 29*, 353–384.

Azim, M. T., & Islam, M. M. (2016). Earthquake preparedness of households in Jeddah, Saudi Arabia: A perceptual study. *Environmental Hazards, 15*(3), 189–208.

Baker, E. J. (2011). Household preparedness for the aftermath of hurricanes in Florida. *Applied Geography, 31*(1), 46–52.

Baker, L. R., & Baker, M. D. (2010). Disaster preparedness among families of children with special health care needs. *Disaster Medicine and Public Health Preparedness, 4*(3), 240–245.

Becker, J. S., Paton, D., Johnston, D. M., & Ronan, K. R. (2012). A model of household preparedness for earthquakes: How individuals make meaning of earthquake information and how this influences preparedness. *Natural Hazards, 64*, 107–137.

Brenkert-Smith, H., Champ, P. A., & Flores, N. (2006). Insights into wildfire mitigation decisions among wildland–urban interface residents. *Society and Natural Resources, 19*(8), 759–768.

Calvello, M., Papa, M. N., Pratschke, J., & Nacchia Crescenzo, M. (2016). Landslide risk perception: A case study in Southern Italy. *Landslides, 13*, 349–360.
Cui, P., Lin, Y.-M., & Chen, C. (2012). Destruction of vegetation due to geo-hazards and its environmental impacts in the Wenchuan earthquake areas. *Ecological Engineering, 44*, 61–69.
Dooley, D., Catalano, R., Mishra, S., & Serxner, S. (1992). Earthquake preparedness: Predictors in a Community Survey 1. *Journal of Applied Social Psychology, 22*(6), 451–470.
Du, Y., Ding, Y., Li, Z., & Cao, G. (2015). The role of hazard vulnerability assessments in disaster preparedness and prevention in China. *Military Medical Research, 2*(1), 1–7.
Ge, Y., Gu, Y., & Deng, W. (2010). Evaluating China's national post-disaster plans: The 2008 Wenchuan earthquake's recovery and reconstruction planning. *International Journal of Disaster Risk Science, 1*, 17–27.
Goltz, J. D., & Bourque, L. B. (2017). Earthquakes and human behavior: A sociological perspective. *International Journal of Disaster Risk Reduction, 21*, 251–265.
Goltz, J. D., Russell, L. A., & Bourque, L. B. (1992). Initial behavioral response to a rapid onset disaster: A case study of the October 1, 1987 Whittier Narrows earthquake. *International Journal of Mass Emergencies & Disasters, 10*(1), 43–69.
Griffin, R. J., Dunwoody, S., & Neuwirth, K. (1999). Proposed model of the relationship of risk information seeking and processing to the development of preventive behaviors. *Environmental Research, 80*(2), S230–S245.
Han, Z., Wang, H., Du, Q., & Zeng, Y. (2017). Natural hazards preparedness in Taiwan: A comparison between households with and without disabled members. *Health Security, 15*(6), 575–581.
Heller, K., Alexander, D. B., Gatz, M., Knight, B. G., & Rose, T. (2005). Social and personal factors as predictors of earthquake preparation: The role of support provision, network discussion, negative affect, age, and education 1. *Journal of Applied Social Psychology, 35*(2), 399–422.
Huang, Y., Chen, W., & Liu, J. (2012). Secondary geological hazard analysis in Beichuan after the Wenchuan earthquake and recommendations for reconstruction. *Environmental Earth Sciences, 66*, 1001–1009.
Koshiba, Y., Hayashibara, N., Suzuki, Y., & Ohtani, H. (2012). Usefulness of fire drills in each university building. *J Environ Saf, 3*, 87–95.
Levac, J., Toal-Sullivan, D., & OSullivan, T. L. (2012). Household emergency preparedness: a literature review. *Journal of Community Health, 37*, 725–733.
Li, S., Zhai, G., Zhou, S., Fan, C., Wu, Y., & Ren, C. (2017). Insight into the earthquake risk information seeking behavior of the victims: Evidence from Songyuan, China. *International Journal of Environmental Research and Public Health, 14*(3), 267.
Liang, Y., Cheng, J., Ruzek, J. I., & Liu, Z. (2019). Posttraumatic stress disorder following the 2008 Wenchuan earthquake: A 10-year systematic review among highly exposed populations in China. *Journal of Affective Disorders, 243*, 327–339.
Lindell, M. K., & Hwang, S. N. (2008). Households' perceived personal risk and responses in a multihazard environment. *Risk Analysis: An International Journal, 28*(2), 539–556.
Lindell, M. K., & Perry, R. W. (2012). The protective action decision model: Theoretical modifications and additional evidence. *Risk Analysis: An International Journal, 32*(4), 616–632.
Lindell, M. K., Prater, C. S., Wu, H. C., Huang, S. K., Johnston, D. M., Becker, J. S., & Shiroshita, H. (2016). Immediate behavioural responses to earthquakes in Christchurch, New Zealand, and Hitachi, Japan. *Disasters, 40*(1), 85–111.
Lindell, M. K., & Whitney, D. J. (2000). Correlates of household seismic hazard adjustment adoption. *Risk Analysis, 20*(1), 13–26.
Lo, A. Y., & Cheung, L. T. (2015). Seismic risk perception in the aftermath of Wenchuan earthquakes in southwestern China. *Natural Hazards, 78*, 1979–1996.
Mabuku, M. P., Senzanje, A., Mudhara, M., Jewitt, G., & Mulwafu, W. (2018). Rural households' flood preparedness and social determinants in Mwandi district of Zambia and Eastern Zambezi Region of Namibia. *International Journal of Disaster Risk Reduction, 28*, 284–297.

Martins, V. N., Nigg, J., Louis-Charles, H. M., & Kendra, J. M. (2019). Household preparedness in an imminent disaster threat scenario: The case of superstorm sandy in New York City. *International Journal of Disaster Risk Reduction, 34*, 316–325.

McNeill, I. M., Dunlop, P. D., Heath, J. B., Skinner, T. C., & Morrison, D. L. (2013). Expecting the unexpected: Predicting physiological and psychological wildfire preparedness from perceived risk, responsibility, and obstacles. *Risk Analysis, 33*(10), 1829–1843.

Miceli, R., Sotgiu, I., & Settanni, M. (2008). Disaster preparedness and perception of flood risk: A study in an alpine valley in Italy. *Journal of Environmental Psychology, 28*(2), 164–173.

Mishra, S., & Suar, D. (2007). Do lessons people learn determine disaster cognition and preparedness? *Psychology and Developing Societies, 19*(2), 143–159.

Mulilis, J. P., & Lippa, R. (1990). Behavioral change in earthquake preparedness due to negative threat appeals: A test of protection motivation theory. *Journal of Applied Social Psychology, 20*(8), 619–638.

Paul, B. K., & Bhuiyan, R. H. (2010). Urban earthquake hazard: Perceived seismic risk and preparedness in Dhaka City, Bangladesh. *Disasters, 34*(2), 337–359.

Rüstemli, A., & Karanci, A. N. (1999). Correlates of earthquake cognitions and preparedness behavior in a victimized population. *The Journal of Social Psychology, 139*(1), 91–101.

Shah, A. A., Shaw, R., Ye, J., Abid, M., Amir, S. M., Pervez, A. K., & Naz, S. (2019). Current capacities, preparedness and needs of local institutions in dealing with disaster risk reduction in Khyber Pakhtunkhwa, Pakistan. *International Journal of Disaster Risk Reduction, 34*, 165–172.

Shaw, R., Shiwaku Hirohide Kobayashi, K., & Kobayashi, M. (2004). Linking experience, education, perception and earthquake preparedness. *Disaster Prevention and Management: An International Journal, 13*(1), 39–49.

Siegrist, M., & Gutscher, H. (2006). Flooding risks: A comparison of lay people's perceptions and expert's assessments in Switzerland. *Risk Analysis, 26*(4), 971–979.

Siriwardena, M., Malalgoda, C., Thayaparan, M., Amaratunga, D., & Keraminiyage, K. (2013). Disaster resilient built environment: Role of lifelong learning and the implications for higher education. *International Journal of Strategic Property Management, 17*(2), 174–187.

Slovic, P. (1993). Perceived risk, trust, and democracy. *Risk Analysis, 13*(6), 675–682.

Slovic, P. (1999). Trust, emotion, sex, politics, and science: Surveying the risk-assessment battlefield. *Risk Analysis, 19*, 689–701.

Soffer, Y., Goldberg, A., Adini, B., Cohen, R., Ben-Ezra, M., Palgi, Y., & Bar-Dayan, Y. (2011). The relationship between demographic/educational parameters and perceptions, knowledge and earthquake mitigation in Israel. *Disasters, 35*(1), 36–44.

Spittal, M. J., McClure, J., Siegert, R. J., & Walkey, F. H. (2008). Predictors of two types of earthquake preparation: Survival activities and mitigation activities. *Environment and Behavior, 40*(6), 798–817.

Tekeli-Yeşil, S., Dedeoğlu, N., Braun-Fahrlaender, C., & Tanner, M. (2010). Factors motivating individuals to take precautionary action for an expected earthquake in Istanbul. *Risk Analysis: An International Journal, 30*(8), 1181–1195.

Wang, S., Tang, W., Qi, D., Li, J., Wang, E., Lin, Z., & Duffield, C. F. (2017). Understanding the role of built environment resilience to natural disasters: Lessons learned from the Wenchuan earthquake. *Journal of Performance of Constructed Facilities, 31*(5), 04017058.

Wei, H.-H., Sim, T., & Han, Z. (2019). Confidence in authorities, neighborhood cohesion and natural hazards preparedness in Taiwan. *International Journal of Disaster Risk Reduction, 40*, 101265.

Wu, G., Han, Z., Xu, W., & Gong, Y. (2018). Mapping individuals' earthquake preparedness in China. *Natural Hazards and Earth System Sciences, 18*(5), 1315–1325.

Xu, D., Peng, L., Liu, S., & Wang, X. (2018). Influences of risk perception and sense of place on landslide disaster preparedness in southwestern China. *International Journal of Disaster Risk Science, 9*, 167–180.

Zhang, L., Liu, X., Li, Y., Liu, Y., Liu, Z., Lin, J., Shen, J., Tang, X., Zhang, Y., & Liang, W. (2012). Emergency medical rescue efforts after a major earthquake: Lessons from the 2008 Wenchuan earthquake. *The Lancet, 379*(9818), 853–861.

Chapter 3
Trust and Stakeholders' Assistance in Households' Earthquake Preparedness Behavior

Abstract In this chapter, the rural residents affected by Wenchuan earthquake, Ya 'an earthquake and Yibin earthquake were taken as the research objects, and 674 valid questionnaires were obtained through field household surveys. A Multinominal Logit Model (MNL) was constructed to explore the influence of villagers' trust in the disaster relief ability of stakeholders and the help they can get from stakeholders on their preparedness behavior. The results show that the less trust the villagers have on the government and the community, and the more help they can get from the outside while preparing measures, the more inclined they are to take the disaster preparedness measures. Furthermore, the education level of villagers in earthquake-stricken areas has significant positive impacts on people's earthquake preparedness behavior. People who are not born in rural areas are more likely to take earthquake preparedness measures. In addition, male, young and married villagers are more likely to take earthquake preparedness measures in their daily lives. This study enriches the theory of rural disaster prevention and mitigation, and provides reference for the practice of disaster prevention and mitigation in earthquake-stricken rural areas.

Keywords Villager earthquake preparedness behaviour · MNL model · Trust · Help · Stakeholders

3.1 Introduction

China is one of the countries with the most serious earthquake disasters in the world, with many wide distributed, highly intense earthquakes, which causes serious disaster consequences (Arken Imirbaki, 2018a, 2018b). Three recent earthquakes in Sichuan Province in China have caused huge damage. Wenchuan M8.0 earthquake in 2008, Lushan M7.0 earthquake in Ya 'an in 2013 and Changning M6.0 earthquake in Yibin in 2019 have caused more than 70,000 deaths, with losses exceeding one trillion yuan, especially in rural areas. It is clear to see that in Sichuan region, earthquakes happen frequently and it is important to be prepared for residents for the next coming earthquake.

Y. Ao and H. Bahmani, *Navigating Complexity: Understanding Human Responses to Multifaceted Disasters*, https://doi.org/10.1007/978-981-99-8207-3_3

Existing research shows that scientific and reasonable disaster preparedness behavior can reduce the negative impact of earthquake disasters (Li et al., 2017b). Xu et al. systematically analyzed the influence relationship between risk perception and residents' disaster preparedness behavior (Dingde Xu et al., 2019). Zhou believed that improving residents' livelihood resilience and adjusting residents' livelihood strategies were effective means to deal with disaster risks, and discussed the relationship between residents' livelihood resilience and livelihood strategies from the perspective of residents' livelihood resilience (Zhou et al., 2021a). Ma explored the correlation between community resilience and residents' disaster preparedness by establishing a Tobit regression model (Ma et al., 2021a). However, existing studies mostly focus on the driving factors of disaster-preparedness behaviors among urban residents, while few studies consider such factors among rural residents. (Lian et al., 2021a), while the research on rural areas is less.

Furthermore, effective disaster preparedness behavior also needs the participation of other stakeholders, including the government, communities, families and individuals (Han et al., 2020). Existing research shows that perceived stakeholder characteristics, such as trust, responsibility and help affect people's judgment of disasters, and thus affect people's disaster preparedness behavior (Han et al., 2021; Wei et al., 2016a). Among the perceived characteristics of stakeholders, the public's trust in the ability of relevant government departments and social organizations to cope with disasters has received great attention in recent years (Deyoung & Peters, 2016; Wei et al., 2019a, 2019b). The help from different stakeholders that local residents can get when preparing for earthquake disaster is another emerging factor that affects their disaster preparedness behavior (Wei et al., 2019a, 2019b). However, the current research results are not always consistent on these two factors. Therefore, it is necessary to explore the impact of stakeholders on residents' disaster preparedness behavior in China.

Therefore, this study takes the villagers in the three rural earthquake-stricken areas in Sichuan Province as the case areas, and controls the socio-demographic variables of the villagers in the disaster areas to explore whether the villagers' trust in stakeholders' disaster relief ability and the assistance they can get from stakeholders affect their earthquake preparedness behavior. The contribution of this study is to systematically investigate the disaster situation and influence factors of rural residents, reveals the differences between the different individuals of disaster prevention and disaster prevention act the role of stakeholders in residents, confirmed the importance of education and the earthquake disaster prevention aid for the rural disaster prevention planning and construction provides practical guidance and advice.

The significance of this study is to supplement and improve the influencing factors of rural farmers' earthquake preparedness behavior, further enrich the theoretical framework of disaster prevention and mitigation, provide theoretical support for the formulation of disaster prevention and mitigation guidelines, and promote the development of disaster prevention and mitigation in China. At the same time, the awareness of disaster preparedness has been effectively spread to residents and farmers have been advocated to take effective disaster preparedness behavior.

3.2 Literature Review

There is a growing body of empirical research that explores the relationship between preparedness and demographic factors, including residents' age, gender, education level, earthquake experience and so on. Different ages and genders have differences in disaster preparedness behavior. With the increase of age, the probability of residents taking disaster preparedness measures will decrease (Tang & Feng, 2018; Wu et al., 2018). The study found that men had higher levels of preparedness than women (Chen et al., 2021; Wang et al., 2021). Some studies believe that education level has a positive impact on disaster preparedness (Mabuku et al., 2018; Zheng & Wu, 2020). For example, Hoffmann et al. (Hoffmann & Muttarak, 2017) found that education can increase disaster preparedness actions, and made it clear that education is the basic mechanism to promote disaster preparedness. Atreya et al. (2017)also confirmed that people with higher education levels were more active in disaster preparedness. In addition, there is a significant correlation between residents' disaster preparedness behavior and their disaster experience (Bronfman et al., 2016). People who have experienced an earthquake are more likely to take disaster preparedness measures than people who have not. Another study finds out that migrants are more prepared than locals (Green et al., 2021).

Stakeholders involved in disaster control and prevention have significant influence on residents' disaster preparedness behaviour (Kim & Jae, 2020). Stakeholders refer to the individuals and groups that have important interests in an organization's decisions or activities, or all the individuals and groups that are influenced by an organization in realizing its goals. Stakeholder theory is one of the dominant approaches for analyzing the normative obligations of those engaged in business (Hasnas, 2013), and was widely applied in the study of earthquake preparedness behaviour (Wei et al., 2016b). Stakeholders in earthquake preparedness are individuals or groups that have an important influence on residents' disaster preparedness behaviors (Deng et al., 2015; Wu et al., 2020). In most cases, the government and its relevant parts act as stakeholders, and the stakeholder characteristics include the trust in varied stakeholders, feeling of responsibility, etc. (Wu et al., 2018) In the event of an earthquake, the participation of stakeholders is very important in disaster response decision-making (Coppola, 2018), and stakeholders can support the resilience of buildings and infrastructure, the delivery of health and human services, and the restoration of transport and transportation systems (Taeby & Zhang, 2019). Existing research shows that people's trust in stakeholders' disaster relief ability (Cheng & Tsou, 2018; Deyoung & Peters, 2016) and the help they can get from stakeholders (Wei et al., 2019a, 2019b) for preparing are two factors that significantly affect their earthquake preparedness behaviors.

Stakeholders have an important influence on residents' disaster preparedness behavior, but there are different results about this aspect of the research. One study has found that the public confidence in local governments' ability to respond to disasters enhances their willingness to prepare for disasters, but has no significant impact on their actual preparedness behaviour (Basolo et al., 2009). Another evidence shows that residents' confidence in government disaster relief cannot predict the degree of disaster preparedness measures they take (Deyoung & Peters, 2016). In Chile, researchers found that trust in authorities was a strong predictor of environmental hazards risk perception (Bronfman et al., 2016). However, lessons from the Netherlands showed that trust in government reduced the public's risk perception of flooding, and in turn, discouraged individual's preparedness intention (Terpstra, 2011). Similarly, scholars also find out that Chinese residents' trust in the government's ability of disaster prevention and mitigation can reduce their judgment on the degree of disaster impact, thus reducing the disaster preparedness behavior (Han et al., 2017a, 2017b; Han et al., 2017a, 2017b); and American residents' confidence in Federal Emergency Management Agency (FEMA) is also positively correlated with their probability of taking disaster preparedness measures (Kim & Oh, 2015). Similarly, Facing different natural disasters, the research conclusions on the influence of the help and support that residents can get from stakeholders on their disaster preparedness behavior are not completely consistent. For example, the emergency information that individuals can get through social networks such as family, friends and neighbors (Rooney & White, 2017) as well as learning about other people's disaster experiences can help individuals understand the consequences of disasters and thus facilitate their disaster preparedness behaviour (Becker et al., 2017). However, some studies have found that if a person thinks that the help they can get is enough to resist threats, the probability that they take the initiative to prepare for disasters will be reduced (Mulilis et al., 2010). Therefore, more related empirical research is very necessary to provide context-specific recommendations, especially for areas like rural China that have not received much attention until recently.

3.3 Methods

3.3.1 Questionnaire Design

Based on the literature review, the questionnaire designed in this study consists of three parts, namely, social and demographic variables (control variables), villagers' trust in disaster relief ability of stakeholders and help from stakeholders variables (explanatory variables) and disaster preparedness behavior variables (dependent variables).

Control variables: Demographic variables studied by Wei et al. (2019a, 2019b) and Goltz and Bourque (2017) include gender, age, education, birthplace and other factors. In addition, the type of housing structure (Liu et al., 2007), geographical location (Sim et al., 2021) and residence year (Probst et al., 2009) also have an impact on disaster preparedness behavior. Combined with the pre-investigation, this study increased the housing type index, and finally designed eight sociodemographic variables (shown in Table 3.1):

Table 3.1 Selection and explanation of social demographic variables

	Variable	Variable declaration
Demographic characteristic	City and county	1 = Wenchuan County; 2 = Lushan County; 3 = Changning County
	Gender	1 = Man; 2 = Woman
	Age	The corresponding numerical value is the corresponding age For example: 25 = 25 years old
	Is the current place of residence the birthplace?	1 = No; 2 = Yes
	Education Level	1 = Uneducated; 2 = Primary School; 3 = Junior High School; 4 = Senior High School; 5 = University and above
Architectural features	Year of residence	The corresponding value is the corresponding year For example, 2008 = 2008
	Residential structure type	1 = Bamboo-grass structure, 2 = Stone-wood structure; 3 = Masonry structure; 4 = Brick-concrete structure; 5 = Steel–concrete structure
	Type of house	1 = Own house; 2 = Rent house; 3 = Other

Explanatory variables: In previous studies, different stakeholders are considered as residents, peers, government officials, media and other forms (Apatu et al., 2015). Bo et al. (Fan & Zhan, 2013) conducted a systematic stakeholder analysis on the government, military, non-governmental organizations, enterprises, victims and media. And Basolo's (Basolo et al., 2009) study found that the ability of local government to deal with disasters is positively correlated with the willingness of residents to prepare for disasters. Therefore, this study takes the government as a typical stakeholder. Babcicky and Seebauer (2017) believe that more forms of social capital or social network should be added, such as kinship and connection with external resources. In general, individuals can access all kinds of urgent information through social networks such as family, friends and neighbors (Rooney & White, 2017), therefore, this study also considers the role of family members, relatives, friends and neighbors as stakeholders in residents' disaster preparedness behavior. To sum up this study mainly considers the following six kinds of social stakeholders: (Imirbaki, 2018a, 2018b) Government; (Li et al., 2017a) Family members; (Xu et al., 2019) Relatives; (Zhou et al., 2021b) Friends; (Ma et al., 2021b) Neighbors; (Lian et al., 2021b) Other social relations.

Trust in disaster relief ability and available help in disaster preparedness measures are considered as two independent variables. The question is set to: How much do you trust the ability of six different groups of stakeholders to respond to disasters. For each kind of stakeholders (Imirbaki, 2018a, 2018b) Government; (Li et al., 2017a) Family members; (Xu et al., 2019) Relatives; (Zhou et al., 2021b) Friends; (Ma et al., 2021b) Neighbors; (Lian et al., 2021b) Other social relations. the responses are measured by Likert 5 sub-scale, ranging from 1 to 5, representing no confidence at all to very confident. As such, the range of measurement value is 6–30. Another question is: the degree of help that can be obtained from six different stakeholders in earthquake preparedness. Similarly, for each kind of stakeholders, Likert 5 subscale is used, with 1 representing no help and 5 representing all-round help, so the range of measurement value is 6–30.

In addition, risk perception has a certain relationship with residents' disaster preparedness behaviour (Qing et al., 2021; Xu et al., 2018a, 2018b), this study sets a question of perceived disaster preparedness degree: How do you think you are prepared for the earthquake with the help of Likert 5 scale. 1 means completely unprepared, 5 means completely ready.

Dependent variables: According to the existing research, the most common disaster preparedness behaviors are preparing emergency disaster kits (Doyle et al., 2018; Lindell & Perry, 2000), buying insurance (Xu et al., 2018a, 2018b; Xu et al., 2018a, 2018b), making escape plans (Sudo et al., 2019) and learning knowledge (Yong et al., 2020) etc. Based on the characteristics of villagers in Sichuan earthquake-stricken areas, this study considers six specific earthquake preparedness measures: (Imirbaki, 2018a, 2018b) purchasing disaster insurance; (Li et al., 2017a) preparing valuables for carrying; (Xu et al., 2019) preparing sufficient food, medicine and other storage materials; (Zhou et al., 2021b) Participate in evacuation drills; (Ma et al., 2021b) making an escape plan; (Lian et al., 2021b) learning disaster prevention knowledge. Respondents have selected the above six disaster preparedness activities according to the actual disaster preparedness measures. The research design questionnaire is included in the appendix.

3.3.2 *Sample Selection and Data Collection*

Sichuan Province is located in the hinterland of southwest China, with an area of 486,000 square kilometers. In Southwest of China, especially in Sichuan Province, There are high frequency and magnitude of earthquakes, causing great damages. At 14: 28 on May 12th, 2008, an earthquake measured 8.0 on the Richter scale occurred in Wenchuan County, Sichuan Province and it left 89,000 people dead and missing (Guo, 2009); At 21: 19 on August 8, 2017, an earthquake measured 7.0 on the Richter scale occurred in Jiuzhaigou County, Aba Prefecture, Sichuan Province, caused 20 people's lives and injured about 500 others (Liang, 2019); At 22: 55 on June 17, 2019, an earthquake measured 6.0 on the Richter scale occurred in Changning County, Yibin City, Sichuan Province, killed 13 compatriots (Jia, 2019). Therefore, this study takes the villagers in rural areas seriously affected by these three earthquakes as the research objects, and randomly selects three sample villages in each earthquake-stricken area, counting totally nine sample villages, namely: Younian Village, Yuzixi Village and Xingwenping Village in Wenchuan County; Renjia Village, Shuanghe Village and Caoping Village in Lushan County, Ya 'an City; Bijia Village, Jinyu Village and Longtou Village in Changning County, Yibin City. The geographical locations of the sample villages are shown in Fig. 3.1, the spatial distribution of respondents is shown in Fig. 3.2

The questionnaire survey was carried out by three groups, each with six people, during January 5th and January 10th, 2020. The researchers randomly selected households to carry out the questionnaire survey. In order to obtain the accurate geographical location of the respondents' households, the researchers used Ovi map software to locate the geographical location with the consent of the respondents. A total of 714 questionnaires were collected during this survey period, of which 674 were valid, with an effective rate of 94.40%. The number of valid returned questionnaires are shown in Table 3.2. The socio-demographic information of interviewees is shown in Table 3.3:

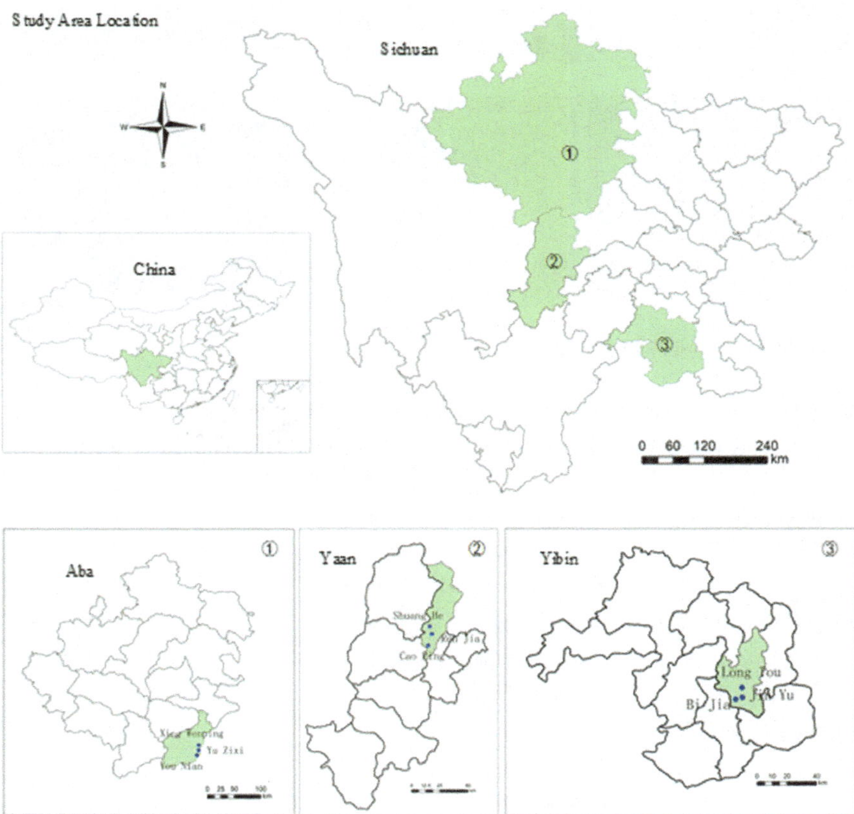

Fig. 3.1 Location of sample villages

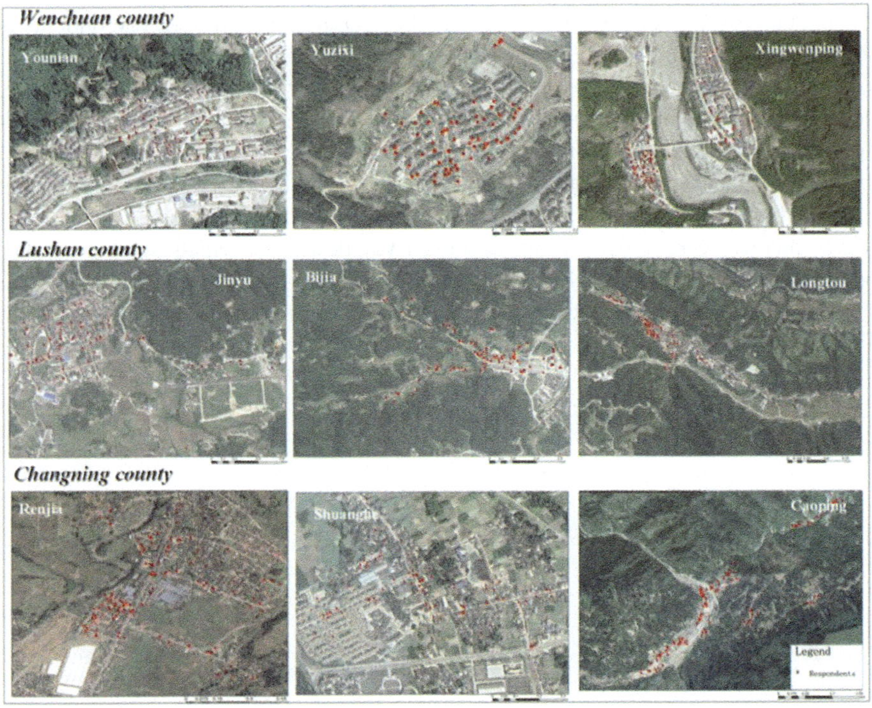

Fig. 3.2 Spatial distribution of respondents

Table 3.2 Number and proportion of questionnaires collected by villages

City and county where the sample village is located	Sample village	Number of valid returned questionnaires	Percentage%
Wenchuan County, Aba Autonomous Prefecture	Younian village	78	11.57
	Yuzixi village	71	10.53
	Xingwenping village	65	9.64
Lushan County, Ya 'an City	Renjia village	76	11.28
	Shuanghe village	83	12.31
	Caoping village	68	10.09
Changning County, Yibin City	Bijia village	78	11.57
	Jinyu village	71	10.53
	Longtou village	84	12.46
Total		674	100.00

Table 3.3 Socio-demographic information of respondents

Variable	Variable definition	Frequency	Percentage%
City and county	Yibin	210	31.2
	Ya ' an	254	37.7
	Wenchuan	210	31.2
Gender	Man	311	46.1
	Woman	363	53.9
Age	35 below	152	22.6
	36 ~ 60	380	56.4
	61 above	142	21.1
Birthplace	Not local	441	65.4
	Local	233	34.6
Academic degree	Primary school and below	340	50.4
	Junior school	223	33.1
	Senior high school	72	10.7
	University and above	39	5.8

3.3.3 Model Specification

Multinominal Logit Model (MNL) is widely used in the research of multiple choices, mainly through the calculation of utility functions to determine the item to obtain the probability of individual different choices. Different disaster preparedness behaviors as discrete variables with general models will cause deviation. Therefore, this study chooses MNL to establish the model, and explores the relationship between villagers' trust in the disaster relief ability of stakeholder, the disaster preparedness help available from stakeholder and the daily disaster preparedness behaviors of the topic, by controlling the demographic variables of respondents.

Suppose that the nth respondent chooses the effect of the i-th disaster preparedness behavior as Uni, Jn is the scheme set, then $i \in Jn$, $Uni = Vni + \varepsilon ni$, and $Vni = \beta' Xnk$. Among them, εni is the random error term; Xnk is the k-th factor which affects the nth disaster preparedness behavior; β' is the parameter to be estimated. Then the probability that the nth respondent chooses the i-th disaster preparedness behavior is:

$$\begin{aligned} P_n(i) &= Prob\left(U_{ni} \geq U_{nj}, j \in J_n, i \neq j\right) = Prob(V_{ni} + \varepsilon_{ni}, j \in J_n, i \neq j) \\ &= Prob\left[V_{ni} + \varepsilon_{ni} \geq \max_{j \in J_n}\left(V_{nj} + \varepsilon_{nj}\right)\right] \end{aligned} \tag{3.1}$$

If each random term ε_{ni} obeys independent identical distribution, then:

$$f(\varepsilon_1, \varepsilon_2, \ldots, \varepsilon_n) = \prod_n g(\varepsilon_n) \tag{3.2}$$

where $g(\varepsilon_n)$ is the distribution function corresponding to the nth respondent. Assuming that $g(\varepsilon_n)$ obeys the double exponential distribution, the probability of choosing the i-th disaster preparedness behavior in Jn is:

$$\begin{aligned} p_{in} &= \frac{\exp(V_{in})}{\sum_{j \in J_n} \exp(V_{jn})} = \frac{1}{\sum_{j \in J_n} \exp(V_{jn}) - \sum_{j \in J_n} \exp(V_{in})} \\ &= \frac{\exp(\beta / X_{nk})}{\sum_{j \in J_n} \exp(\beta' X_{nk})} \end{aligned} \tag{3.3}$$

3.4 Results and Discussion

3.4.1 Reliability and Validity Test

SPSS software was used to test the reliability and validity of the questionnaire, As shown in the Table 3.4, Cronbach's Alpha of all scales was greater than 0.7, indicating that the scales had good reliability.

As shown in the Table 3.5, through KMO and Bartlett's Test of Sphericity, the KMO value of the scale was greater than 0.7, and the P value of Bartlett's Test of Sphericity was lower than 0.05, indicating that the scale had good validity.

Table 3.4 Scale reliability test

Variables	Items	Cronbach's alpha
Trust in stakeholders	6	0.904
Help available for disaster preparedness	6	0.865
Disaster preparedness behaviour	6	0.732

Table 3.5 Scale validity test

KMO	Bartlett's test of sphericity		
	Approx. Chi-Square	df	Sig
0.886	7038.737	171	0.00

Table 3.6 Multiple collinearity and independent variable likelihood ratio test

Variable	Model fitting condition	Likelihood ratio test			Collinearity test	
	The-2 log-likelihood of the reduced model	Chi Square	Degree of freedom	P-value	Tolerance	VIF
Intercept distance	3300.279	0	0			
City and county	3328.695	28.417	12	0.005	0.858	1.166
Gender	3321.374	21.096	6	0.002	0.901	1.11
Age	3312.988	12.709	6	0.048	0.627	1.595
Birthplace	3319.238	18.96	6	0.004	0.93	1.075
Level of education	3318.778	17.499	6	0.013	0.63	1.586
Years of residence	3320.605	20.327	6	0.002	0.986	1.014
Structure type	3339.503	39.224	24	0.026	0.93	1.075
Type of house	3314.546	12.981	6	0.043	0.972	1.028
Perceptual preparation	3348.043	99.226	6	0.000	0.846	1.181
Trust in disaster relief ability	3317.278	16.999	6	0.009	0.674	1.483
Help available for disaster preparedness	3316.801	16.522	6	0.011	0.663	1.507

3.4.2 Collinearity Analysis

In this study, Variance Inflation Factor (VIF) is used to test multicollinearity. When VIF value is greater than 10, it is considered that there is strong multicollinearity among variables, which will seriously affect the model fitting (Wu & Pan, 2014). After inspection, the VIF values of the variables in this study are all less than 2, which indicates that there is no multicollinearity among the variables, and model analysis can be carried out. The results of multicollinearity detection are shown in Table 3.6

3.4.3 Model Result Fitting

MNL model was established by SPSS to fit the above data. Options for disaster preparedness are defined as: purchasing disaster insurance, preparing valuables, preparing sufficient materials, participating in disaster drills, making an escape plan, learning disaster prevention knowledge, and nothing as the reference group. The model fitting significance P is 0.008 (< 0.05) and Nagellkerke R2 is 0.235, which indicates that the model fitting effect is good. It can be seen from Table 3.5 that the likelihood ratio test P values of all variables are less than 0.05, which indicates that these variables have a significant impact on the choice of disaster preparedness behavior of villagers in earthquake-stricken areas in the selected sample villages. The fitting results of MNL model are shown in Table 3.7.

3.4.4 Socio-demographic Variables

As shown in Table 3.7. Compared with the villagers living in Yibin, the villagers in Ya 'an (− 1.040) and Wenchuan (− 0.507) were less likely to take part in disaster preparedness drills, and there was no significant difference in other measures (P > 0.05); In addition, other variables have significant influence on disaster preparedness behavior. Gender has significant difference in disaster preparedness behavior, compared with women, men are more likely to purchase disaster insurance (0.786), prepare valuables (0.827), prepare sufficient materials (0.695), participate in evacuation drills (0.775), make an escape plan (0.812) and learn disaster prevention knowledge (0.686); All the six measures: purchase disaster insurance (− 0.592), prepare valuables (− 0.517), prepare sufficient materials (− 0.646), participate in evacuation drills (− 0.545), make an escape plan (− 0.316), and learn disaster prevention knowledge (− 0.464) are negatively correlated with age; Non-native-born residents are more inclined to prepare valuables (0.822), prepare sufficient materials (0.771), participate in evacuation drills (0.808), make an escape plan (0.701), and learn disaster prevention knowledge (0.657). Among them, the probability of preparing valuables is the highest, and the probability of learning disaster prevention knowledge is the lowest; The education level of villagers in earthquake-stricken areas is significantly positively correlated with purchasing of disaster insurance (0.324), preparing valuables (0.322), preparing sufficient materials (0.298), participate in evacuation drills (0.223), making an escape plan (0.207), and learning disaster prevention knowledge (0.292); Length of residence had a positive impact on participation in disaster drills (0.031), development of escape plans (0.037) and learning disaster prevention knowledge (0.029); The seismic capability of rural buildings is negatively correlated with residents' disaster preparedness behavior, compared with steel–concrete structure, villagers living in masonry structures are more likely to purchase disaster insurance (2.039), prepare valuables (1.672), prepare sufficient materials (1.022), participate in evacuation drills (1.290), make an escape plan (1.364), and learn disaster prevention knowledge (1.182).

Table 3.7 MNL model parameter estimation (ref = reference group; β = intercept)

	Purchasing disaster insurance		Preparing valuables		Preparing sufficient materials		Participate in evacuation drills		making an escape plan		Learning disaster prevention knowledge	
	β	P-value	β	P-value	β	P-value	β	P-value	β	P-value	β	P-value
Intercept distance	– 43.616	0.125	– 7.786	0.685	– 5.492	0.576	– 62.402	0.020	– 75.370	0.002	– 58.826	0.007
City and county (Yibin = ref)												
Wenchuan	– 0.041	0.908	0.500	0.141	– 0.032	0.921	– 0.507	0.041	0.108	0.689	– 0.046	0.861
Ya ‘ an	– 0.572	0.107	– 0.326	0.374	– 0.55	0.105	– 1.04	0.000	– 0.302	0.263	– 0.176	0.489
Male (Female = ref)	0.786	0.004	0.827	0.002	0.695	0.007	0.775	0.000	0.812	0.000	0.686	0.001
Age	– 0.592	0.017	– 0.517	0.036	– 0.646	0.008	– 0.545	0.005	– 0.316	0.097	– 0.464	0.011
Not born locally (Local birth = ref)	0.450	0.109	0.822	0.002	0.771	0.003	0.808	0.000	0.701	0.001	0.657	0.002
Level of education	0.324	0.040	0.322	0.034	0.298	0.049	0.223	0.075	0.207	0.092	0.292	0.013
Years of residence	0.021	0.133	0.003	0.734	0.002	0.664	0.031	0.020	0.037	0.002	0.029	0.007
Type of house (other = ref)												
Own house	0.056	0.941	– 0.311	0.638	0.522	0.543	0.692	0.296	0.74	0.272	0.643	0.298
Rent a house	– 2.167	0.093	– 0.317	0.688	– 0.032	0.974	– 0.24	0.763	0.349	0.654	0.032	0.965
Type of structure (steel concrete structure = ref)												

(continued)

Table 3.7 (continued)

	Purchasing disaster insurance		Preparing valuables		Preparing sufficient materials		Participate in evacuation drills		making an escape plan		Learning disaster prevention knowledge	
	β	P-value	β	P-value	β	P-value	β	P-value	β	P-value	β	P-value
Bamboo and grass structure	0.693	0.592	1.37	0.207	0.674	0.603	0.788	0.423	– 0.202	0.873	0.770	0.406
Stone and wood structure	0.089	0.939	– 0.744	0.542	0.169	0.857	– 19.672		– 1.539	0.173	– 0.974	0.265
Masonry structure	2.039	0.000	1.672	0.001	1.022	0.052	1.290	0.004	1.364	0.001	1.182	0.004
Brick– concrete structure	0.315	0.301	0.395	0.178	0.284	0.318	– 0.151	0.516	0.110	0.628	0.061	0.782
Perceptual preparation	1.039	0.000	0.979	0.000	1.260	0.000	0.716	0.000	0.810	0.000	0.764	0.000
Trust in disaster relief ability	0.230	0.728	0.335	0.208	1.040	0.117	– 0.882	0.048	– 0.784	0.009	– 0.855	0.068
Help available for disaster preparedness	0.088	0.894	0.036	0.956	– 0.508	0.441	1.200	0.017	1.085	0.027	1.112	0.018

3.4.5 Explanatory Variables

As shown in Table 3.7, there was a significant positive correlation between the perceived level of earthquake preparedness and the actual probability of taking measures such as purchasing disaster insurance (1.039), preparing valuables (0.979), preparing sufficient materials (1.260), participate in evacuation drills (0.716), making an escape plan (0.810), and learning disaster prevention and reduction knowledge (0.764). That is, the self-perception of earthquake preparedness level of rural residents is consistent with the actual disaster preparedness behavior, and there is no obvious deviation in perception; There is a significant negative correlation between the degree of trust of villagers regarding stakeholders' disaster relief ability and the probability of participating in evacuation drills (− 0.822), making an escape plan (− 0.784), and it has no significant effect on other disaster preparedness behaviors ($P > 0.05$); The degree of help that villagers in earthquake-stricken areas can get from different stakeholders in disaster preparedness is significantly positively correlated with the probability of participating in evacuation drills (1.200), making an escape plan (1.085), and learning disaster prevention knowledge (1.112). The more help rural residents can get from stakeholders in earthquake preparedness, the higher the degree of disaster preparedness actions they take.

In addition, the degree of help that villagers in earthquake-stricken areas can get from different stakeholders in earthquake preparedness measures has significant positive impacts on their preparedness measures., the more help rural residents can get from six different stakeholder groups in earthquake preparedness measures, the higher the degree of disaster preparedness actions they would take. This is consistent with the previous conclusion that when faced with disasters, the degree of social support has a positive impact on the perception of residents in disaster areas, thus promoting them to take disaster preparedness actions (Han et al., 2017a, 2017b). In an emergency, people depend on each other to get help and information (Perry & Lindell, 2003), which will promote the transformation of preventive measures into concrete actions (Kim & Kang, 2010).

3.5 Discussion

In this paper, we analyze the disaster preparedness behavior of rural residents using typical disaster areas in Sichuan Province. We classify six specific disaster preparedness behaviors and explore their influencing factors.

There was no significant difference in the residents' disaster preparedness behavior in the three regions except the participating in evacuation drills. This is because Ya 'an, Wenchuan, Yibin, these three areas have experienced many earthquakes, these residents had similar earthquake experiences. Disaster experience is regarded as an important driving factor of disaster preparedness behaviour (Atreya et al., 2017) and one of the key predictors of disaster preparedness behaviour (Hoffmann & Muttarak, 2017). Therefore, we believe that similar disaster experiences can motivate residents to take similar disaster preparedness behaviors, Becker et al. (2017)found that disaster experiences, life accidents, and other people's disaster descriptions can help individuals understand the consequences of disasters and facilitate disaster-preparedness interactions in communities.

There are significant differences between individuals with different demographic characteristics. Men are more active in disaster preparedness than women. Because women have more family activities, they are less aware of crisis, so their cognition of earthquake disaster ability is obviously lower than that of men, and the corresponding possibility of disaster preparedness is also lower (Su et al., 2007; Yue & Ou, 2005). Older people are less likely to take disaster preparedness measures, and people with higher levels of education are more likely to take disaster preparedness measures, this is consistent with the conclusion of Wu et al. (2018). While it is not possible to improve disaster preparedness by changing the nature of the public, disaster preparedness education can be more targeted to women, the elderly, the less educated population and other groups. Non-native residents are more likely to take disaster a range of preparedness measures, This is consistent with the conclusion of Paul et al. (2010), Compared with native people, non-native people have a weaker sense of belonging. Yang et al. pointed out the connection between belonging and sense of security. Non-natives have a sense of crisis due to their weak sense of belonging, so they are more likely to take disaster prevention measures.

The characteristics of houses can also affect residents' disaster preparedness behavior, the younger the age of the house, the more likely the villagers of the house to prepare for disaster. Zhou An et al. (Probst et al., 2009) also found that residential year significantly affected residents' disaster preparedness level. Villagers living in masonry structures were more likely to adopt a range of disaster preparedness measures than those living in steel–concrete structures, According to the statistics of Ming et al. (2017), in the earthquake, masonry structure and civil structure of the damage rate of the damage rate is relatively high. The seismic performance of buildings is negatively correlated with villagers' disaster preparedness measures.

This study found that trust in government is negatively correlated with individual disaster preparedness behavior. By surveying survivors of the Yushu earthquake, Han et al. found that trust in government tends to decrease the respondents' perceived consequences of and reported preparedness for future potential earthquakes (Han et al., 2020). However, there are many conflicting results on the relationship between government trust and disaster preparedness, Through a survey of residents in North Carolina, Deyoung et al. (2016) found that there was no significant relationship between trust in the government and individual disaster preparedness behavior. Some scholars have also found that American residents' confidence in Federal Emergency Management Agency (FEMA) is also positively correlated with their probability of taking disaster preparedness measures (Kim & Kang, 2010). Wang and Han (Wang & Han, 2018) also found that in the United States, where individualism prevails, the more the general public trusts the government, the higher the disaster preparedness level is. Although it is positively correlated with the actual emergency preparedness level, it is not significant. He believes this is a result of cultural differences. The general public in China have a much higher degree of trust in government than the citizens in western countries (Li, 2016), about disasters and emergency management, China has no any policy made it clear that the public should share in emergency situations to protect our own responsibility, but on the contrary, the government often give a person a kind of impression, the government can save you in disaster (Han et al., 2020), as a result, Chinese residents are highly dependent on the government, which makes them less prepared for disasters. Therefore, we need to recognize that disaster reduction is not entirely a matter of the government, nor can the government cover all aspects of disaster reduction (Wang & Han, 2018). While emphasizing the leading role of the government, we must also advocate the participation of social forces.

3.6 Conclusions

To understand how residents in rural areas of China choose various earthquake preparedness behavior to reduce damages, it is desired to understand how these behaviors are related to their perception of the help they can get in preparation and their trust in the different stakeholders in reducing and preventing earthquake damages. Since there is no consistent conclusion regarding these two explanatory variables, it is necessary to carry out related research in rural earthquake-stricken areas with control variables like socio-demographic variables. Through theoretical and empirical research, this paper investigates several villages in Sichuan province that have experienced major earthquakes. MNL model is used to explore and analyze the influencing factors of rural residents' disaster preparedness behavior, especially the influence of stakeholders on residents' disaster preparedness behavior. The results of this study show that stakeholders play a very important role in residents' disaster preparedness behavior, which is manifested in that trust and available help to stakeholders will reduce residents' willingness to take disaster preparedness behavior. In addition, the willingness of residents to take disaster preparedness behavior is affected by age, gender, educational background, housing type and other factors.

The main findings of the study areas are as follows:

(1) Among the sociodemographic variables, males and young villagers are more inclined to actively take disaster preparedness measures. The villagers' education level and residence years significantly affect their disaster preparedness activities.
(2) The villagers' trust in disaster relief ability of different stakeholders has significant negative impacts on their disaster preparedness behavior, this conclusion applies to Chinese mainland, in addition, the relationship between trust and disaster preparedness may be different in different regions and cultures. On the contrary, the degree of disaster preparedness assistance that villagers can get

from stakeholders significantly affects their daily disaster preparedness behavior positively.

This study further confirmed the importance of earthquake education and disaster preparedness assistance. Due to the differences in residence, age, education level and social experience, different groups of people face different risks during the earthquake. Therefore, it is necessary to formulate targeted publicity and education measures according to the different characteristics of the publicity objects (Peng & Huang, 2020). Some recommendations are given in this context. Organizing regular earthquake education and disaster preparedness measure publicity through village speakers; Setting up early warning devices and equipment in places with large crowds such as squares or commissary; Posting comprehensive earthquake preparedness behaviors and implementation methods on community bulletin boards, and establishing community disaster preparedness work boxes to obtain specific disaster preparedness assistance that residents want, etc. In addition, because the villagers' high trust in the disaster relief ability of stakeholders will reduce the villagers' daily vigilance in disaster prevention, it is necessary to further objectively emphasize the uncertainty of earthquake disasters and the importance of self-prevention in daily publicity and education, so as to promote disaster prevention and mitigation for all people, while relying on the government, we must also advocate the participation of social forces, not only social organizations, but also enterprises and the public.

To summarize, this research sheds lights on the rural residents earthquake preparedness behavior, and provide practical guidance for rural disaster prevention planning and construction, especially from the stakeholder perspective.

While empirical results for rural areas in Sichuan have been found in this research, there are limitations. These may be taken up in subsequent research work:

(1) The scope of the study is limited. The selected villages are all from Sichuan province, and most of them are near the epicenter. Although these locations are representative to some extent, they cannot represent other earthquake-affected areas in China, and the general applicability of their conclusions needs further study. This paper has some limitations in regional distribution.
(2) This study did not deeply explore the specific impact path and intermediary effect between each influencing factor and residents' disaster preparedness behavior, it will be further studied in the future research.

Appendix

Questionnaire survey on disaster preparedness behavior of residents in earthquake-stricken areas

Hello, thank you for participating in this questionnaire survey of residents' disaster preparedness behavior. This survey will not reveal any personal information about you, but will only be used for scientific research. Please feel free to fill in. Please sign underneath to confirm your participation afterwards.

I. Social demographic characteristics:

1. Village:
2. Gender: ○ Male ○ Female
3. Age:
4. Whether the current place of residence is the place of birth: ○ No ○ Yes
5. Education level: ○ No primary school ○ Primary school ○ Junior school ○ Senior high school ○ University and above
6. Year of residence:
7. Residential structure type: ○ Bamboo and grass structure ○ Stone and wood structure
   ○ Masonry structure ○ Brick-concrete structure ○ Steel-concrete structure
8. Type of house: ○ Own house ○ Rent a house ○ Other.

II. Explanatory variable:

① Stakeholders (1. Strongly disagree 2. Disagree 3. Generally agree 4. Agree 5. Strongly agree)

See Table 3.8.

② Perceptual preparation

You think you're ready for an earthquake: **1.** Strongly disagree **2.** Disagree **3.** Generally agree **4.** Agree **5.** Strongly agree

III. Dependent variable: investigation of disaster preparedness behavior

See Table 3.9.

Table 3.8 Disaster preparedness beliefs

Questions	Option				
	Strongly disagree	Disagree	Generally agree	Agree	Strongly agree
You have great trust in the disaster relief ability of the government/district community					
You have great trust in your familys' disaster relief ability					
You have great trust in your relatives' disaster relief ability					
You have great trust in your friends' disaster relief ability					
You have a lot of faith in your neighbors' ability to help					

(continued)

Table 3.8 (continued)

You have a lot of faith in his ability to work with other communities					
Government/community help is available for earthquake preparedness					
Get help from family members for earthquake preparedness					
Get help from relatives for earthquake preparedness					
Get help from friends for earthquake preparedness					
Get help from your neighbors for earthquake preparedness					
Get help from other social relations for earthquake preparedness					

Table 3.9 Disaster preparedness behavior

Which of the following disaster preparedness behaviors have you adopted (Useful painting√)	
Buy disaster insurance	○
Prepare valuables for carrying	○
Sufficient water, food, medicine and other storage materials	○
Participate in disaster relief exercises or drills	○
Make an escape plan/route	○
Learn the knowledge of disaster prevention and mitigation	○

References

Apatu, E., Gregg, C., Lindell, M. K., Hillhouse, J., & Wang, L. (2015). Survivors perceptions of stakeholders and the 2009 South Pacific tsunami. *Disaster Prevention and Management, 24*(5), 596–609. https://doi.org/10.1108/dpm-11-2014-0230

Atreya, A., Czajkowski, J., Botzen, W., Bustamante, G., Campbell, K., Collier, B., Montgomeru, M., et al. (2017). Adoption of flood preparedness actions: A household level study in rural communities in Tabasco, Mexico. *International Journal of Disaster Risk Reduction, 24*, 428–438. https://doi.org/10.1016/j.ijdrr.2017.05.025

Babcicky, P., & Seebauer, S. (2017). The two faces of social capital in private flood mitigation: Opposing effects on risk perception, self-efficacy and coping capacity. *Journal of Risk Research, 20*(8), 1017–1037. https://doi.org/10.1080/13669877.2016.1147489

Basolo, V., Steinberg, L. J., Burby, R. J., Levine, J., Cruz, A. M., & Huang, C. (2009). The effects of confidence in government and information on perceived and actual preparedness for disasters. *Environment and Behavior, 41*(3), 338–364. https://doi.org/10.1177/0013916508317222

Becker, J. S., Paton, D., Johnston, D. M., Ronan, K. R., & Mcclure, J. (2017). The role of prior experience in informing and motivating earthquake preparedness. *International Journal of Disaster Risk Reduction, 22*, 179–193.

Bronfman, N. C., Cisternas, P. C., López-Vázquez, E., & Cifuentes, L. A. (2016). Trust and risk perception of natural hazards: Implications for risk preparedness in Chile. *Natural Hazards, 81*, 307–327.

Bronfman, C. N., Cisternas, P. C., López-Vázquez, E., & Cifuentes, L. A. (2016). Trust and risk perception of natural hazards: Implications for risk preparedness in Chile. *Natural Hazards, 81*(1), 307–327

Chen, Y., He, S., Xiao, C., & Tan, C. (2021). Evaluation and influencing factors of residents' disaster response ability. *Science of Disaster, 36*(02), 228–234.

Cheng, H. T., & Tsou, K. W. (2018). Mitigation policy acceptance model: An analysis of individual decision making process toward residential seismic strengthening. *International Journal of Environmental Research and Public Health, 15*(9), Article 1883. https://doi.org/10.3390/ijerph15091883

Coppola, D. (2018). *Strategic approach to capacity development for implementation of the Sendai framework for disaster risk reduction: A vision of risk-informed sustainable development by 2030.*

Deng, S., Liu, X., & Jiang, X. (2015). Construction of emergency case knowledge base based on stakeholder theory *Journal of Books and Information* (03), 1–8.

Deyoung, S., & Peters, M. (2016). My community, my preparedness: The role of sense of place, community, and confidence in government in disaster readiness. *International Journal of Mass Emergencies and Disasters, 34*(2), 250–282.

Doyle, E., John, M. C., Potter, S. H., Becker, J. S., Johnston, D. M., Lindell, M. K., Coomer, M. A., et al. (2018). Motivations to prepare after the 2013 Cook Strait Earthquake, N.Z. *International Journal of Disaster Risk Reduction, 31*, 637–649.

Fan, B., & Zhan, H. (2013). Collaborative research on emergency response based on stakeholder theory. *Journal of Theoretical Inquiry* (05), 150–153. https://doi.org/10.16354/j.cnki.23-1013/d.2013.05.038

Goltz, J. D., & Bourque, L. B. (2017). Earthquakes and human behavior: A sociological perspective. *International Journal of Disaster Risk Reduction, 21*, 251–265.

Green, D., Linley, M., Whitney, J., & Sano, Y. (2021). Factors affecting household disaster preparedness among foreign residents in Japan. *Social Science Japan Journal, 24*(1), 185–208. https://doi.org/10.1093/ssjj/jyaa026

Guo, X. (2009). Analysis on the characteristics and causes of the wenchuan earthquake. *Journal of Earthquake Engineering and Engineering Vibration, 29*(06), 74–87. https://doi.org/10.13197/j.eeev.2009.06.008

Han, Z., Lu, X., EI, H., & Yan, J. (2017). The effects of trust in government on earthquake survivors' risk perception and preparedness in China. *Natural Hazards: Journal of the International Society for the Prevention and Mitigation of Natural Hazards, 86*(1), 437–452.

Han, Z., Wang, H., Du, Q., & Zeng, Y. (2017). Natural hazards preparedness in Taiwan: A comparison between households with and without disabled members. *Health Security*, 575.

Han, Z., Wang, L., & Cui, K. (2020). Trust in stakeholders and social support: Risk perception and preparedness by the Wenchuan earthquake survivors. *Environmental Hazards* (5), 1–14.

Han, Z., Wang, L., & Cui, K. (2021). Trust in stakeholders and social support: Risk perception and preparedness by the Wenchuan earthquake survivors. *Environmental Hazards-Human and Policy Dimensions, 20*(2), 132–145. https://doi.org/10.1080/17477891.2020.1725410

Hasnas, J. (2013). Whither stakeholder theory? A guide for the perplexed revisited. *Journal of Business Ethics, 112*(1), 47–57. https://doi.org/10.1007/s10551-012-1231-8

Hoffmann, R., & Muttarak, R. (2017). Learn from the past, prepare for the future: Impacts of education and experience on disaster preparedness in the Philippines and Thailand. *World Development.*

Imirbaki, A. (2018). Report of the Law Enforcement Inspection Team of the Standing Committee of the National People's Congress on Inspection of the Implementation of the Law of the People's Republic of China on Protecting against and Mitigating Earthquake Disasters—delivered at the

Sixth Session of the Standing Committee of the 13th National People's Congress on October 24, 2018 *Chinas National Peoples Congress* (23), 15–19.

Imirbaki, A. (2018b). Report of the law enforcement inspection team of the standing committee of the National People's Congress on Inspection of the implementation of the law of the People's Republic of China on Protecting against and mitigating Earthquake disasters-delivered at the sixth session of the Standing Committee of the 13th National People's Congress on October 24, 2018. *Chinas National Peoples Congress, 37*(23), 15–19.

Jia, L. (2019). Concern about the 6.0 magnitude earthquake in Changning, Sichuan province. *China's Disaster Reduction* (13), 38–41.

Kim, J., & Oh, S. S. (2015). Confidence, knowledge, and compliance with emergency evacuation. *Journal of Risk Research, 18*(1), 111–126. https://doi.org/10.1080/13669877.2014.880728

Kim, K. Y., & Jae, M. M. (2020). A study of the effects of government disaster management performance and trust in government on public engagement in social disaster preparedness [재난관리 정부성과 인식과 정부신뢰가 일반시민의 사회재난 예방참여에 미치는 영향]. *The Korean Journal of Local Government Studies, 24*(1), 187–213. https://doi.org/10.20484/klog.24.1.9

Kim, Y.-C., & Kang, J. (2010). Communication, neighbourhood belonging and household hurricane preparedness. *Disasters, 34*(2), 470–488. https://doi.org/10.1111/j.1467-7717.2009.01138.x

Li, L. (2016). Reassessing trust in the Central Government: Evidence from five national surveys. *The China Quarterly* (225), 100–121.

Li, S., Zhai, G., Zhou, S., Fan, C., Wu, Y., & Ren, C. (2017a). Insight into the earthquake risk information seeking behavior of the victims: Evidence from Songyuan, China. *International Journal of Environmental Research and Public Health, 14*(3), 267.

Li, S., Zhai, G., Zhou, S., Fan, C., Wu, Y., & Ren, C. (2017b). Insight into the earthquake risk information seeking behavior of the victims: Evidence from Songyuan, China. *International Journal of Environmental Research and Public Health, 14*(3), Article 267. https://doi.org/10.3390/ijerph14030267

Lian, P., Zhuo, Z., Qi, Y., Xu, D., & Deng, X. (2021a). The impacts of training on farmers' preparedness behaviors of earthquake disaster-evidence from earthquake-prone settlements in Rural China. *Agriculture-Basel, 11*(8), Article 726. https://doi.org/10.3390/agriculture11080726

Lian, P., Zhuo, Z., Qi, Y., Xu, D., & Deng, X. (2021b). The impacts of training on farmers' preparedness behaviors of earthquake disaster—Evidence from earthquake-prone settlements in rural China. *Agriculture, 11*(8), 726.

Liang, C. (2019). *Seismogenic structure and genetic mechanism of the Jiuzhaigou Ms7.0 earthquake in Sichuan province* [master's thesis, Chengdu Univerisity of Technology].

Lindell, M. K., & Perry, R. W. (2000). Household adjustment to earthquake hazard: A review of research. *Environment and Behavior, 32*(4), 461–501.

Liu, Y., Guo, Z., Yang, Y., Dong, X., & Ye, B. (2007). Current situation of rural housing safety and disaster prevention suggestions in Southern Fujian. *Journal of Huaqiao University* (01), 63–67.

Ma, Z., Guo, S., Deng, X., & Xu, D. (2021a). Community resilience and resident's disaster preparedness: Evidence from China's earthquake-stricken areas. *Natural Hazards, 108*(1), 567–591. https://doi.org/10.1007/s11069-021-04695-9

Ma, Z., Guo, S., Deng, X., & Xu, D. (2021b). Community resilience and resident's disaster preparedness: Evidence from China's earthquake-stricken areas. *Natural Hazards, 108*, 567–591.

Mabuku, M. P., Senzanje, A., Mudhara, M., Jewitt, G., & Mulwafu, W. (2018). Rural households' flood preparedness and social determinants in Mwandi district of Zambia and Eastern Zambezi Region of Namibia. *International Journal of Disaster Risk Reduction, 28*, 284–297. https://doi.org/10.1016/j.ijdrr.2018.03.014

Ming, X., Zhou, Y., Lu, Y., & He, J. (2017). Building characteristics and seismic capability evaluation in northwest Yunnan. *Seismic Study, 40*(04), 646–654.

Mulilis, J. P., Duval, T. S., & Bovalino, K. (2010). Tornado preparedness of students, nonstudent renters, and nonstudent owners: Issues of PrE Theory1. *Journal of Applied Social Psychology, 30*(6), 1310–1329.

Paul, B. K., & Bhuiyan, R. H. (2010). Urban earthquake hazard: Perceived seismic risk and preparedness in Dhaka City, Bangladesh. *Disasters, 34*(2), 337–359. https://doi.org/10.1111/j.1467-7717.2009.01132.x

Peng, T., & Huang, W. (2020). Strengthen the measures of popularization of earthquake popular science knowledge. *China Newspaper Industry* (02), 48–49. https://doi.org/10.13854/j.cnki.cni.2020.02.022

Perry, R. W., & Lindell, M. K. (2003). Preparedness for emergency response: Guidelines for the emergency planning process. *Disasters, 27*(4), 336–350. https://doi.org/10.1111/j.0361-3666.2003.00237.x

Probst, J. C., Baek, J.-D., & Laditka, S. B. (2009). Characteristics and recruitment paths of certified nursing assistants in rural and urban nursing homes. *The Journal of Rural Health*, 267–275.

Qing, C., Guo, S., Deng, X., & Xu, D. (2021). Farmers' disaster preparedness and quality of life in earthquake-prone areas: The mediating role of risk perception. *International Journal of Disaster Risk Reduction, 59*, Article 102252. https://doi.org/10.1016/j.ijdrr.2021.102252

Rooney, C., & White, G. W. (2017). Consumer perspective: Narrative analysis of a disaster preparedness and emergency response survey from persons with mobility impairments. *Journal of Disability Policy Studies, 17*(4), 206–215.

Sim, T., Han, Z., Guo, C., Lau, J., Yu, J., & Su, G. (2021). Disaster preparedness, perceived community resilience, and place of rural villages in northwest China. *Natural Hazards, 108*(1), 907–923. https://doi.org/10.1007/s11069-021-04712-x

Su, J., Wu, G., Ju, l., & Chen, Q. (2007). A survey on the cognition of natural disaster and suggestions on disaster reduction education among College students in Beijing. *Science of Disaster* (03), 100–104.

Sudo, N., Mashiro, G., Beppu, S., & Hakamata, R. (2019). A Training program to enhance disaster preparedness of group companies in the Tokyo Metropolitan Area. *International Journal of Environmental Research and Public Health, 16*(23).

Taeby, M., & Zhang, L. (2019). Exploring stakeholder views on disaster resilience practices of residential communities in South Florida. *Natural Hazards Review, 20*(1), 04018028.04018021-04018028.04018018.

Tang, J. S., & Feng, J. Y. (2018). Residents' disaster preparedness after the Meinong Taiwan Earthquake: A test of protection motivation theory. *International Journal of Environmental Research and Public Health, 15*(7), Article 1434. https://doi.org/10.3390/ijerph15071434

Terpstra, T. (2011). Emotions, trust, and perceived risk: Affective and cognitive routes to flood preparedness behavior. *Risk Analysis, 31*.

Wang, H., & Han, Z. (2018). Why people do not prepare for earthquake emergency—A trial investigation from chengdu grassroots cadres *Risk Disaster Crisis Research* (01), 210–227.

Wang, Z. Y., Han, Z. Q., Liu, L., & Yu, S. B. (2021). Place attachment and household disaster preparedness: Examining the mediation role of self-efficacy. *International Journal of Environmental Research and Public Health, 18*(11), Article 5565. https://doi.org/10.3390/ijerph18115565

Wei, H.-H., Sim, T., & Han, Z. (2019). Confidence in authorities, neighborhood cohesion and natural hazards preparedness in Taiwan. *International Journal of Disaster Risk Reduction, 40*, Article 101265. https://doi.org/10.1016/j.ijdrr.2019.101265

Wei, H. H., Sim, T., & Han, Z. (2019b). Confidence in authorities, neighborhood cohesion and natural hazards preparedness in Taiwan. *International Journal of Disaster Risk Reduction, 40*, 101265.

Wei, J., Wang, F., & Lindell, M. K. (2016a). The evolution of stakeholders' perceptions of disaster: A model of information flow. *Journal of the Association for Information Science and Technology, 67*(2), 441–453. https://doi.org/10.1002/asi.23386

Wei, J., Wang, F., & Lindell, M. K. (2016b). *The evolution of stakeholders' perceptions of disaster: A model of information flow*. John Wiley & Sons, Inc.

Wu, G., Han, Z., Xu, W., & Gong, Y. (2018). Mapping individuals' earthquake preparedness in China. *Natural Hazards and Earth System Sciences, 18*(5), 1315–1325. https://doi.org/10.5194/nhess-18-1315-2018

Wu, H. C., Greer, A., & Murphy, H. (2020). Perceived stakeholder information credibility and hazard adjustments: A case of induced seismic activities in Oklahoma. *Natural Hazards Review, 21*(3), Article 04020017. https://doi.org/10.1061/(asce)nh.1527-6996.0000378

Wu, S., & Pan, F. (2014). *SPSS statistical analysis.* Tsinghua University Press.

Xu, D., Liu, E., Wang, X., Hong, T., & Liu, S. (2018). Rural households' livelihood capital, risk perception, and willingness to purchase earthquake disaster insurance: Evidence from Southwestern China. *International Journal of Environmental Research & Public Health, 15*(7), 1319.

Xu, D., Liu, Y., Deng, X., Qing, C., Zhuang, L., Yong, Z., & Huang, K. (2019). Earthquake disaster risk perception process model for rural households: A pilot study from southwestern China. *International Journal of Environmental Research and Public Health, 16*(22), 4512.

Xu, D., Peng, L., Liu, S., & Wang, X. (2018b). Influences of risk perception and sense of place on landslide disaster preparedness in Southwestern China. *International Journal of Disaster Risk Science, 9*(2), 167–180. https://doi.org/10.1007/s13753-018-0170-0

Yong, Z., Zhuang, L., Liu, Y., Deng, X., & Xu, D. (2020). Differences in the disaster-preparedness behaviors of the general public and professionals: Evidence from Sichuan Province, China. *International Journal of Environmental Research and Public Health, 17*(14), Article 5254. https://doi.org/10.3390/ijerph17145254

Yue, L., & Ou, G. (2005). Comparative study on residents' awareness of mountain hazards. *Science of Disaster* (03), 117–120.

Zheng, Y., & Wu, G. (2020). Analysis of factors affecting public earthquake emergency preparedness. *Technology for Earthquake Disaster Prevention, 15*(03), 591–600.

Zhou, W., Guo, S., Deng, X., & Xu, D. (2021a). Livelihood resilience and strategies of rural residents of earthquake-threatened areas in Sichuan Province, China. *Natural Hazards, 106*(1), 255–275. https://doi.org/10.1007/s11069-020-04460-4

Zhou, W., Guo, S., Deng, X., & Xu, D. (2021b). Livelihood resilience and strategies of rural residents of earthquake-threatened areas in Sichuan Province, China. *Natural Hazards, 106*, 255–275.

Chapter 4
Flood-Prone Rural China: Famers' Livelihood Capital and Strategy Choices

Abstract Following the sustainable development analysis framework proposed by the Department for International Development (DFID), this chapter collects empirical data of 360 rural households in six sample villages in the Jialing River Basin of Sichuan Province, China through a village-to-household field questionnaire and applies the Multinominal Logit Model (MNL) to explore the influence of farmer households' capital on livelihood strategy choice. Research results show that: (1) In human capital category, the education level of the household head has a significant positive impact on the livelihood strategies of farmers' families; (2) In physical capital category, farmer households with larger space have more funds to choose among flood adaptation strategies; (3) In natural capital category, house location and the sale of family property for cash have the greatest negative impact on farmers' livelihood strategies; (4) Rural households with more credit opportunities in financial capital are more willing to obtain emergency relief funds; (5) Farmers' families helped by the village for a long time will probably not choose to move to avoid floods, but are more likely to choose buying flood insurance. This study provides an empirical reference for effective short and long term prevention and mitigation strategies design and application in rural in flood-prone areas.

Keywords Farmers · Household · Flood disaster · Capital · Livelihood strategy

4.1 Introduction

According to the International Emergency Disaster Database (EM-DAT), floods rank the first among all natural disasters in terms of frequency, total population, area and direct economic losses. China is one of the countries with the highest flood frequency and flood range in the world, and the economic losses caused by floods are extremely high. According to the statistical distribution of flood in China from 2000 to 2015, Sichuan is a region that suffers from floods frequently. Mianyang, Dazhou and Nanchong in Jialing River Basin have a wide range of exposure to flood disasters, and all of them have suffered from severe flood disasters. Dazhou

Y. Ao and H. Bahmani, *Navigating Complexity: Understanding Human Responses to Multifaceted Disasters*, https://doi.org/10.1007/978-981-99-8207-3_4

suffered the "9.3" catastrophic flood in 2004, which directly caused an economic loss of 6.1 billion yuan. From July to August, 2020, Nanchong suffered a severe flood, resulting in a direct economic loss of 755 million yuan. In 2020, Mianyang suffered the biggest rainstorm and flood since the founding of the People's Republic of China. The intensity, duration and severity of the process are rare in history. The continuous torrential rain caused frequent flood hazards in Mianyang, resulting in a direct economic loss of 2.968 billion yuan. Rural areas in China face more flood disasters and farmer households are the most basic disaster-bearing units. The sudden flood disasters have shown that the lack of flood disaster knowledge and capital reserves of farmers have become important issues. Choosing appropriate livelihood strategies during the disasters and in the long-term adaptation will greatly improve resilience and reduce famers' vulnerability to future disasters.

Following the sustainable development analysis framework of the Department for International Development (DFID), this study obtained the research data of 360 farmer households in six sample villages in Jialing River Basin through the questionnaire survey in villages and households, and used the Multinominal Logit Model (MNL) to explore the influence mechanism of livelihood capital on farmer households' livelihood strategies in flood-prone areas step by step. The structure of the rest of this paper is as follows: Sect. 4.2 lists the literature results and Sect. 4.3 introduces the research area and research methods, and Sect. 4.4 analyzes the empirical results. Section 4.5 presents the conclusions and suggestions.

4.2 Literature Review

The literature review has revealed that the most critical factor to alleviate livelihood stress and improve resilience to disasters of farm households is to improve the livelihood capital of farm households (Mavhura, 2017). From the perspective of livelihood capital, human, physical and social capital all have an impact on farmers' changing livelihood strategies in the face of climate extremes (Atube et al., 2021). Livelihood strategies are not only central to sustainable livelihoods, but can also provide important guidance for addressing the problems caused by disasters (Haeffner et al., 2018).

Cai (2010) executed a quantitative study on the influencing factors of the restoration and reconstruction of farmers' livelihood system in four poverty-stricken counties in Wenchuan areas, and the results show that the imbalance in the sub-capital of the livelihood capital of farm households is extremely high, resulting in the low livelihood of farm households in natural disasters. In the southwest minority areas, Zhuang et al. (2010) find that due to the increased frequency of natural disasters, the incidence of poverty among low-income farmers increased, and most of the poor were trapped in the poverty afterwards. Cao et al. (2016) analyze the coupling model and coordination degree of ecological vulnerability and economic poverty in 14 contiguous areas with special difficulties, and confirmed that the ecological vulnerability and economic poverty in the region coexist, and attention should be paid to the protection of the ecological environment. Yang et al. (2016) analyze the financial policies implemented in various regions aiming in helping different types of farmers to get rid of poverty, and there are significant differences in scalability and sustainability of such policies and financial poverty alleviation has an important impact on the livelihood development of poor farmers.

Regarding livelihood capital influencing factors, Ma et al. (2018) use a multi-class logistic regression model to quantitatively simulate the relationship between farmers' livelihood capital and livelihood strategies in Qingpu District, Shanghai under the background of rapid urbanization in China, and find that the average annual income of farmers is the key factor affecting the choice of farmers' livelihood strategies. It is suggested to help farmers optimize their livelihood strategies and develop sustainable livelihoods. Yang (2018a, 2018b) takes the desertification land closed to protection zones as the object, and empirically analyzes the relationship between farmers' livelihood capital and livelihood strategies using the logistic regression model. The results show that farmers with much financial and social capital choose a variety of livelihood strategies. Su et al. (2009) take the livelihood capital of farmers in Ganzhou District, Zhangye City as the research object to study the relationship between farmers' livelihood capital and their livelihood strategies, and find that farmers with more natural capital tend to choose agricultural livelihood strategies, while the ones with more financial capital prefer off-farm livelihood strategies. Zhao (2017) conducts a regression analysis on the relationship between the livelihood capital and livelihood strategies before and after the relocation of rural households for poverty alleviation. The results show that the relocation change the livelihood capital of farmers, and the livelihood strategies gradually tend towards non-agricultural livelihood strategies.

Furthermore, the livelihood strategies of residents will be changed constantly according to the adjustments of policies, institutions, external environment and personal livelihood capital (Zhou et al., 2021a, 2021b). For example, when rural residents face risks or impacts from natural disasters, famine, or ecological degradation, they often change their livelihood strategies according to their own capital (Amir et al., 2020). There has been much research on livelihood strategies and their drivers in the context of disasters. Mentamo and Geda (2016) analysis showed that the education level of the head of the household, access to credit, participation in a food for work programme and the land size owned by households were the key predictors of livelihood diversification. Rahut et al. (2018) analyzed strategies for diversifying rural livelihoods and their impacts on household welfare, pointing out the importance of diversifying livelihoods to reduce poverty and increase household income. Hriday Lal Koirala et al. used PRA method to conduct field research and quantitative analysis of the livelihood capital status of farmers in The Merramzi Basin, the impact of livelihood capital on livelihood strategies and the sensitivity characteristics of different livelihood strategies to livelihood capitals (Su et al., 2016).

It can been seen that existing research is limited in the sense that they focus on the relationship between livelihood capital and general livelihood strategies (Wei et al., 2019). And they only examine several types of livelihood capital, and do not subdivide the impact of farmers' household livelihood capital indicators on the choice of livelihood strategies for farmers (Qin et al., 2021). Furthermore, the choices made to cope with short-term disaster and adapt to natural disasters after being hit by natural disasters in the long run are not separated (Zhou et al., 2021a, 2021b). Therefore, it is very necessary to study the relationship between livelihood capital and livelihood strategies in more detail with a systematic approach (Shah et al., 2000).

4.3 Methodology

4.3.1 *Questionnaire Design*

In this study, 360 rural households in six sample villages with frequent flood disasters in the Jialing River Basin of Sichuan province were interviewed to conduct empirical research using questionnaires. The main purpose is two-folds: the choice of livelihood strategies of households in flood-prone areas and the specific livelihood capital indicators of households in the sample area.

(1) Independent variable: The selection of livelihood capital indicators is based on the principles of professionalism, experience, authenticity, feasibility and representativeness, according to the five-dimension criteria of livelihood capital proposed by the Department for International Development (DFID). These variables are divided into Human capital, Natural capital, Financial capital, Social capital and Physical capital on the first level. Furthermore, based on the context of the research area and referring to the domestic and foreign literature (Adelekan & Asiyanbi, 2016; Collier & Babich, 2019; Dalu & Shackleton, 2018; Lee & Kim, 2021; Memon et al., 2020; Nakayachi, 2015; Qin et al., 2021; Sadri et al., 2018; Shah et al., 2000; Wei et al., 2019; Yang et al., 2021; Zhou, Ma, et al., 2021a, 2021b), sub-indicators of each first-level indicators are identified and categorized (see Table 4.1).

It is found that the age of the head of household, family size, illness status and education level will affect the family's choice of disaster mitigation strategies, and will also have a significant impact on vulnerability factors such as sensitivity and adaptability. The family building structure is selected as the index of material capital, because it is of great significance to maintain building safety and family development (Wei et al., 2019). Most of the damages and losses caused by flood impact occur in the housing structure, and the age of housing increases the vulnerability of flood impact (Dalu & Shackleton, 2018). Studies have also shown that families with a relatively high number of income members may have diversified income combinations (Memon et al., 2020). High income and income diversification can also improve families' ability to cope with and recover from floods, and reduce their social vulnerability to floods (Dalu & Shackleton, 2018). Credit opportunities provide uninsured families with a means to manage disaster losses, but after a serious incident, access to credit may be fragile (Collier & Babich, 2019). Flood prevention is affected by community awareness and land use area. When the elements of the family at risk are exposed to disasters, the more times they are exposed, the more vulnerable they are to the power and influence of disasters, and the more vulnerable they are (Adelekan &

Table 4.1 Independent variable index, source and likelihood ratio test

First-level indicator	Secondary indicators	Indicator meaning	Indicator source	Collinearity test	
				Tolerance	VIF
H Human capital	H1 Age of household head	0.2 = 18 years and under 0.6 = 18 to 30 years old 1 = 31 to 50 years old 0.8 = 51 to 60 years old 0.4 = 61 years and over	(Bhattacharjee & Behera, 2018)	0.997	1.003
	H2 Education level household head	0.2 = Ll literacy; 0.4 = Primary school 0.6 = Junior high school 0.8 = High school and secondary school; 1 = University and above	(Kong et al., 2021)	0.997	1.003
	H3 Family illness	1 = Yes; 0 = No	(Saqib et al., 2016)	0.997	1.003
	H4 Total family size	1 = Less than two people 0.8 = Two to four people 0.6 = Four to six people 0.4 = Six to eight people 0.2 = Eight or more people	(Bhattacharjee & Behera, 2018)	0.998	1.002
P Physical capital	P1 House area	0.25 = Less than 100 square meters; 0.5 = 100 to 150 square meters 0.75 = 150 to 200 square meters 1 = More than 200 square meters	(Bhattacharjee & Behera, 2018)	0.953	1.050

(continued)

Table 4.1 (continued)

First-level indicator	Secondary indicators	Indicator meaning	Indicator source	Collinearity test	
				Tolerance	VIF
	P2 House age	1 = Less than 10 years 0.8 = 10 to 20 years 0.6 = 20 to 30 years; 0.4 = 30 to 40 years; 0.2 = More than 40 years	(Adelekan & Asiyanbi, 2016)	0.992	1.008
	P3 House structure	1 = Reinforced concrete; 0.8 = Brick concrete; 0.6 = Cob house 0.4 = wooden house 0.2 = thatched cottage	(Zhao, 2011)	0.972	1.029
	P4 Household livestock value	0.2 = Blow 1 thousand yuan 0.4 = 1 to 2 thousand yuan 0.6 = 2 to 3 thousand yuan 0.8 = 3 to 4 thousand yuan 1 = 4 thousand yuan and above	(Hao et al., 2014)]	0.978	1.022
	P5 Value of household items	0.2 = Blow 10 thousand yuan 0.4 = 10 to 50 thousand yuan 0.6 = 50–100 thousand yuan 0.8 = 100–150 thousand yuan 1 = 15 thousand yuan and above	(Hao et al., 2014)	0.960	1.042

(continued)

Table 4.1 (continued)

First-level indicator	Secondary indicators	Indicator meaning	Indicator source	Collinearity test	
				Tolerance	VIF
N Natural capital	N1 Own land area	0.2 = 0 to 1 mu; 0.4 = 1 to 2 mu 0.6 = 2 to 3 mu; 0.8 = 3 to 4 mu 1 = 4 mu and above	(Adelekan & Asiyanbi, 2016)	0.974	1.027
	N2 Family location (the distance between the house and the river)	0.2 = Below 0.5 km 0.4 = 0.5 to 1 km 0.6 = 1 to 1.5 km 0.8 = 1.5 to 2 km 1 = More than 2 km	(Wang et al., 2012)	1.000	1.000
	N3 Drain condition	1 = Yes; 0 = No	(Danso & Addo, 2016)	0.974	1.027
F Financial capital	F1 Number of households with income	0.2 = One person; 0.4 = Two people 0.6 = Three people;0.8 = Four people 1 = Five or more people	(Atreya et al., 2017)	0.923	1.083
	F2 Average annual household income	0.2 = Blow 10 thousand yuan 0.4 = 10 to 20 thousand yuan 0.6 = 20 to 30 thousand yuan 0.8 = 30 to 40 thousand yuan 1 = 40 thousand yuan and more	(Yuan, 2019)	0.772	1.295
	F3 Credit opportunity	1 = Yes; 0 = No	(Zhao, 2011)	0.926	1.080

(continued)

Table 4.1 (continued)

First-level indicator	Secondary indicators	Indicator meaning	Indicator source	Collinearity test	
				Tolerance	VIF
	F4 Borrowing opportunity	1 = Yes; 0 = No	(Zhao, 2011)	0.930	1.075
S Social capital	S1 Community help during disasters	1 = Yes; 0 = No	(Bhattacharjee & Behera, 2018)	0.462	2.163
	S2 Helped by neighbors during disasters	1 = Yes; 0 = No	(Ma, 2017)	0.622	1.608
	S3 Trust in village managers	1 = Trust all; 0.75 = Mostly trust 0.5 = Half trust; 0.25 = Few trust 0 = Hardly trust	(Zhao, 2011)	0.490	2.039
	S4 Occupation in the village group	1 = Yes; 0 = No	(Wu et al., 2019)	0.669	1.495

Asiyanbi, 2016). Therefore, the distance between houses and rivers is taken as the index of natural capital. In the event of a disaster, trusting neighbors and communities has a significant impact on reducing the experience of disaster conflicts (Adelekan & Asiyanbi, 2016). Those families who have received recovery assistance from neighbors, stronger personal networks and higher levels of social capital recover faster (Sadri et al., 2018). Similarly, in terms of social capital, the public's trust in the management organization after being severely damaged by major disasters will also affect the public's preparedness for disasters (Nakayachi, 2015).

When there are many independent variables in the practical problems studied, there may be a certain degree of correlation between two or more independent variables, which is called multicollinearity. If the collinearity trend of independent variables is obvious, the fitting effect of the model will be seriously affected (Yang et al., 2021, 2022, 2020a, 2020b). In this study, the Variance inflation factor (VIF) is used to test multicollinearity. When the VIF value is greater than 10, it is considered that the variables have strong multicollinearity, which is unacceptable (Yang & Yang, 2021). The VIF values of the independent variables in this research are all less than 3, indicating that there is no multicollinearity among the independent variables. The test results are shown in Table 4.1.

Dependent variable: The farmer's family can choose short-term and long-term livelihood strategies. During the flood, the short-term livelihood strategies of farmers' families include: relying on government and social organizations' relief funds, borrowing money from banks or loan companies, selling livestock for cash, using past savings, borrowing money from relatives and friends, and taking temporary jobs in nearby areas that are not affected by the disaster to earn money. According to the existing literature and the characteristics of rural residents in Sichuan, this study sums up six farmers' families' choice of livelihood strategies for flood response in short-term regarding their livelihood capitals (Ao et al., 2021). The choices of long-term livelihood adaptation strategies of farmers' families in flood-prone areas include: increasing the height of houses, improving drainage ditches, participating in flood emergency training, purchasing flood insurance, changing the type and date of crop planting, choosing to move, and increasing agricultural irrigation measures. In this study, seven farmers' families' livelihood capital choices of livelihood strategies for flood adaptation are selected.

In short-term flood coping strategies, after the disaster, first of all, farmers' families will have no financial income, and some families will use their accumulated savings to ensure their daily expenses and recover the losses caused by the flood disaster (Bhattacharjee & Behera, 2018). The government and social welfare organizations will often provide materials and financial assistance to the disaster areas (Liang & Cao, 2015), so as to help farmers get through the flood difficulties as soon as possible. Farmers can also rely on this source of funds to relieve the pressure of temporary lack of financial income due to the flood. Floods have a great impact on farmers' families, and will cause great harm to their houses and crops. When the past savings can't support the family's recovery, the head of household who is short of funds will choose to borrow some funds from relatives or friends whose families are relatively rich or haven't been hit by floods to quickly recover the losses caused by floods (Bhattacharjee & Behera, 2018), or immediately borrow money from banks and loan companies as a response measure to tide over the floods (Mondal et al., 2020). Severe flood impacts often force farmers to sell livestock to maintain their basic consumption. Even wealthy farmers have to reduce livestock to cope with floods (Helgeson et al., 2013). When floods occur, farmers sell livestock to reduce their feed expenses, so as to obtain cash. And it is an important risk treatment mechanism to earn money by taking temporary jobs in nearby areas that are not affected by disasters (Mondal et al., 2020). Floods reduce farmers' agricultural labor, and the labor force is idle. Choosing to stay nearby without being affected by floods can make up for the loss of crop income caused by floods, thus protecting family income.

In the long-term flood adaptation strategy, because of the perennial flooding, families will raise the foundation of newly repaired houses appropriately according to the situation of previous floods (Shah et al., 2000), so as to avoid flooding the houses where families live again, and take precautions, or family heads will choose to build or repair the drainage measures near the houses (Abebe et al., 2019), so as to ensure that the houses can escape the disaster when the floods come. The impact of flood on the crops of farmers' families is enormous, which makes the families whose income is agriculture lose their economic resources. Families will choose to increase and improve agricultural irrigation measures to alleviate the impact of flood on crops and minimize the possible impact of flood (Schad et al., 2012). Based on the analysis of the occurrence time of perennial floods and the prediction results of floods, the farmers in flood-prone areas will change the types of crops into waterlogging-resistant crops to keep their crops able to withstand the impact of floods, or choose to advance or postpone the planting date of crops to avoid the invasion of floods, and ensure the family's economic benefits by changing crops. Because of the perennial nature and great destructiveness of floods, families suffering from disasters choose to buy some natural disaster accident insurance and rely on insurance premiums to make up for the losses caused by floods (Bao et al., 2021). This is a livelihood strategy choice of scientific disaster prevention. The destructive and destructive nature of floods has a great impact on farmers. Farmers can improve their awareness of disaster prevention and self-protection skills by participating in flood training (Xu et al., 2018). Choosing this long-term livelihood prevention strategy can reduce losses. If suffering from long-term flooding, some families simply abandon their original homes and rebuild their homes on high ground that will not be invaded by flooding (King et al., 2014), so as to fundamentally solve the possible flooding problem, but the cost of such a choice is enormous.

4.3.2 Model Specification

Multinominal Logit Model (MNL) is often used in the study of the behavior choice (Ao et al., 2020; Guo et al., 2020; Wang & Ross, 2018; Yang, 2018a, 2018b). This model can calculate the probability of farmers' families choosing different livelihood strategies through a utility function. This study takes individual rural residents as the object and explores the influence of livelihood capital on rural residents' choice of different livelihood strategies by establishing an MNL model.

Suppose that the nth respondent chooses the effect of the i-th disaster preparedness behavior as Uni, Jn is the scheme set, then i ∈ Jn, Uni = Vni + εni, and Vni = β'Xnk. Among them, εni is the random error term; Xnk is the K factor which affects the nth disaster preparedness behavior; β' is the parameter to be estimated. Then the probability that the nth respondent chooses the i-th disaster preparedness behavior is:

$$
\begin{aligned}
P_n(i) &= \mathrm{Prob}\left(U_{ni} \geq U_{nj}, j \in J_n, i \neq j\right) = \mathrm{Prob}(V_{ni} + \varepsilon_{ni}, j \in J_n, i \neq j) \\
&= \mathrm{Prob}\left[V_{ni} + \varepsilon_{ni} \geq \max_{j \in J_n}\left(V_{nj} + \varepsilon_{nj}\right)\right]
\end{aligned}
\tag{4.1}
$$

If each random term ε_{ni} obeys independent identical distribution, then $f(\varepsilon_1, \varepsilon_2, \ldots, \varepsilon_n) = \prod_n g(\varepsilon_n)$ Where $g(\varepsilon_n)$ is the distribution function corresponding to the nth respondent. Assuming that $g(\varepsilon_n)$ obeys the double exponential distribution, the probability of choosing the i-th disaster preparedness behavior in Jn is:

$$
p_{in} = \frac{\exp(V_{in})}{\Sigma_{j \in J_n} \exp\left(V_{jn}\right)} = \frac{1}{\Sigma_{j \in J_n} \exp\left(V_{jn}\right) - \Sigma_{j \in J_n} \exp(V_{in})} = \frac{\exp(\beta / X_{nk})}{\Sigma_{j \in J_n} \exp(\beta / X_{nk})}
\tag{4.2}
$$

4.3.3 Sample Selection and Data Collection

Based on the daily precipitation records of Sichuan Province from 1961 to 2017, it is known that Sichuan Province has experienced significant climate change (Chen et al., 2010; Li et al., 2019). Mianyang, Dazhou and Nanchong in eastern Sichuan belong to high-risk areas, high-sensitivity areas and high-vulnerability areas to storms and floods, and suffer from floods all year round (Du & Dong, 2016). Based on the research purpose and the relevant literature, it is found that the research sample points of flood disaster are basically communities along rivers or low-lying areas (Roy et al., 2020). Therefore, when selecting sample villages, the villages with frequent floods along rivers are selected as sample points. Six sample villages are determined, namely Pengjiaxiang Village and Fucheng Village in Fujiang River Basin of Mianyang City, Baoshamiao Village and Diankouzhai Village in Jialing River Basin of Nanchong City, Shizi Village and Xikou Village in Qujiang River Basin of Dazhou City. The geographical distribution of the sample villages is shown in Fig. 4.1.

Questionnaire data collection of influencing factors of household livelihood strategy choice in flood-prone areas is mainly divided into two stages: preparation and formal implementation. In the research preparation stage, the team recruited students from rural areas for the Chengdu University of Technology as the researchers and explained the research purpose, research methods, research contents, itinerary planning and time arrangement face to face. After all the preparatory work is completed, the research team is divided into three groups, taking the villagers' activity center as

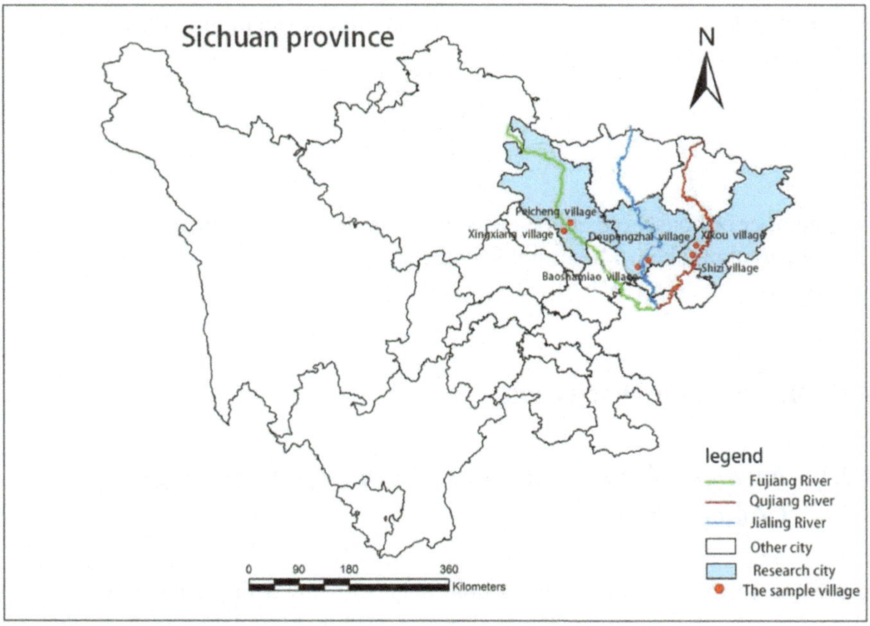

Fig. 4.1 Geographical distribution of sample villages

the starting point, to visit the villagers in the sample village and random villagers are selected to conduct household survey, so as to ensure that the interviewees are evenly distributed in the village and do not repeat. Under the condition of certain error and confidence, the sample size will change with the overall size. The larger the total, the less obvious the change, while the smaller the total, the obvious change, but the change between them is not linear. Therefore, the larger the sample size, the better. The commonly used formula for calculating the minimum sample size is shown in (4.3), so the minimum sample size should be 269 when the confidence interval is 90% and the sampling error is 0.05.

$$\mathrm{n} = \frac{\mathrm{z}^2\sigma^2}{d^2} \tag{4.3}$$

n: sample size
z: standard score of confidence interval; the value of z of 90% confidence interval is 1.64
σ: population standard deviation, and generally 0.5
d: sample error

The number of households in each village sampling survey is determined according to Formula (4.4) (Yamane, 1965), to determine the minimum number of sampling copies. The sampling number of sample villages is shown in Table 4.2. A total of 360 questionnaires were distributed this time, and 325 questionnaires were recovered, with an effective recovery rate of 90.28%. It shows that the questionnaires collected in this survey meet the minimum sampling requirements, and further research can be carried out.

$$n = \frac{N}{1 + Ne^2} \tag{4.4}$$

n: number of households to be investigated in the sample village
N: the total number of households in the sample village
e: accuracy is set to 15% (0.15)

4.4 Results and Discussion

4.4.1 *Farmers' Household Short-Term Coping Strategy Choice Analysis*

According to the data of the household survey in the village, the proportion of farmer households in sample villages choosing livelihood strategies for flood disaster response is shown in Fig. 4.2. Flood response strategies have the most options to use

Table 4.2 Sample number of sample villages

Sample villages		Total households	Number of sample households	Number of valid survey	Sampling rate (%)
Fujiang River Basin of Mianyang City	Pengjiaxiang Village	726	42	42	5.79
	Fucheng Village	1155	46	43	4.36
Jialing River Basin of Nanchong City	Cloak Stronghold	854	58	42	6.79
	Baosha Temple	483	59	41	12.21
Qujiang River Basin of Dazhou City	Xikou Village	760	60	43	7.89
	Shizi Village	1080	60	42	5.55

past savings. Pengjiaxiang village and Fucheng village in the Fujiang River Basin of Mianyang account for 76.19% and 76.60% respectively. In the Jialing River Basin of Nanchong, 81.36% of Baosha Temple village and 98.25% of Cloak Stronghold village. Shizi village and Xikou village in Dazhou Qujiang River Basin account for 91.53% and 93.33% respectively. It can be seen from this that most farmers have a certain amount of capital reserves to cope with the emergency needs of floods, however, in the short-term capital strategy to deal with flood disasters, the probability of choosing to borrow from banks or loan companies is low, which is due to the shortage of financial capital stock of farmers in the sample area.

The proportions of farmer households choosing flood disaster response strategies is shown in Fig. 4.3. 8.75% of them chose 0, that is, only a few families did not take corresponding financial strategies to deal with the flood risk when the flood came, the proportions of adopting 1, 2, 3, 4, 5 and 6 coping strategies were 21.60%, 26.54%, 21.30%, 16.05%, 7.1% and 1.00% respectively. Farmers choose 1–4 coping strategies when dealing with flood risk, accounting for 83.15% of farmers in the Mianyang Fujiang area, 86.21% in the Nanchong Jialing river area and 86.55% in the Dazhou Qujiang river area. It shows that most farmers in the study area will choose a variety of strategies to deal with the short-term financial difficulties when the flood comes, and families also have corresponding living capital storage to provide conditions for implementing the corresponding strategies.

4.4.2 Farmers' Household Long-Term Adaption Strategy Choice Analysis

The proportion of farmer households choosing flood disaster adaptation livelihood strategy is shown in Fig. 4.4, and the most selected strategy to adapt to floods is to

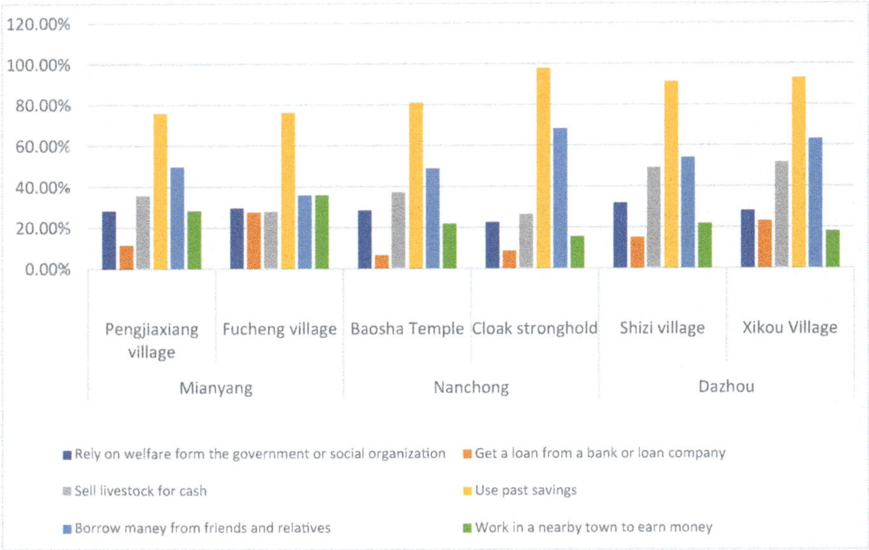

Fig. 4.2 Proportion of farmer households in sample villages choosing livelihood strategies for flood disaster response

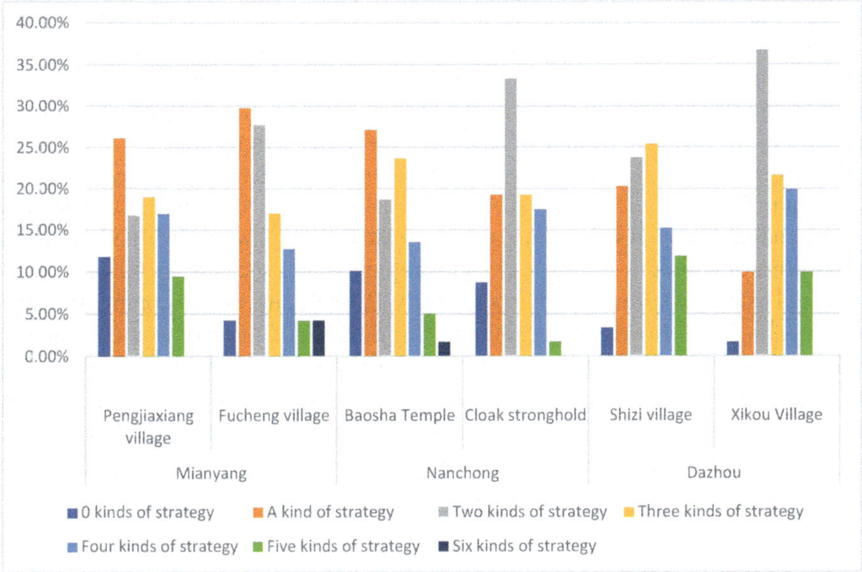

Fig. 4.3 Proportion of types of flood disaster response strategies selected by farmer households in sample villages

participate in flood emergency training. Pengjiaxiang village and Fucheng village in the Fujiang River Basin of Mianyang account for 66.67% and 65.96% respectively. In the Jialing River Basin of Nanchong, 67.80% of Baosha Temple village and 56.14% of Cloak Stronghold village. Shizi village and Xikou village in Dazhou Qujiang River Basin account for 59.32% and 30% respectively. It can be seen that most farmer families in Jialing River Basin have been invaded by floods for a long time, and families will choose corresponding community flood emergency training to learn more emergency knowledge of flood escape or property protection, so as to reduce the losses caused by disasters. However, among all the strategies, the probability of choosing changing crop types and planting dates as the strategy to adapt to floods is low. This is because farmers' habitual planting of crops and planting time has been deeply rooted in the sample area. Although farmers' habits from ancient times to today face the challenge of flood, most families will still solve the farming problems caused eventually (Roy et al., 2020).

The survey results reveal that farmers who choose no strategies to adapt to floods account for less than 1%. Only one family in Fucheng village in Mianyang and one family in Cloak Stronghold village in Nanchong did not adopt any flood adaptation strategy. It can be seen that only a very small number of farmers' livelihood capital stock is at a low level, and the accumulated livelihood capital is small. The proportions of adopting 1, 2, 3, 4, 5, 6 and 7 flood adaptation strategies are 9%, 22.22%, 28.40%, 19.75%, 12.96%, 5.25% and 1.85% respectively. The proportion of farmers who choose 2–5 strategies is relatively high. The proportion of farmers in the Fujiang River area of Mianyang is 78.65%, that in the Jialing River area of Nanchong is 84.48%, and that in Qujiang River of Dazhou is 85.71%. Most farmers in the study

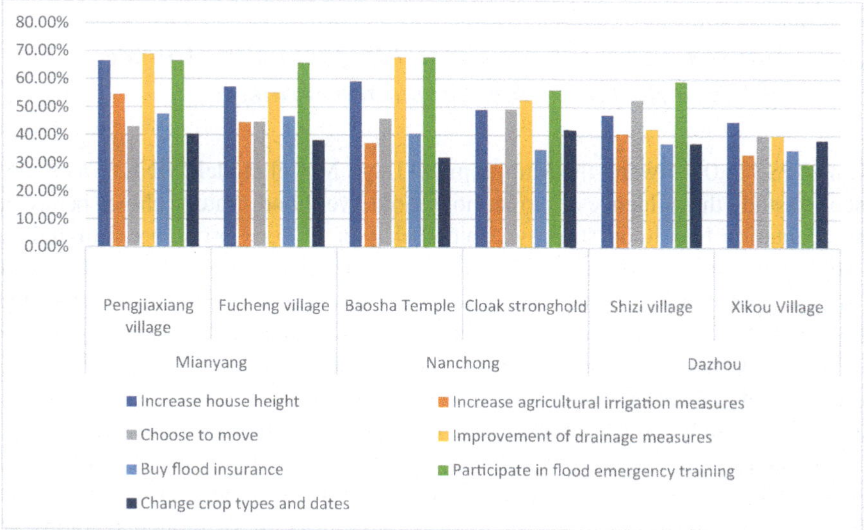

Fig. 4.4 Proportion of households in sample villages choosing flood disaster adaptation strategy

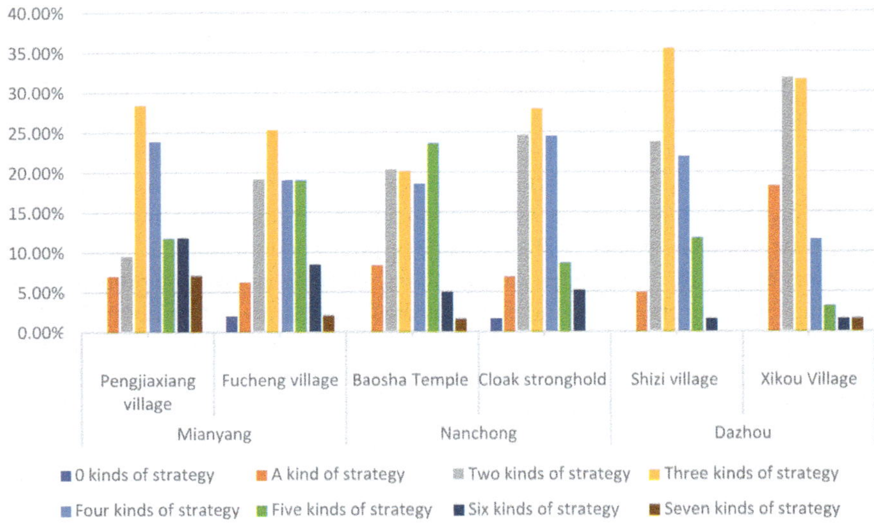

Fig. 4.5 Proportion of types of flood disaster adaptation strategies selected by farmer households in sample villages

area will choose a variety of livelihood strategies to adapt to the perennial floods, and families also have corresponding livelihood capital storage to provide conditions for implementing adaptation strategies. The number of livelihood strategies types selected by farmers is shown in Fig. 4.5.

4.4.3 Household Livelihood Capital Impact on the Choice of Flood Response and Adaptation Strategies

Using SPSS23.0 software, the Multinominal Logit Model model (MNL) was established to study the influence of farmer household livelihood strategy choice behavior and adaptation livelihood strategy choice behavior in flood-prone areas of Jialing River Basin. The proportion of coping flood without any livelihood strategy behavior is 8.75%, and this part of the questionnaire is screened out. The coping with the livelihood strategy behavior analysis uses the past savings as the reference group, and model fitting significance p is 0.000(< 0.050) and Nagellkerke R^2 is 0.378. The proportion of adaptation strategy model without any strategy is less than 1%, so this part of the questionnaire is screened out. The selection of perfect drainage measures is set as the reference group in adaptation strategy, and the model fitting significance p is 0.000(< 0.050) and Nagellkerke R^2 is 0.443. It can be judged that the fitting effect of the model is good, and the fitting results of the MNL model are shown in Table 4.3.

Table 4.3 MNL model parameter estimation

MNL Model Fitting Results of Farmers' Household Livelihood Capital on the Choice of Livelihood Strategies in Response to Floods										
Index	Rely on Welfare from the Government or Social Organization		Get a Loan from a Bank or Loan Company		Sell Livestock for Cash		Borrow Money from Friends and Relatives		Work in a Nearby Town to Earn Money	
	B	p-Value	B	p-Value	B	p-Value	B	p-Value	B	p-Value
Intercept	– 0.369	0.716	– 1.636	0.167	– 1.45	0.116	– 1.121	0.183	– 1.631	0.171
Human capital										
Age of head of household	– 0.325	0.004	– 1.081	0.041	0.544	0.005	0.162	0.71	– 1.775	0.007
Education level of head of household	0.199	0.824	2.15	0.034	– 0.407	0.616	0.217	0.768	1.952	0.044
Family illness	1.882	0.039	– 0.534	0.022	0.126	0.562	0.147	0.456	– 0.395	0.15
Total family size	1.842	0.037	– 1.325	0.035	0.337	0.537	– 0.018	0.972	– 0.651	0.35
Natural capital										
Own land area	– 0.502	0.393	0.24	0.718	– 0.029	0.953	– 0.06	0.893	– 0.384	0.549
Family location (the distance between the house and the river)	– 0.601	0.014	– 0.769	0.199	– 2.199	0	– 0.007	0.986	– 1.877	0.002
Drain condition	0.817	0.001	0.405	0.03	0.103	0.005	– 0.238	0.256	– 0.336	0.267

(continued)

Table 4.3 (continued)

MNL Model Fitting Results of Farmers' Household Livelihood Capital on the Choice of Livelihood Strategies in Response to Floods

Index	Rely on Welfare from the Government or Social Organization		Get a Loan from a Bank or Loan Company		Sell Livestock for Cash		Borrow Money from Friends and Relatives		Work in a Nearby Town to Earn Money	
	B	p-Value	B	p-Value	B	p-Value	B	p-Value	B	p-Value
Physical capital										
House area	– 0.246	0.761	– 1.022	0.267	– 2.133	0.023	– 0.509	0.423	– 0.741	0.406
House age	– 0.015	0.98	– 0.54	0.456	1.63	0.017	– 0.187	0.707	0.932	0.169
House structure	– 0.246	0.761	– 1.022	0.267	1.03	0	– 0.509	0.423	– 0.741	0.406
Household livestock value	– 0.03	0.935	– 0.597	0.184	0.304	0.337	0.179	0.544	– 0.245	0.568
Value of household items	0.05	0.959	– 0.009	0.993	0.621	0.473	– 0.113	0.891	– 1.519	0.189
Financial capital										
Average annual household income	– 0.814	0.168	– 0.423	0.514	– 0.476	0.355	– 0.211	0.645	0.278	0.65
Average annual household income	0.583	0.353	0.356	0.622	0.025	0.963	0.229	0.64	– 0.224	0.755
Credit opportunity	0.031	0.928	1.147	0.001	0.072	0.81	0.367	0.162	0.726	0.037
Borrowing opportunity	0.208	0.437	0.033	0.913	0.262	0.272	0.671	0.003	0.275	0.357

(continued)

Table 4.3 (continued)

MNL Model Fitting Results of Farmers' Household Livelihood Capital on the Choice of Livelihood Strategies in Response to Floods

Index	Rely on Welfare from the Government or Social Organization		Get a Loan from a Bank or Loan Company		Sell Livestock for Cash		Borrow Money from Friends and Relatives		Work in a Nearby Town to Earn Money	
	B	p-Value	B	p-Value	B	p-Value	B	p-Value	B	p-Value
Social capital										
Community help during disasters	0.458	0.016	0.308	0.134	0.712	0	0.519	0.003	0.805	0
Helped by neighbors during disasters	– 0.079	0.788	0.552	0.141	– 0.114	0.668	– 0.298	0.219	– 0.064	0.843
Trust in village managers	– 0.157	0.591	0.225	0.518	0.143	0.599	0.297	0.241	0.272	0.427
Occupation in the village group	0.817	0	0.405	0.018	0.229	0.247	– 0.146	0.455	– 0.012	0.958

MNL Model Fitting Results of Farmers' Household Livelihood Capital on the Choice of Livelihood Strategies in Flood Disaster Adaptation

Index	Increase House Height		Choose to Move		Increase Agricultural Irrigation Measures		Change Crop Types and Dates		Buy Flood Insurance		Participate in Flood Emergency Training	
	B	p-Value	B	p-Value	B	p-Value	B	p-Value	B	p-Value	B	p-Value
Intercept	0.109	0.001	– 0.138	0	0.143	0.001	– 0.066	0.003	0.149	0.002	0.054	0
Human capital												
Age of head of household	– 0.055	0.839	0.057	0.847	– 0.159	0.545	– 0.243	0.4	– 0.111	0.636	– 0.117	0.633

(continued)

Table 4.3 (continued)

MNL Model Fitting Results of Farmers' Household Livelihood Capital on the Choice of Livelihood Strategies in Flood Disaster Adaptation												
Index	Increase House Height		Choose to Move		Increase Agricultural Irrigation Measures		Change Crop Types and Dates		Buy Flood Insurance		Participate in Flood Emergency Training	
	B	p-Value	B	p-Value	B	p-Value	B	p-Value	B	p-Value	B	p-Value
Education level of head of household	0.036	0.001	0.142	0.205	0.162	0.002	0.107	0.013	0.118	0	0.11	0.021
Family illness	–0.056	0.813	–0.272	0.015	–0.221	0.298	–0.321	0.207	0.258	0.017	–0.241	0.235
Total family size	–0.019	0.924	–0.226	0.304	–0.199	0.28	–0.011	0.957	–0.117	0.485	–0.134	0.44
Natural capital												
Own land area	–0.193	0.22	–0.15	0.01	0.332	0.005	0.176	0.002	–0.063	0.619	–0.091	0.499
Family location (the distance between the house and the river)	–0.454	0.001	–0.634	0.004	0.186	0.687	0.311	0.564	–0.184	0.013	0.597	0.171
Drain condition	–0.144	0.664	–0.18	0.024	0.087	0	–0.011	0.977	–0.055	0.852	–0.019	0.949
Physical capital												
House area	0.327	0.326	–0.198	0.003	0.579	0.047	1.023	0.004	0.38	0.034	0.562	0.03
House age	0.08	0.016	–0.043	0.62	0.071	0.292	0.038	0.622	0.078	0.206	0.045	0.481
House structure	–0.058	0.041	0.281	0.11	0.052	0.73	0.188	0.264	0.145	0.297	0.148	0.299
Household livestock value	–0.034	0.868	–0.056	0.801	0.06	0.05	0.195	0.014	–0.003	0.988	–0.09	0.62

(continued)

Table 4.3 (continued)

MNL Model Fitting Results of Farmers' Household Livelihood Capital on the Choice of Livelihood Strategies in Flood Disaster Adaptation												
Index	Increase House Height		Choose to Move		Increase Agricultural Irrigation Measures		Change Crop Types and Dates		Buy Flood Insurance		Participate in Flood Emergency Training	
	B	p-Value	B	p-Value	B	p-Value	B	p-Value	B	p-Value	B	p-Value
Value of household items	– 0.162	0.34	– 0.22	0.228	– 0.12	0.457	0.133	0.447	0.037	0.015	– 0.058	0.7
Financial capita												
Average annual household income	0.761	0.031	0.208	0.279	0.056	0.012	0.233	0.081	0.655	0.005	0.131	0.141
Average annual household income	– 0.204	0.554	0.685	0.052	– 0.124	0.706	0.172	0.625	0.52	0.003	– 0.37	0.016
Credit opportunity	– 0.279	0.577	1.249	0.018	– 0.015	0.974	– 0.169	0.76	0.355	0.014	– 0.02	0.962
Borrowing opportunity	0.197	0.649	– 0.405	0.415	– 0.048	0.908	0.109	0.818	– 0.101	0.788	0.555	0.163
Social capital												
Community help during disasters	– 0.251	0.67	– 0.547	0.032	– 0.049	0.929	1.105	0.057	– 0.006	0.991	0.062	0.001
Helped by neighbors during disasters	– 0.095	0.831	– 0.448	0.014	0.653	0.131	– 0.496	0.303	– 0.265	0.486	– 0.524	0.186
Trust in village managers	0.111	0.843	– 0.085	0	– 0.059	0.919	– 0.053	0.93	0.24	0.016	– 0.117	0.827
Occupation in the village group	0.222	0.018	– 0.209	0.746	0.557	0.292	0.157	0.81	0.257	0.018	0.505	0.032

According to the fitting results of the MNL model, the choice behavior of farmer households' livelihood strategy in flood-prone areas is affected by livelihood capital, and households with sufficient livelihood capital reserves have a stronger ability to cope with and adapt to flood risks.

In human capital, compared with the coping strategy of families using their past savings to solve the lack of livelihood during the flood, the choice of family heads to work and earn money in nearby areas without disaster ($p = 0.007$, $B = -1.755$), relying on government or social organizations' relief funds ($p = 0.004$, $B = -0.325$) and loans from banks or loan companies ($p = 0.041$, $B = -1.081$) is significantly negatively correlated with age, while the choice of changing sellers to produce livestock for cash ($p = 0.005$, $B = 0.544$) is significantly positively correlated with age, which is consistent with the conclusion of Coninx (Helgeson et al., 2013) on the choice of coping strategies of farmers' families when the flood disaster comes. In addition, it also shows that after the disaster, older farmers are more inclined to choose to rely on themselves than the government or banks. Borrowing from banks or loan companies ($p = 0.034$, $B = 2.150$) is significantly positively correlated with the education level of household heads, which is consistent with the research conclusion of Bhattacharjee (Yang & Yang, 2021). Compared with the choice of flood adaptation livelihood strategy to repair drainage facilities, the education level of the household head has a significant positive impact on the increase of house height ($p = 0.001$, $B = 0.036$), improvement of agricultural irrigation measures ($p = 0.002$, $B = 0.162$) and purchase of flood insurance ($p = 0.000$, $B = 0.110$), which is consistent with the research conclusion of Bhattacharjee (Yang & Yang, 2021).

In physical capital, compared with the coping strategy of families using their past savings to solve the lack of livelihood during the flood, the choice behavior of selling family property or livestock for cash ($p = 0.023$, $B = -2.133$) is negatively correlated with the construction area of the family house, which is consistent with the research conclusion of Xun (Abebe et al., 2019). Compared with choosing to repair drainage facilities in the flood livelihood strategy, buying flood protection ($p = 0.034$, $B = 0.380$), participating in flood emergency training ($p = 0.030$, $B = 0.562$), increasing agricultural irrigation measures ($p = 0.047$, $B = 0.579$), and changing crop planting date or planting ($p = 0.004$, $B = 1.023$) are positively correlated with the building area of family houses. Yang (Schad et al., 2012) also believes that when the housing construction area of farmer households is larger, they will have sufficient funds to choose more livelihood strategies. However, choosing to move ($p = 0.003$, $B = -0.198$) is negatively related to the building area of family houses, because the cost of implementing this strategy will be higher.

In natural capital, compared with using past savings, the exchange of property for cash ($p = 0.000$, $B = -2.199$) is significantly negatively correlated with the distance between the house and the river and lake, which is consistent with the research conclusion of Baker (Bao et al., 2021). The closer the distance to the river, the greater the loss and damage caused by the flood, and selling the property can quickly solve the economic needs. In addition, farmers with drainage ditches are more likely to choose to rely on government or social organizations for relief funds ($p = 0.001$, $B = 0.817$), borrowing from banks or loan companies ($p = 0.03$, $B =$

0.405), selling property and livestock in exchange for cash (p = 0.005, B = 0.103). Compared with the adaptive livelihood strategy of improving drainage facilities, the choice of increasing house height (p = 0.001, B = − 0.454) and moving (p = 0.004, B = − 0.634) are significantly negatively correlated with the distance from the house to the river and lake, which is consistent with the research conclusions of Farman et al. (Xu et al., 2018) and Gioli et al. (King et al., 2014). Farmers with drainage ditches are more likely to choose to increase agricultural irrigation measures (p = 0.00, B = 0.087), and farmers with drainage ditches will choose a variety of flood disaster response livelihood strategies, but the choice of flood disaster adaptation livelihood strategies is limited.

In financial capital, compared with using past savings to solve the lack of livelihood during the flood, there is a significant positive correlation between credit opportunities and borrowing from banks or loan companies (p = 0.001, B = 1.147), which can be explained by the fact that households have a loan history, indicating that they have a good understanding of the bank's model and other conditions, and can also be able to do so when disasters occur. Guo (Wang & Ross, 2018) also draws consistent research conclusions. Compared with the livelihood adaptation strategy of choosing improved drainage facilities, the average annual household income has the greatest positive impact on increasing the height of the house (p = 0.031, B = 0.761). Rehman (Yang, 2018a, 2018b) also showed that family income was significantly positively correlated with the adaptive strategies made by families.

In social capital, compared with using past savings to solve the lack of livelihood during the flood, employment in villages has a positive impact on relying on the government or social company relief (p = 0.000, B = 0.817), which is consistent with the study of Bhattacharjee (Bhattacharjee & Behera, 2018). Compared with choosing the adaptive livelihood strategy of improving drainage facilities, having family members as cadres in the village and community has the greatest positive impact on choosing to participate in flood emergency training (p = 0.032, B = 0.505), followed by choosing to buy flood insurance (p = 0.018, B = 0.257). Wu et al. (2019) in the research on the influencing factors of herdsmen's livelihood capital on livelihood strategies also confirmed that the family's employment as village cadres has a significant impact on the choice of herdsmen's livelihood strategies.

4.5 Conclusions

Facing the climate change challenges, short-term and long-term livelihood strategies are of great importance to help farmers to be resilient in disaster-prone areas. Previous research only study the impact of various livelihood capital on disaster preparedness and post disaster recovery when farmers respond to disasters. The purpose of this study is to subdivide each livelihood capital into sub capitals and analyze their impacts on both short and long-term livelihood strategies. We aim to provide suggestions based on the analysis accordingly. This study applies the Department for International Development (DFID) farmer household sustainable livelihood analysis framework to conduct field investigation of household surveys in six sample villages of Dazhou Qujiang, Mianyang Fujiang and Nanchong Jialing River in Sichuan. Based on the analysis of the current situation of household livelihood capital, the factors affecting the household's choice of livelihood strategies to cope with floods and adapt to floods, this study constructs the household livelihood capital evaluation index system. Using the collected data and Multinominal Logit Model (MNL) analysis, the research draws the following conclusions: Through investigation and analysis, it can be seen that most of the farmer households in the flood-prone areas in the six sample villages have a certain amount of capital reserves to deal with flood disasters, but the proportion of responding to disasters by borrowing from banks or loan companies is relatively low, And farmer households generally choose 1–4 strategies to deal with floods. Among the livelihood strategies that farmers choose to adapt to flood disasters, participating in flood emergency training accounted for the largest proportion of the sample villages, while the proportion of choosing to change the type and date of crop planting was the least. The coping strategies adopted by farmers when adapting to flood risks with 2–5 species as the main type, and less than 1% of farmers in flood-prone areas did not choose any flood-adaptive livelihood strategies. According to the fitting results of the MNL model, the choice behavior of farmers' livelihood strategy in flood-prone areas is affected by the status of farmers' livelihood capital. Farmers' families with sufficient reserves of livelihood capital have a stronger ability to cope with and adapt to flood risk.

In terms of human capital, farmers with higher education can choose more scientific and effective livelihood strategies to deal with and adapt to floods, however, the choice of livelihood strategies for older farmers is limited. Therefore, local governments should increase local investment in education to enable farmers and families to have more choices of livelihood strategies to deal with and adapt to floods. At the same time, banks should appropriately relax the age limit for applying for loans to the elderly who meet the credit conditions, so that the elderly can receive financial assistance in the event of disasters.

Considering physical capital, households with larger built-up areas tend to adopt rich adaptation strategies (increasing agricultural irrigation, changing crop planting types and dates, purchasing flood insurance, participating in flood emergency training), when the housing area of farmers' families is larger, their economy is richer and they have more funds to make livelihood strategies to adapt to floods. We can increase the material capital by increasing family fixed assets, appropriate compensation from the government, improving housing conditions and increasing the source of income of farmers. At the same time, we can strengthen the demolition and reconstruction of old and dilapidated houses in this area and improve the ability of houses to cope with the impact of floods.

Talking about natural capital, families closer to rivers and lakes will choose measures that can be quickly realized (selling livestock, working), while families with drainage ditches will choose richer flood response strategies. It is needed to improve the grass-roots transportation network and the family location advantages by increasing the cultivated land area of farmers' families. The output value of cultivated land per hectare and land-use efficiency should be improved. We need to intensify efforts to build flood control and waterlogging prevention measures, and strengthen the construction of rural drainage facilities and flood control dams for rivers.

For financial capital, farmers and families with more credit opportunities are more willing to choose loans to deal with floods. Improving the microfinance system to facilitate farmers' family loans is therefore necessary. At the same time, we should improve the employment rate of the rural population and increase the average annual income of farmers' families. Sufficient financial capital reserves can enable households to have more and more effective livelihood strategies to choose in the process of flood response and adaptation.

Regarding social capital, with the help of the village or neighbors for a long time, farmers' families will not need to choose to move to adapt to the flood disaster. They can increase the publicity and support of farmers' cooperative organizations, integrate village social relationship network resources, and enhance the collective ability to resist the flood risk. Farmers and families shall be encouraged to strengthen contact with relatives and friends. Relevant local government departments shall timely and truly inform local farmers of the real-time flood information and the flood control measures taken, so as to enhance the trust of local farmers in the flood control capacity of the media and the government.

This study also plays an important role in other parts of the world. It expands the research area on the impact of both short-term and long-term flood-response strategies and adaptation strategies of farmers under different sub-divided livelihood capitals, and enriches therefore the theoretical analysis framework of livelihood capital and livelihood strategies in disaster-prone areas. Other areas can follow the same analysis framework to customize their analysis. It also provides theoretical reference for further research and analysis on sustainable development of farm households in similar regions.

This research has several limitations that should be addressed in the future research. On the one hand, the selected areas are only the flood-prone rural areas in Sichuan. Whether they are representative for the whole country needs further investigation in flood-prone rural areas in other provinces. On the other hand, although the sample size meets the minimum sample requirements, obtaining more samples will make the data analysis more accurate. Last but not least, although several rural areas in different regions were selected, samples from different regions were not compared, so it is necessary to further explore whether there is spatial heterogeneity in each variable in future studies.

References

Abebe, Y. A., Ghorbani, A., Nikolic, I., Vojinovic, Z., & Sanchez, A. (2019). Flood risk management in Sint Maarten—A coupled agent-based and flood modelling method. *Journal of Environmental Management, 248*, 109317.

Adelekan, I. O., & Asiyanbi, A. P. (2016). Flood risk perception in flood-affected communities in Lagos, Nigeria. *Natural Hazards, 80*, 445–469.
Amir, S., Saqib, Z., Khan, M. I., Ali, A., Khan, M. A., Bokhari, S. A., & Zaman ul, H. (2020). Determinants of farmers' adaptation to climate change in rain-fed agriculture of Pakistan. *Arabian Journal of Geosciences 13*, 1025.
Ao, Y., Zhang, H., Yang, L., Wang, Y., Martek, I., & Wang, G. (2021). Impacts of earthquake knowledge and risk perception on earthquake preparedness of rural residents. *Natural Hazards, 107*, 1287–1310.
Ao, Y., Zhang, Y., Wang, Y., Chen, Y., & Yang, L. (2020). Influences of rural built environment on travel mode choice of rural residents: The case of rural Sichuan. *Journal of Transport Geography, 85*, 102708.
Atreya, A., Czajkowski, J., Botzen, W., Bustamante, G., Campbell, K., Collier, B., & Montgomery, M. (2017). Adoption of flood preparedness actions: A household level study in rural communities in Tabasco, Mexico. *International Journal of Disaster Risk Reduction, 24*, 428–438.
Atube, F., Malinga, G. M., Martine, N., Okello, D. M., & Okello-Uma, I. (2021). Determinants of smallholder farmers' adaptation strategies to the effects of climate change: Evidence from northern Uganda. *Agriculture & Food Security, 10*, 6.
Bao, X., Zhang, F., Deng, X., & Xu, D. (2021). Can trust motivate farmers to purchase natural disaster insurance? Evidence from earthquake-stricken areas of Sichuan, China. *Agriculture, 11*, 783.
Bhattacharjee, K., & Behera, B. (2018). Determinants of household vulnerability and adaptation to floods: Empirical evidence from the Indian State of West Bengal. *International Journal of Disaster Risk Reduction, 31*, 758–769.
Cai, Z. (2010). Analysis on livelihood capital of peasant households in poverty-stricken villages in Wenchuan earthquake area. *Chin Rural Econ, 12*, 55–67.
Cao, S., Wang, Y., Duan, F., Zhao, W., Wang, Z., & Fang, N. (2016). Coupling relationship between ecological environmental vulnerability and economic poverty in poor areas of China: An empirical analysis of 714 poor counties in contiguous poor areas. *Chinese Journal of Applied Ecology, 27*, 2614–2622.
Chen, C., Pang, Y., & Zhang, Y. (2010). The characteristics of climate change in Sichuan Basin in recent 50 years. *Journal of Southwest University, 32*, 115–120.
Collier, B. L., & Babich, V. O. (2019). Financing recovery after disasters: Explaining community credit market responses to severe events. *The Journal of Risk and Insurance, 86*, 479–520.
Dalu, M. T. B., & Shackleton, C. M. (2018). The potential use of natural resources in urban informal settlements as substitutes for financial capital during flooding emergencies. *Physics and Chemistry of the Earth, 104*, 18–27.
Danso, S. Y., & Addo, I. Y. (2016). Coping strategies of households affected by flooding: A case study of Sekondi-Takoradi Metropolis in Ghana. *Urban Water Journal, 14*, 539–545.
Du, H., & Dong, T. (2016). Risk assessment of rainstorm and flood in main flood season in Sichuan-Yunnan region. *Bulletin Soil Water Conservation 36*, 27–31+35.
Guo, Y., Yang, L., Huang, W., & Guo, Y. (2020). Traffic safety perception, attitude, and feeder mode choice of metro commute: Evidence from Shenzhen. *International Journal of Environmental Research and Public Health, 17*, 9402.
Haeffner, M., Baggio, J. A., & Galvin, K. (2018). Investigating environmental migration and other rural drought adaptation strategies in Baja California Sur, Mexico. *Regional Environmental Change, 18*, 1495–1507.
Hao, W., Yang, D., Zhang, J., Li, W., & Wang, Z. (2014). A study on the relationship between sustainable livelihood capital and livelihood strategy of farmers and Herdsmen: A case study of Nyingchi, Tibet. *Journal of Arid Land Resource Environment, 28*, 37–41.
Helgeson, J. F., Dietz, S., & Hochrainer-Stigler, S. (2013). Vulnerability to weather disasters: The choice of coping strategies in Rural Uganda. *Ecology and Society, 18*, 13.

King, D., Bird, D., Haynes, K., Boon, H., Cottrell, A., Millar, J., Okada, T., Box, P., Keogh, D., & Thomas, M. (2014). Voluntary relocation as an adaptation strategy to extreme weather events. *International Journal of Disaster Risk Reduction, 8*, 83–90.

Kong, L., Li, Y., Wang, M., & Zheng, T. (2021). Research on livelihood capital and livelihood strategy of peasant households under project poverty alleviation—Based on investigation data of Shule County, Xinjiang. *Agric Resour Reg China, 42*, 134–142.

Lee, D. W., & Kim, H. Y. (2021). The effect of social capital on disaster conflicts in local communities: Focusing on disaster victims. *International Journal of Disaster Risk Reduction, 63*, 102445.

Li, J., Zhao, Y., & Iqbal, J. (2019). Variation patterns of extreme precipitation and relation to ocean-atmospheric climate in Sichuan province China from 1961 to 2017. *Theoretical and Applied Climatology, 137*, 3009–3026.

Liang, Y., & Cao, R. (2015). Employment assistance policies of Chinese government play positive roles! The impact of post-earthquake employment assistance policies on the health-related quality of life of Chinese earthquake populations. *Social Indicators Research, 120*, 835–857.

Ma, L. (2017). *Study on flood vulnerability and sustainable livelihood strategies of farmers in Poyang Lake Region*. Master's Thesis, Jiangxi Normal University, Nanchang, China.

Ma, C., Liu, L., Yuan, C., & Ren, G. (2018). The differentiation characteristics of farmers' livelihood capital and its impact on livelihood strategy in rapid urbanization area: A case study of Qingpu District, Shanghai. *Study Agricultural Modernization, 39*, 316–324.

Mavhura, E. (2017). Applying a systems-thinking approach to community resilience analysis using rural livelihoods: The case of Muzarabani district, Zimbabwe. *International Journal on Disaster Risk Reduction, 25*, 248–258.

Memon, M. H., Ali, M., & Khalil, S. (2020). Determinants of income diversification in flood-prone rural Pakistan. *International Journal of Disaster Risk Reduction, 50*, 101914.

Mentamo, M., & Geda, N. R. (2016). Livelihood diversification under severe food insecurity scenario among smallholder farmers in Kadida Gamela District, Southern Ethiopia. *Kontakt, 18*, e258–e264.

Mondal, M. S. H., Murayama, T., & Nishikizawa, S. (2021). Determinants of household-level coping strategies and recoveries from riverine flood disasters: Empirical evidence from the right bank of Teesta River, Bangladesh. *Climate, 9*, 4.

Nakayachi, K. (2015). Examining public trust in risk-managing organizations after a major disaster. *Risk Analysis, 35*, 57–67.

Qin, Y., Shi, X., Li, X., & Yan, J. (2021). Geographical indication agricultural products, livelihood capital, and resilience to meteorological disasters: Evidence from kiwifruit farmers in China. *Environmental Science and Pollution Research, 28*, 65832–65847.

Rahut, D. B., Mottaleb, K. A., & Ali, A. (2018). Rural livelihood diversification strategies and household welfare in Bhutan. *European Journal of Development Research, 30*, 718–748.

Roy, P., Pal, S. C., Chakrabortty, R., Chowdhuri, I., Malik, S., & Das, B. (2020). Threats of climate and land use change on future flood susceptibility. *Journal of Cleaner Production, 272*, 122757.

Sadri, A. M., Ukkusuri, S. V., Lee, S., Clawson, R., Aldrich, D., Nelson, M. S., Seipel, J., & Kelly, D. (2018). The role of social capital, personal networks, and emergency responders in post-disaster recovery and resilience: A study of rural communities in Indiana. *Natural Hazards, 90*, 1377–1406.

Saqib, S. E., Ahmad, M. M., Panezai, S., & Rana, I. A. (2016). An empirical assessment of farmers' risk attitudes in flood-prone areas of Pakistan. *International Journal of Disaster Risk Reduction, 18*, 107–114.

Schad, I., Schmitter, P., Saint-Macary, C., Neef, A., Lamers, M., Nguyen, L., Hilger, T., & Hoffmann, V. (2012). Why do people not learn from flood disasters? Evidence from Vietnam's northwestern mountains. *Natural Hazards, 62*, 221–241.

Shah, A. A., Ye, J., Abid, M., & Ullah, R. (2000). Determinants of flood risk mitigation strategies at household level: A case of Khyber Pakhtunkhwa (KP) province, Pakistan. *Natural Hazards, 88*, 415–430.

Su, F., Pu, X., Xu, Z., & Wang, L. (2009). Study on the relationship between livelihood capital and livelihood Strategy—A case study of Ganzhou District in Zhangye City, China. *Population, Resources and Environment, 19*, 119–125.

Su, Y., Deng, W., Zhang, J., Hriday Lal, K., & Paudel Khatiwada, S. (2016). Relationship between livelihood capital and livelihood strategies of rural households in Melamchi Basin of Central Mountainous Area in Nepal. *Mt Res, 34*, 645–653.

Wang, F., & Ross, C. L. (2018). Machine learning travel mode choices: Comparing the performance of an extreme gradient boosting model with a multinomial logit model. *Transportation Research Record, 2672*, 35–45.

Wang, L., Wang, C., & Li, X. (2012). Household differentiation based on livelihood asset quantification: A case study of 471 households in Bailin Village, Shapingba District, Chongqing. *Geogr Res, 31*, 945–954.

Wei, B., Su, G., Li, Y., & Ma, Y. (2019). Livelihood strategies of rural households in Ning'er Earthquake-Stricken Areas, Yunnan Province, China. *Sustainability, 11*, 5905.

Wu, T., Wu, Y., Wang, D., & Lin, H. (2019). Study on the impact of livelihood capital on livelihood strategy of herders in Sanjiangyuan Region. *Jounal of Grass Ind, 28*, 12–21.

Xu, D., Peng, L., Liu, S., & Wang, X. (2018). Influences of risk perception and sense of place on landslide disaster preparedness in Southwestern China. *International Journal of Disaster Risk Science, 9*, 167–180.

Yamane, T. (1965). Statistics: An introductory analysis. *Harper Row, 60*, 678–680.

Yang, L. (2018b). Modeling the mobility choices of older people in a transit-oriented city: Policy insights. *Habitat International, 76*, 10–18.

Yang, L., Ao, Y., Ke, J., Lu, Y., & Liang, Y. (2021). To walk or not to walk? Examining non-linear effects of streetscape greenery on walking propensity of older adults. *Journal of Transport Geography, 94*, 103099.

Yang, L., Chau, K. W., Szeto, W. Y., Cui, X., & Wang, X. (2020a). Accessibility to transit, by transit, and property prices: Spatially varying relationships. *Transportation Research Part D: Transport and Environment, 85*, 102387.

Yang, L., Chu, X., Gou, Z., Yang, H., Lu, Y., & Huang, W. (2020b). Accessibility and proximity effects of bus rapid transit on housing prices: Heterogeneity across price quantiles and space. *Journal of Transport Geography, 88*, 102850.

Yang, L., Liang, Y., He, B., Lu, Y., & Gou, Z. (2022). COVID-19 effects on property markets: The pandemic decreases the implicit price of metro accessibility. *Tunnelling and Underground Space Technology, 125*, 104528.

Yang, X. (2018a). *Study on the relationship between livelihood capital and livelihood strategy of farmers in desertification land restricted reserve: A case study of six counties in Hexi Corridor of Gansu Province*. Lanzhou University.

Yang, Y., Wang, H., & He, W. (2016). Research on the mode of Targeted financial poverty alleviation in China under the "four-element structure." *Tibet Financ, 10*, 47–53.

Yuan, Q. (2019). *Study on sustainable livelihood of farmers in Zhangjiajie under the background of rural ecotourism development*. Master's Thesis, Jishou University, Xiangxi Tujia and Miao Autonomous Prefecture, China.

Zhao, Y. (2017). *Study on sustainable Livelihood of poverty alleviation migrants in inhospitable areas*. master, Shanxi University.

Zhao, X. (2011). Impact of livelihood capital on life satisfaction of farmers and herdsmen: A case study of Gannan Plateau. *Geographical Research, 30*, 687–698.

Zhou, W. F., Guo, S. L., Deng, X., & Xu, D. D. (2021a). Livelihood resilience and strategies of rural residents of earthquake-threatened areas in Sichuan Province, China. *Natural Hazards, 106*, 255–275.

Zhou, W., Ma, Z., Guo, S., Deng, X., & Xu, D. (2021b). Livelihood capital, evacuation and relocation willingness of residents in earthquake-stricken areas of rural China. *Safety Science, 141*, 105350.

Zhuang, T., Zhang, H., & Yang, J. (2010). Study on the impact of natural disasters on rural poverty in Southwest China—Based on the analysis of 67 villages in 21 national poverty counties. *Rural Econ, 3*, 52–56.

Chapter 5
Early Stage COVID-19 Impact on Chinese Residents' Mental Health

Abstract This chapter aims to investigate the effect of coronavirus disease 2019 (COVID-19) on the Chinese public's mental health during its early stage. We collected our data through an online questionnaire survey. Specifically, we adopted the Impact of Event Scale-Revised (IES-R) and State-Trait Anxiety Inventory (STAI) to assess symptomatic responses to exposure to traumatic life events and public anxiety, respectively, in the COVID-19 pandemic outbreak. Then, we evaluated the differences in the scores among various socio-demographic groups using Kruskal-Wakkis H tests and t-tests and analyzed the IES-R, state anxiety (SA) score, and trait anxiety (TA) score using the Pearson correlation analysis. Finally, we conducted a path analysis to determine the mediating role of post-traumatic stress disorder symptoms (measured by the IES-R) in the relationship between TA and SA. The results show that the average of the SA and TA scores were 48.0 ± 10.4 and 38.0 ± 8.2, respectively; the respondents who suffered from mild, moderate, and severe psychological impacts because of the health crisis accounted for 21.9%, 5.2%, and 13.1%, respectively; farmers have the highest IES-R score than others; people with the highest income have the lowest SA level; a significant positive correlation existed between the IES-R and STAI scores; and TA produces both direct and indirect (through the IES-R) effects on SA. Overall, the general Chinese public exhibited much higher anxiety levels than normal in the early days of the pandemic outbreak. Accordingly, we strongly recommend psychological counseling and intervention support to mitigate the adverse psychological impacts of such an event.

Keywords Impact of event scale-revised · Public anxiety · State-Trait anxiety inventory · COVID-19 · Post-traumatic stress disorder · Questionnaire survey · China

Y. Ao and H. Bahmani, *Navigating Complexity: Understanding Human Responses to Multifaceted Disasters*, https://doi.org/10.1007/978-981-99-8207-3_5

5.1 Introduction

Since the 1950s, the number and effect of natural and human-related disasters have considerably increased (Ao et al., 2021; Li et al., 2020, 2021). There have been many epidemics in recent decades. On March 12, 2003, the World Health Organization (WHO) issued an unprecedented global alert. In particular, it declared an outbreak of severe acute respiratory syndrome (SARS), which became the first severe human-to-human disease in the twenty-first century (Wu et al., 2005). The virus spread rapidly in Asia, causing over 8,000 confirmed cases and at least 774 deaths. Its inherent scientific uncertainty also led to epidemic fear and suffering in various communities (Cheng & Wong, 2005). Similarly, in 2012, the Middle East respiratory syndrome (MERS) coronavirus emerged as a devastating public health threat (Khan et al., 2017). The epidemic caused human panic and financial loss and affected the mental health of infectious disease survivors and medical workers.

At the end of December 2019, Wuhan, the capital of Hubei Province (China), witnessed many cases of unknown viral pneumonia. On January 7, 2020, the Chinese Center for Disease Control and Prevention identified and isolated a new coronavirus disease. The novel coronavirus was named SARS coronavirus 2 (SARS-CoV-2). SARS-CoV-2 can spread through close contact between people. The novel coronavirus disease 2019 (COVID-19) outbreak spread rapidly and affected all regions of China and many other countries of the world. On January 30, 2020, the WHO Emergency Committee announced that the COVID-19 outbreak had become a public health emergency of international concern (Lee, 2020). In March 2020, the deadly virus swiftly swept across the whole world. Surely, the rapid spread of the pandemic has highly affected the physical and mental health of the global community. Therefore, the mental health of the public during this health crisis warrants scholarly and medical attention.

Public health emergencies may adversely affect the mental health of the public (Cannito et al., 2021; Chirumbolo et al., 2021; Di Crosta et al., 2020, 2021). This effect can manifest into different results, such as burnout, post-traumatic stress disorder (PTSD, a recurrent mental & physical distress happening after dangerous or catastrophic, transient, or long-lasting situations), anxiety, and depressive symptoms (Daniali & Flaten, 2021; Fontalba-Navas et al., 2017; Lancee et al., 2008; Probst et al., 2020; Wheaton et al., 2021; Xu et al., 2020). Previous studies consistently found that PTSD increases the prevalence of infectious diseases among survivors (Ro et al., 2017). Therefore, the timely detection of the symptoms of patients and psychological intervention will help reduce the psychological impact of health crises on the public and consequentially benefit the whole society.

To understand the mental health status of the Chinese public in the early COVID-19 stage, this study used the Impact of Event Scale-Revised (IES-R) and the State-Trait Anxiety Inventory (STAI) to collect data through an online questionnaire survey. These two scales determine the extent of the COVID-19 effect on the public's mental health during the tumultuous time. On the basis of our collected data, we analyze the relationship between the IES-R, SA, and TA scores of the public to provide theoretical support for their subsequent psychological intervention, treatment, and recovery. Overall, this study aims to provide a valuable basis for individuals, government agencies, and medical personnel to take timely and effective measures to alleviate the psychological anxiety of the public developed in the COVID-19 outbreak. We also hope that these measures will reduce the lasting psychological adverse effects caused by such a health crisis.

The contributions of this paper include (1) examining the effect of COVID-19 on mental health during the early days of the pandemic; (2) comparing the COVID-19 effect on people with different socio-demographic characteristics; (3) examining the association of the IES-R, SA, and TA scores; and (4) determining that the relationship between TA and SA is mediated by symptomatic responses to exposure to traumatic life events (measured by the IES-R), which contributes to a better understanding of the connection between the IES-R, SA, and TA scores.

The remainder of this paper is organized as follows. Section 5.2 introduces the data collected for analysis. Section 5.3 presents the empirical results. Section 5.4 discusses the findings of this study. The final section (Sect. 5.5) winds up the paper and identifies avenues for further research.

5.2 Materials and Methods

5.2.1 Measures

5.2.1.1 IES-R

The Impact of Event Scale-Revised (IES-R), originally proposed by Weiss and Marmar (1997), is a self-report measure to evaluate subjective distress caused by traumatic events. IES-R respondents are required to determine a certain stressful life event and demonstrate the magnitude of the bother in the past week (measured by Likert-type scores). The IES-R is one of the most popular measures of subjective distress assessment around the world. It has been translated into many languages for international use. Its internal consistency and test–retest reliability have been extensively confirmed in numerous studies (Ripon et al., 2020; Wang et al., 2020a, 2020b; Weiss, 2007).

The IES-R is the enhanced version of the Impact of Event Scale (IES), a simple yet compelling self-report measure initially put forward by Horowitz et al. (1979). The original IES, which has been widely employed to measure PTSD symptoms among various populations, comprises 15 items for the diagnosis of PTSD and 2 subscales, namely intrusion (7 items) and avoidance (8 items). The IES-R adds 7 items to the original IES and creates a new subscale, namely hyperarousal. As such, it contains a total of 22 items ($= 7 + 8 + 7$) and 3 subscales, namely intrusion, avoidance, and hyperarousal (Wagner & Waters, 2014; Zhang et al., 2014).

The scoring method of the IES-R for each item is 0 = Not at all, 1 = A little bit, 2 = Moderately, 3 = Quite a bit, and 4 = Extremely, which is slightly from that of the IES (0, 1, 3, 5). Therefore, the total IES-R score ranges from 0 to 88 (= 22 × 4). According to the IES-R score, people can usually be categorized into four groups: 0–23 (normal), 24–32 (mild psychological impact), 33–36 (moderate psychological impact), and ≥ 37 (severe psychological impact) (Alkhamees et al., 2020).

Because the COVID-19 pandemic was still ongoing at the period of data collection, Item 1 ("Any reminder brought back feelings about it") and Item 12 ("I was aware that I still had a lot of feelings about it, but I didn't deal with them") were excluded in this study. Therefore, in this study, a total of 20 IES-R items are used, and the maximum IES-R score is 80. We slightly modified the cut-off values according to the maximum score comparison (80 for the revised version vs. 88 for the original) and thus categorized people into four groups: 0–20 (normal), 21–29 (mild psychological impact), 30–33 (moderate psychological impact), and ≥ 34 (severe psychological impact).

5.2.1.2 STAI

This study uses the STAI to evaluate the state and trait anxieties of the public. The STAI is a self-reporting tool used to assess the level of anxiety associated with a situation and is made up of the state anxiety (SA) inventory and the trait anxiety (TA) inventory. A higher score means a higher anxiety level (Dennis et al., 2013).

The SA scale is used to assess the current state of anxiety. It measures how respondents feel "now," using items that measure subjective feelings, namely fear, tension, worry, and autonomic nervous system activation/arousal. By contrast, the TA scale is used to evaluate relatively stable aspects of "anxiety propensity," including general states of calm, confidence, and security (Julian, 2011). The STAI has a total of 40 items, including 20 SA ones and 20 TA ones. For SA, ten items capture negative emotions, and the other ten describe positive emotions. For TA, eleven items capture negative emotions, and the other nine describe positive emotions.

The STAI uses a 4-point Likert-type score (ranging from 1 to 4) for all 40 items. Therefore, the SA and TA scores are 20–80. According to their STAI score, people can be categorized into four groups: no anxiety (= 20), mild anxiety (21–39), moderate anxiety (40–59), and severe anxiety (60–80) (Liu et al., 2020).

5.2.2 Data

Many socio-demographic factors are found to affect the transmission and spread of infectious diseases, such as influenza (Liu, 2020). Recently, research on the influence of COVID-19 shows that gender is a consistent predictor of psychological outcomes. Specifically, females exhibited a higher level of psychological distress than males, indicating moderate levels of anxiety. Concerning age groups, although the results subtly varied across studies, the young (18–30 years old) and older adults ($\geq$ 60 years old) commonly reported the highest level of psychological distress (Mazza et al., 2020). Moreover, the poor, the disabled, the marginalized, those on insecure employment, and other vulnerable groups had the greatest risk of infection and indirect consequences (Lee & Morling, 2020). Therefore, in addition to IES-R and STAI scores, our survey records many socio-demographic factors, such as gender, marriage, occupation, education attainment, annual household income, and recent location.

A pilot survey was conducted on February 8, 2020. We meticulously revised the questionnaire according to the feedback from the pilot survey. We distributed its final version on February 10, 2020, through an online questionnaire survey platform ("Questionnaire Star," Changsha Ranxing Information Technology Co., Ltd., Changsha, China). Until 00:00 on February 19, we collected 956 questionnaires. After eliminating 200 invalid questionnaires (e.g., uncompleted questionnaires and respondents aged < 17 years), we retained 736 valid ones for subsequent analyses (effective rate = 76.98%). The respondents covered a wide range of locations: 33 provincial administrative regions in China. More information regarding this online survey can be found in our previous publication (Ao et al., 2020). Table 5.1 shows the demographic information of the 736 respondents.

5.2.3 Methods

We employed the SPSS 24.0 statistical software for the following statistical analysis: First, we determined the Cronbach's α and KMO to verify the internal consistency and structural validity of the IES-R and STAI scores. Next, we sorted out the obtained socio-demographic information by frequency and descriptive statistics. Then, we used Kruskal–Wakkis H tests and t-tests to identify the score differences among different socio-demographic groups. Finally, we investigated the correlations among IES-R, SA, and TA through a correlation analysis and determined that TA affects SA both directly and indirectly (through IES-R, acting as mediator variable) using a path analysis (also known as causal modeling, latent variable models, and analysis of covariance structures).

5.3 Results

The Cronbach's α values of the IES-R, SA, and TA were determined to be 0.916, 0.919, and 0.862, respectively. The KMO values of the IES-R, SA, and TA were 0.943, 0.933, and 0.921, respectively. Therefore, we can conclude that our questionnaire data has excellent internal consistency and structural validity.

Table 5.1 Demographic information of the respondents

Variable	Number of observations	Percentage (%)	Variable	Number of observations	Percentage (%)
Gender			Education attainment		
Male	307	41.7	Middle school or below	30	4.1
Female	429	58.3	High school/ technical secondary school/ vocational high school	39	5.3
Age (years)			Junior college	104	14.1
Under 25	224	30.4	Undergraduate	299	40.6
26–30	141	19.2	Master	170	23.1
31–40	223	30.3	Doctor	94	12.8
41–50	113	15.4	Annual household income (RMB)		
51–60	31	4.2	Under 30,000	61	8.3
Above 60	4	0.5	30,000–50,000	92	12.5
Marital status			60,000–100,000	164	22.3
Unmarried	315	42.8	110,000–150,000	122	16.6
Married	406	55.2	160,000–200,000	100	13.6
Divorced	12	1.6	210,000–250,000	59	8
Widowed	3	0.4	260,000–300,000	52	7.1
Occupation			Over 300,000	86	11.7
Ordinary/ enterprise employee	238	32.3	Residential location		
Government/ public institution servant	215	29.2	Rural	122	16.6
Farmer	9	1.2	Township	136	18.5
College student	190	25.8	City	478	64.9
Others	84	11.4			

5.3.1 IES-R Score Analysis

Table 5.2 shows the IES-R scores of the respondents. The IES-R score for all respondents was 18.8 ± 12.4. Those with scores above and below the mean accounted for 48% and 52%, respectively.

The IES-R score was 19.4 ± 12.6 for males and 18.4 ± 12.2 for females. Males and females were found to only have minor differences. Although previous empirical

Table 5.2 IES-R score of the respondents

Characteristic	Category	Mean	Standard deviation
All respondents		18.8	12.4
Gender	Male	19.4	12.6
	Female	18.4	12.2
	t-statistic	− 1.077	
	p-value	0.281	
Age (years)	18–25	17.6	12.7
	26–30	19.4	12.3
	31–40	19.4	12.0
	41–50	18.6	12.9
	51–60	22.2	11.0
	61–70	15.8	13.6
	H-statistic	8.95	
	p-value	0.176	
Marital status	Unmarried	19.0	13.5
	Married	18.2	11.5
	Divorced	20.6	12.4
	Widowed	13.0	5.0
	H-statistic	1.258	
	p-value	0.739	
Current location	Rural	18.3	11.1
	Town	18.7	12.5
	City	19.0	12.7
	H-statistic	0.052	
	p-value	0.974	
Education attainment	Junior secondary or below	18.8	12.5
	High school	21.9	13.6
	Junior college	19.5	10.9
	Undergraduate	18.1	12.5
	Master	19.9	12.3
	Doctor	17.2	13.0
	H-statistic	9.181	
	p-value	0.102	
Annual household income (RMB)	Under 30,000	18.6	13.8
	30,000–50,000	21.1	13.2
	60,000–100,000	19.3	13.1
	110,000–150,000	18.1	11.2

(continued)

Table 5.2 (continued)

Characteristic	Category	Mean	Standard deviation
	160,000–200,000	17.9	12.6
	210,000–250,000	19.1	11.2
	260,000–300,000	20.9	12.0
	Over 300,000	16.5	10.9
	H-statistic	9.508	
	p-value	0.218	
Occupation	Ordinary employee/enterprise employee	18.8	11.7
	Government/public institution servant	19.4	12.3
	Farmer	28.9	8.5
	College student	17.8	12.1
	Others	18.9	13.1
	H-statistic	10.553	
	p-value	0.032	

studies (Pyari et al., 2012) have identified that females are more likely to suffer from PTSD, we do not find consistent outcomes. Future studies with larger sample sizes are, therefore, needed to support or refute our results. In addition, people aged 51–60 years had the highest score (22.2 $\pm$ 11.0). Regarding marital status, divorced people had the highest score (20.6 $\pm$ 12.4). Furthermore, people with the highest academic degree (doctoral degree) had the lowest score (17.2 $\pm$ 13.0). Last but not least, among the occupation groups, farmers scored the highest (28.9 $\pm$ 8.5).

The empirical data also showed that the proportion of people who were not affected by the pandemic outbreak (IES-R score $\leq$ 20) was 59.9% (441 people). The proportion of people with mild (21 $\leq$ IES-R score $\leq$ 29), moderate (30 $\leq$ IES-R score $\leq$ 33), and severe (IES-R score $\geq$ 34) psychological impacts were 21.9% (161 people), 5.2% (38 people), and 13.1% (96 people), respectively. The results showed that the number of people affected by the pandemic outbreak in its early stage accounted for 40.1%. This result indicated that the pandemic had substantially affected the mental health of the masses.

5.3.2 *STAI Score Analysis*

Table 5.3 reveals the SA and TA scores of the respondents. The average SA score of the respondents was 48.0 $\pm$ 10.4, and the average TA score was 38.0 $\pm$ 8.2. The SA score of the Chinese public affected by the health crisis was significantly higher than the individual TA score ($t = 22.66$, $p < 0.01$), and more than 79.6% of the respondents have a high degree of anxiety. This observation reveals that COVID-19 considerably affects the anxiety level of the respondents at its initial stage (the data collection period). Moreover, the detection rates of without anxiety, mild anxiety, moderate anxiety, and severe anxiety symptoms were 0.5%, 19.8%, 68.5%, and 11.1%, respectively (Ao et al., 2020).

In the early stage of the COVID-19 pandemic outbreak, the SA level of the sampled Chinese residents (48.0 ± 10.4) was significantly higher than the national norm (39.7 ± 8.9) ($t = 16.513$, $p < 0.001$). This outcome, indicating that the pandemic significantly affected the public's mental health in its early stage, is as expected. More explanations for STAI scores can be found in Ao et al. (2020).

5.3.3 Analysis of Group Differences

In Tables 5.2 and 5.3, we truly see some differences in the IES-R, SA, and TA scores. A question arises: whether a socio-demographic group has significantly different scores from others? To answer this question, we conducted two-tailed t-tests to assess the statistical significance of the difference in the mean score.

Table 5.4 shows the results of selected socio-demographic group differences in the IES-R score. A significant difference existed in the IES-R score between those with doctoral and high school degrees ($p = 0.039$) and among those with annual family income above 300,000, 30,000–50,000 yuan ($p = 0.013$), and 260,000–300,000 yuan ($p = 0.042$). Besides, the effect of the COVID-19 outbreak on the farmers in its early stage was higher than that on other occupational groups, indicating that they were more vulnerable to the pandemic.

Table 5.3 has shown that people with different annual incomes had different SA levels ($H = 18.963$, $p = 0.008$) but similar TA levels ($H = 8.353$, $p = 0.303$). Therefore, we aim to compare the SA levels of the richest (annual household income > 300,000 yuan) and others.

Table 5.5 shows the results of selected socio-demographic group differences in the SA score. It revealed that a significant difference existed in the SA level, and the SA level of the richest was significantly lower than that of others. There are many possible explanations for this interesting observation. First, the richest residents have more resources, so they are better prepared for the COVID-19 pandemic. In other words, not-so-rich residents (e.g., salesman, car/truck driver, hairdresser, take-away delivery man, small restaurant owner, and small shop owner) may have a limited income and feel helpless in the large-scale closure (for an unknown time) because many of them have house/car loans, children educational expenses, etc. Second, public anxiety increment could come from social isolation due to COVID-19 restrictions. The richest residents often have a strong social network, even in a pandemic context, which makes them exhibit a lower level of SA. Last, subjective expectation matters. Financial resources have a significant influence on how individuals view their future and how much the pandemic will negatively impact their lives. People who are

Table 5.3 STAI scores of the respondents

Characteristic	Category	SA		TA	
		Mean	Standard deviation	Mean	Standard seviation
Gender	Male	48.6	10.6	38.0	7.7
	Female	47.5	10.2	38.0	8.5
	t-statistic	− 1.152		− 0.219	
	p-value	0.249		0.826	
Age (years)	18–25	47.3	11.2	37.9	8.1
	26–30	48.9	10.5	37.5	8.1
	31–40	48.2	9.6	37.8	8.1
	41–50	47.6	9.6	38.7	8.5
	51–60	49.6	11.0	39.3	9.1
	61–70	50.0	9.5	42.5	7.0
	H-statistic	5.848		3.338	
	p-value	0.440		0.765	
Marital status	Unmarried	48.0	11.1	38.1	8.4
	Married	48.0	9.80	37.9	8.0
	Divorced	49.3	9.8	39.6	9.2
	Widowed	40.0	5.0	42.7	6.8
	H-statistic	3.144		1.863	
	p-value	0.37		0.601	
Current location	Rural	47.1	9.8	37.3	7.4
	Town	48.5	10.9	38.3	9.1
	City	48.1	10.4	38.1	8.1
	H-statistic	1.622		1.084	
	p-value	0.444		0.581	
Education attainment	Junior secondary or below	47.8	10.0	36.2	8.6
	High school	49.4	10.5	38.5	7.5
	Junior college	49.2	11.3	37.2	7.2
	Undergraduate	47.9	10.6	38.2	8.3
	Master	47.4	10.0	38.4	8.3
	Doctor	47.7	9.2	37.9	9.0
	H-statistic	1.817		2.706	
	p-value	0.874		0.745	
Annual household income (RMB)	Under 30,000	48.7	11.2	38.0	8.1
	30,000–50,000	48.1	10.4	38.4	8.7
	60,000–100,000	48.1	10.9	36.7	8.5
	110,000–150,000	48.1	10.5	38.4	8.0

(continued)

Table 5.3 (continued)

Characteristic	Category	SA		TA	
		Mean	Standard deviation	Mean	Standard seviation
	160,000–200,000	49.1	9.9	38.5	7.2
	210,000–250,000	48.7	8.0	38.5	8.8
	260,000–300,000	50.8	10.6	38.2	8.7
	Over 300,000	43.9	9.5	38.4	7.6
	H-statistic	18.963		8.353	
	p-value	0.008		0.303	
Occupation	Ordinary employee/ enterprise employee	48.2	10.2	37.3	7.7
	Government/ public institution servant	47.9	9.4	38.9	8.5
	Farmer	52.0	12.0	36.1	9.0
	College student	47.1	11.8	38.4	8.7
	Others	49.5	9.5	37.0	7.1
	H-statistic	4.998		5.877	
	p-value	0.287		0.209	
All respondents		48.0	10.4	38.0	8.2

Source: Ao et al. (2020)

concerned about their financial resources are more prone to fear that the pandemic will ruin their lives. Furthermore, there are other possible explanations for the obtained results. Future studies with rigorous research design can be conducted to investigate this issue to reach a more persuasive conclusion.

5.3.4 Correlation Analysis Between the IES-R, SA, and TA Scores

To explore the relationship between the magnitude of the effect of the pandemic outbreak and the psychological anxiety of the respondents in its early stage, Pearson correlation analyses were conducted to determine the correlation between the IES-R and STAI scores. The results (Table 5.6) show that a significant positive correlation existed between the IES-R and SA scores ($r = 0.629$, $p < 0.01$) and the SA and TA scores ($r = 0.178$, $p < 0.01$). However, the correlation between IES-R and SA scores is statistically insignificant ($r = 0.065$, $p > 0.05$). These findings are in agreement

Table 5.4 Group differences in the IES-R score

Variable	(A) Education	(B) Education	Mean difference (A-B)	Standard error	p-value
IES-R	Doctor	Junior secondary and below	− 1.684	2.589	0.515
		High school	− 4.851*	2.352	0.039
		Junior college	− 2.38	1.757	0.176
		Undergraduate	− 0.951	1.46	0.515
		Master	− 2.78	1.587	0.08
Variable	(C) Annual household income (RMB)	(D) Annual household income (RMB)	Mean difference (C-D)	Standard error	p-value
IES-R	Over 300,000	Under 30,000	− 2.284	2.067	0.269
		30,000–50,000	− 4.633*	1.852	0.013
		60,000–100,000	− 2.839	1.644	0.085
		110,000–150,000	− 1.645	1.738	0.344
		160,000–200,000	− 1.397	1.816	0.442
		210,000–250,000	− 2.631	2.087	0.208
		260,000–300,000	− 4.412*	2.169	0.042
Variable	(E) Occupation	(F) Occupation	Mean difference (E–F)	Standard error	p-value
IES-R	Farmer	Ordinary employees/enterprise employees	10.112*	4.188	0.016
		Government/public institution servant	9.540*	4.196	0.023
		College student	11.068*	4.207	0.009
		Others	9.948*	4.325	0.022

Note: * The mean difference is significant at the 5% level

Table 5.5 Group differences in the SA score

Variable	(I) Annual household income (RMB)	(J) Annual household income (RMB)	Mean difference (I-J)	Standard error	p-value
SA	Over 300,000	Under 30,000	− 4.800*	1.720	0.005
		30,000–50,000	− 4.258*	1.541	0.006
		60,000–100,000	− 4.213*	1.368	0.002
		110,000–150,000	− 4.259*	1.447	0.003
		160,000–200,000	− 5.188*	1.511	0.001
		210,000–250,000	− 4.789*	1.737	0.006
		260,000–300,000	− 6.878*	1.805	0

Note * The mean difference is significant at the 5% level

with our expectations. Besides, there are some correlations between IES-R and STAI scores.

5.3.5 *Path Analysis Between the IES-R, SA, and TA Scores*

The above correlation analysis has revealed the associations between the IES-R, SA, and TA scores. Following the logic of Ao et al. (2020), we can raise a question: is the relationship between TA and SA mediated by symptomatic responses to exposure to traumatic life events (as measured by IES-R)? A visual representation of the overall mediating relationship is presented in Fig. 5.1.

A path analysis (Ceccato et al., 2021) was carried out to answer the above question (testing whether the mediating effect is significant or not) with the help of Software for Statistics and Data Science, STATA 16. Table 5.7 shows the results. It indicates that TA is fairly useful in predicting both IES-R and SA, and IES-R significantly affects SA. In other words, the path analysis outcomes demonstrate that the relationship between TA and SA is truly mediated by symptomatic responses to exposure to COVID-19. The total effect of TA on SA is 0.150 (=0.065 + 0.620 × 0.137).

5.4 Discussion

Interestingly, government/public institution servants and college students have the lowest SA score, although their TA score is quite similar to that of people with other occupations. This may be because their statuses are more stable (e.g., free from unemployment and income reduction) than people with other occupations. Moreover, a significant difference was observed in the SA level among the richest and others.

In the early days of the COVID-19 outbreak, schools in all Chinese cities were required to close for an unknown time. The uncertainty of the return-to-school time and the potential negative effect on academic progress negatively affected students' mental health (Wang et al., 2020a, 2020b). The respondents who had a high school degree scored the highest in the IES-R. During the outbreak, people were forced to stay at their homes because of a "stay at home" executive order issued by the central government, which caused the public to have varying degrees of anxiety. Moreover, previous studies have shown that apart from the national health state, this pandemic has a major effect on the economies of nations and individuals (Cao et al., 2020).

Farmers scored the highest in the IES-R score. They are significantly different from other occupational groups, especially college students. A possible explanation is that compared with college students, farmers' adaptability is weak. It is widely acknowledged that college students are the frontier group of new technologies and new ideas in society. They often have active thinking and can well use the Internet to obtain new information. Therefore, with exposure to new things, college students

Table 5.6 Correlation analysis between the IES-R and STAI

	IES-R				STAI					
	Hyperarousal	Intrusion	Avoidance	Total IES-R score	PE (TA)	NE (TA)	TA score	PE (SA)	NE (SA)	SA score
Hyperarousal (sum of 7 items)	1									
Intrusion (sum of 6 items)	0.755^{**}	1								
Avoidance (sum of 7 items)	0.687^{**}	0.600^{**}	1							
Total IES-R score (sum of 20 items)	0.924^{**}	0.851^{**}	0.880^{**}	1						
PE (TA) (sum of 9 items)	− 0.072	− 0.061	-0.100^{**}	-0.090^{*}	1					
NE (TA) (sum of 11 items)	0.272^{**}	0.212^{**}	0.235^{**}	0.274^{**}	0.258^{**}	1				
TA score (sum of 20 items)	0.078^{*}	0.057	0.038	0.065	0.883^{**}	0.682^{**}	1			
PE (SA) (sum of 10 items)	0.370^{**}	0.353^{**}	0.268^{**}	0.368^{**}	0.227^{**}	0.059	0.201^{**}	1		
NE (SA) (sum of 10 items)	0.686^{**}	0.576^{**}	0.525^{**}	0.673^{**}	-0.077^{*}	0.309^{**}	0.092^{*}	0.366^{**}	1	
SA score (sum of 20 items)	0.638^{**}	0.561^{**}	0.479^{**}	0.629^{**}	0.092^{*}	0.222^{**}	0.178^{**}	0.828^{**}	0.825^{**}	1

Note ** and * mean that correlation is significant at the 1% and 5% levels, respectively. PE (TA): positive emotion of TA. NE (TA): negative emotion of TA. PE (SA): positive emotion of SA. NE (SA): negative emotion of SA

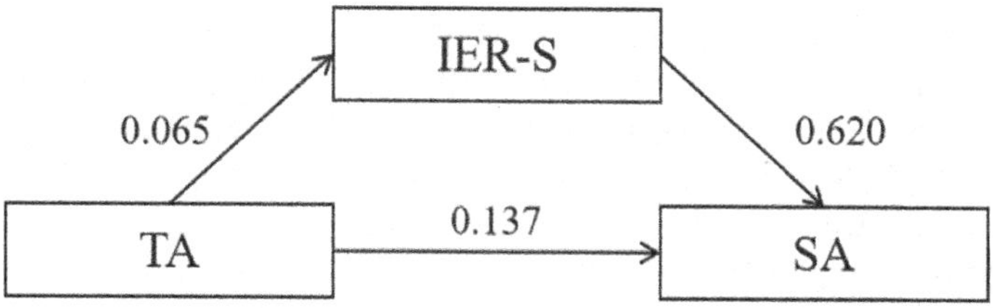

Fig. 5.1 Standardized coefficients of variables in the path analysis

Table 5.7 Path analysis results

Variable	Unstandardized coefficient	Standard error	t-statistic	Standardized coefficient
IES-R				
TA	0.099*	0.06	1.77	0.065
Constant	15.095**	2.16	6.98	
SA				
IES-R	0.520**	0.02	21.88	0.620
TA	0.174**	0.04	4.84	0.137
Constant	31.634**	1.44	22.01	
Number of observations	736			

Note ** and * mean that correlation is significant at the 1% and 10% levels, respectively

tend to accept information quickly and adapt well to changes in the environment. In addition, the farmers who participated in the survey were generally aged between 41–70 years. Many young people coming from rural areas have migrated to cities and towns, leading to an aging population in rural areas (Yang et al., 2021). The pandemic situation mostly affects farmers' work and the agriculture industry. After the SARS epidemic, the Chinese government developed a population-based public health information system. At present, one can report directly to the government at the township level, resulting in quicker administrative responses (Li, 2011). On the basis of the above discussion, in the face of COVID-19, the township community could pay extra attention to the mental health problems reported by farmers and promptly seek help from the higher government in the occurrence of even bigger problems.

PTSD is an anxiety disorder that often develops after a dangerous or disturbing event (Wang & Xu, 2016). The rapid spread of the virus has forced the public to stay at home. Brooks et al. (2020) indicated that people who had been isolated for more than ten days showed significantly higher symptoms of post-traumatic stress than others. PTSD is typically correlated with hypertrophic neurosis, vigilance, and insomnia. Symptoms such as anxiety and depression can also occur. Drug abuse and suicidal thoughts are relatively common. The blockade in Wuhan has greatly slowed down the spread of COVID-19 and put people's lives on pause, but it was short-lived. The continuous presence of the pandemic may have affected suspected individuals of PTSD. If psychological intervention is not promptly provided to patients with suspected PTSD, their psychological trauma may further deepen with extremely serious consequences.

The Pearson correlation analysis results showed that a significant positive correlation exists between the IES-R and SA scores. In addition, the path analysis outcomes revealed that the IES-R servers as a mediator in the relationship between the TA and SA scores. The Pearson correlation and path analysis outcomes jointly indicate that individuals who are easily affected by individual traits (low TA) will be more anxious under the influence of COVID-19 (low SA). As the development of COVID-19 affects more members of the public, the degree of psychological anxiety will increase, and its effect on low-TA people, high school students, and farmers will be more severe. Governments can pay more attention to the psychological state of these people.

5.5 Concluding Remarks

This study explores the COVID-19 pandemic outbreak's effect on the Chinese public's mental health, especially anxiety, in its early stage based on the online survey instrument. It provides a research foundation for public psychological counseling and intervention. Policymakers and public mental health institutions can take effective measures to reduce public psychological anxiety during the pandemic and promote public psychological rehabilitation to help people restore their mental health more quickly after the pandemic.

On the basis of the analysis results, we recommend the following actions during the tumultuous time: (1) Public authorities should promptly establish psychological assistance platforms (e.g., online services and hotline consultation) with a wider reach and account for behavioral reactions of people (Briscese et al., 2020). Specifically, the government should support severely affected families and individuals to respond actively to the pandemic. (2) Government health institutions can improve rural public information service platforms to disseminate real-time updates regarding COVID-19 to farmers. (3) Schools can appropriately upgrade and calibrate preventive measures and emergency response capabilities and promote high school, undergraduate, and graduate students to return to school and continue to attend classes if deemed appropriate. (4) Public authorities can provide some subsidies to families experiencing severe income difficulties and allow the resumption of employment.

This study has the following shortcomings: First, this study has a small sample size and does not involve confirmed and cured patients. Therefore, it is impossible to know the extent of their exposure to COVID-19 events. This feature obviously has a serious effect on people's mental health if they become ill or their family members (or close relatives) become ill and even die. Second, cross-sectional data were used in this study: the questionnaire survey was conducted in February 2020. At present, the pandemic has not completely disappeared. The pandemic has identified in some Chinese cities, such as Beijing, Shanghai, Xiamen, Changsha, and Yangzhou. The struggle of China and other countries against the pandemic may constantly change the psychological impact of their residents. Theremore, tracking the mental health of Chinese (and people in other countries) is essential for making interesting discoveries.

References

Alkhamees, A. A., Alrashed, S. A., Alzunaydi, A. A., Almohimeed, A. S., & Aljohani, M. S. (2020). The psychological impact of COVID-19 pandemic on the general population of Saudi Arabia. *Comprehensive Psychiatry, 102*, 152192.

Ao, Y., Zhang, H., Yang, L., Wang, Y., Martek, I., & Wang, G. (2021). Impacts of earthquake knowledge and risk perception on earthquake preparedness of rural residents. *Natural Hazards, 107*(2), 1287–1310.

Ao, Y., Zhu, H., Meng, F., Wang, Y., Ye, G., Yang, L., Martek, I., et al. (2020). The Impact of social support on public anxiety amidst the COVID-19 pandemic in China. *International Journal of Environmental Research and Public Health, 17*(23). https://doi.org/10.3390/ijerph17239097

Briscese, G., Lacetera, N., Macis, M., & Tonin, M. (2020). *Expectations, reference points, and compliance with COVID-19 social distancing measures.*

Brooks, S. K., Webster, R. K., Smith, L. E., Woodland, L., Wessely, S., Greenberg, N., & Rubin, G. J. (2020). The psychological impact of quarantine and how to reduce it: Rapid review of the evidence. *The Lancet, 395*(10227), 912–920. https://doi.org/10.1016/s0140-6736(20)30460-8

Cannito, L., Anzani, S., Bortolotti, A., Palumbo, R., Ceccato, I., Di Crosta, A., Palumbo, R., et al. (2021). Temporal discounting of money and face masks during the COVID-19 pandemic: The role of hoarding level. *Frontiers in Psychology, 12*, 2215.

Cao, W., Fang, Z., Hou, G., Han, M., Xu, X., Dong, J., & Zheng, J. (2020). The psychological impact of the COVID-19 epidemic on college students in China. *Psychiatry Research, 287*, 112934. https://doi.org/10.1016/j.psychres.2020.112934

Ceccato, I., Palumbo, R., Di Crosta, A., Marchetti, D., La Malva, P., Maiella, R., Di Domenico, A., et al. (2021). "What's next?" Individual differences in expected repercussions of the COVID-19 pandemic. *Personality and Individual Differences, 174*, 110674.

Cheng, S. K. W., & Wong, C. W. (2005). Psychological intervention with sufferers from severe acute respiratory syndrome (SARS): Lessons learnt from empirical findings. *Clinical Psychology & Psychotherapy, 12*(1), 80–86. https://doi.org/10.1002/cpp.429

Chirumbolo, A., Callea, A., & Urbini, F. (2021). The Effect of job insecurity and life uncertainty on everyday consumptions and broader life projects during COVID-19 pandemic. *International Journal of Environmental Research and Public Health, 18*(10), 5363.

Daniali, H., & Flaten, M. A. (2021). What psychological factors make individuals believe they are infected by Coronavirus 2019? [Original Research]. *Frontiers in Psychology, 12*(1377). https://doi.org/10.3389/fpsyg.2021.667722

Dennis, C. L., Coghlan, M., & Vigod, S. (2013). Can we identify mothers at-risk for postpartum anxiety in the immediate postpartum period using the State-Trait Anxiety Inventory? *Journal of Affective Disorders, 150*(3), 1217–1220. https://doi.org/10.1016/j.jad.2013.05.049

Di Crosta, A., Ceccato, I., Marchetti, D., La Malva, P., Maiella, R., Cannito, L., Verrocchio, M. C., et al. (2021). Psychological factors and consumer behavior during the COVID-19 pandemic. *PLoS ONE, 16*(8), e0256095.

Di Crosta, A., Palumbo, R., Marchetti, D., Ceccato, I., La Malva, P., Maiella, R., Verrocchio, M. C., et al. (2020). Individual differences, economic stability, and fear of contagion as risk factors for PTSD symptoms in the COVID-19 emergency. *Frontiers in Psychology, 11*, 2329.

Fontalba-Navas, A., Lucas-Borja, M. E., Gil-Aguilar, V., Arrebola, J. P., Pena-Andreu, J. M., & Perez, J. (2017). Incidence and risk factors for post-traumatic stress disorder in a population affected by a severe flood. *Public Health, 144*, 96–102. https://doi.org/10.1016/j.puhe.2016.12.015

Horowitz, M., Wilner, N., & Alvarez, W. (1979). Impact of event scale: A measure of subjective stress. *Psychosomatic Medicine, 41*(3), 209–218.

Julian, L. J. (2011). Measures of anxiety: State-Trait Anxiety Inventory (STAI), Beck Anxiety Inventory (BAI), and Hospital Anxiety and Depression Scale-Anxiety (HADS-A). *Arthritis Care & Research, 63*(S11), S467–S472. https://doi.org/10.1002/acr.20561

Khan, A. W., Taylor-Robinson, A. W., Nasim, Z., Zahir, F., Ali, A., & Ali, S. (2017). New episode of Middle East respiratory syndrome Coronavirus outbreak in Saudi Arabia: An emerging public health threat. *Public Health, 150*, 149–151. https://doi.org/10.1016/j.puhe.2017.07.016

Lancee, W. J., Maunder, R. G., Goldbloom, D. S., & Coauthors for the Impact of, S. S. (2008). Prevalence of psychiatric disorders among Toronto hospital workers one to two years after the SARS outbreak. *Psychiatric Services (Washington, D. C.), 59*(1), 91–95. https://doi.org/10.1176/ps.2008.59.1.91

Lee, A. (2020). Wuhan novel coronavirus (COVID-19): Why global control is challenging? *Public Health, 179*, A1–A2. https://doi.org/10.1016/j.puhe.2020.02.001

Lee, A., & Morling, J. (2020). COVID19: The need for public health in a time of emergency. *Public Health, 182*, 188–189. https://doi.org/10.1016/j.puhe.2020.03.027

Li, L. (2011). The challenges of healthcare reforms in China. *Public Health, 125*(1), 6–8. https://doi.org/10.1016/j.puhe.2010.10.010

Li, W., Zhu, J., Fu, L., Zhu, Q., Guo, Y., & Gong, Y. (2021). A rapid 3D reproduction system of dam-break floods constrained by post-disaster information. *Environmental Modelling & Software, 139*, 104994.

Li, W., Zhu, J., Fu, L., Zhu, Q., Xie, Y., & Hu, Y. (2020). An augmented representation method of debris flow scenes to improve public perception. *International Journal of Geographical Information Science*, 1–24.

Liu, K., Chen, Y., Wu, D., Lin, R., Wang, Z., & Pan, L. (2020). Effects of progressive muscle relaxation on anxiety and sleep quality in patients with COVID-19. *Complementary Therapies in Clinical Practice, 39*, 101132. https://doi.org/10.1016/j.ctcp.2020.101132

Liu, L. (2020). Emerging study on the transmission of the Novel Coronavirus (COVID-19) from urban perspective: Evidence from China. *Cities, 103*. https://doi.org/10.1016/j.cities.2020.102759

Mazza, C., Ricci, E., Biondi, S., Colasanti, M., Ferracuti, S., Napoli, C., & Roma, P. (2020). A nationwide survey of psychological distress among Italian people during the COVID-19 pandemic: Immediate psychological responses and associated factors. *International Journal of Environmental Research and Public Health* (No. 9).

Probst, T., Humer, E., Stippl, P., & Pieh, C. (2020). Being a psychotherapist in times of the novel Coronavirus disease: Stress-level, job anxiety, and fear of coronavirus disease infection in more than 1,500 psychotherapists in Austria [Original Research]. *Frontiers in Psychology, 11*(2557). https://doi.org/10.3389/fpsyg.2020.559100

Pyari, T., Kutty, R. V., & Sarma, P. (2012). Risk factors of post-traumatic stress disorder in tsunami survivors of Kanyakumari District, Tamil Nadu, India. *Indian Journal of Psychiatry, 54*(1), 48.

Ripon, R. K., Mim, S. S., Puente, A. E., Hossain, S., Babor, M. M. H., Sohan, S. A., & Islam, N. (2020). COVID-19: Psychological effects on a COVID-19 quarantined population in Bangladesh. *Heliyon, 6*(11), e05481.

Ro, J. S., Lee, J. S., Kang, S. C., & Jung, H. M. (2017). Worry experienced during the 2015 Middle East Respiratory Syndrome (MERS) pandemic in Korea. *PLoS ONE, 12*(3), e0173234. https://doi.org/10.1371/journal.pone.0173234

Wagner, S. L., & Waters, C. (2014). An initial investigation of the factor-analytic structure of the impact of event scale-revised with a volunteer firefighter sample. *Journal of Loss and Trauma, 19*(6), 568–583. https://doi.org/10.1080/15325024.2013.810443

Wang, C., Pan, R., Wan, X., Tan, Y., Xu, L., Ho, C. S., & Ho, R. C. (2020a). Immediate psychological responses and associated factors during the initial stage of the 2019 coronavirus disease (COVID-19) epidemic among the general population in China. *International Journal of Environmental Research and Public Health, 17*(5), 1729.

Wang, C., Pan, R., Wan, X., Tan, Y., Xu, L., Ho, C. S., & Ho, R. C. (2020). Immediate psychological responses and associated factors during the initial stage of the 2019 Coronavirus Disease (COVID-19) epidemic among the general population in China. *International Journal of Environmental Research Public Health, 17*(5). https://doi.org/10.3390/ijerph17051729

Wang, Z., & Xu, J. (2016). A cross-sectional study on risk factors of posttraumatic stress disorder in Shidu parents of the Sichuan Earthquake. *Journal of Child and Family Studies, 25*(9), 2915–2923. https://doi.org/10.1007/s10826-016-0454-1

Weiss, D., & Marmar, C. (1997). The impact of event scale—Revised. In J. P. Wilson, & T. M. Keane (Ed.), *Assessing psychological trauma and PTSD: A handbook for practitioners*. Guildford Press.

Weiss, D. S. (2007). The impact of event scale: revised. In *Cross-cultural assessment of psychological trauma and PTSD* (pp. 219–238). Springer.

Wheaton, M. G., Prikhidko, A., & Messner, G. R. (2021). Is fear of COVID-19 contagious? The effects of emotion contagion and social media use on anxiety in response to the coronavirus pandemic [Brief Research Report]. *Frontiers in Psychology, 11*(3594). https://doi.org/10.3389/fpsyg.2020.567379

Wu, K. K., Chan, S. K., & Ma, T. M. (2005). Posttraumatic stress, anxiety, and depression in survivors of severe acute respiratory syndrome (SARS). *Journal of Traumatic Stress: Official Publication of the International Society for Traumatic Stress Studies, 18*(1), 39–42.

Xu, J., Ou, J., Luo, S., Wang, Z., Chang, E., Novak, C., Wang, Y., et al. (2020). Perceived social support protects lonely people against COVID-19 Anxiety: A three-wave longitudinal study in China [Original Research]. *Frontiers in Psychology, 11*(2759). https://doi.org/10.3389/fpsyg.2020.566965

Yang, L., Ao, Y., Ke, J., Lu, Y., & Liang, Y. (2021). To walk or not to walk? Examining non-linear effects of streetscape greenery on walking propensity of older adults. *Journal of Transport Geography, 94*, 103099.

Zhang, M. W., Ho, C. S., Fang, P., Lu, Y., & Ho, R. C. (2014). Methodology of developing a smartphone application for crisis research and its clinical application. *Technology and Health Care, 22*(4), 547–559. https://doi.org/10.3233/THC-140819

Chapter 6
Social Support Impact on Public Anxiety During the COVID-19 Pandemic in China

Abstract This chapter investigated the psychological anxiety of the Chinese public and its relationship with social support during the early stage of the COVID-19 outbreak, thereby providing empirical support for interventions on improving the public's mental health. On the basis of an online questionnaire survey conducted on February 10th–18th, 2020, this study shows that 19.8% of the respondents suffered mild anxiety, 68.5% suffered moderate anxiety, and 11.1% suffered severe anxiety. Significant differences are reported in state anxiety between people with different household incomes. Social support and trait anxiety were significantly correlated. Social support and state anxiety were negatively correlated. There were significant differences in trait anxiety and state anxiety between different social support groups. Social support affected state anxiety both directly and indirectly (through the mediation of trait anxiety). Therefore, during the COVID-19 outbreak, increasing public support for society can effectively reduce public psychological anxiety.

Keywords Social support · Public anxiety · State · Trait anxiety inventory · Social support rating scale · COVID-19 · China

6.1 Introduction

The COVID-19 pandemic, which is characterized by fast and multi-channelled transmission and high infectivity, hit Wuhan, China, in December 2019 (Cheng et al., 2020; Sun et al., 2020a, 2020b; Zhou et al., 2020). It spread quickly throughout China and beyond (Deng & Peng, 2020; Rahman et al., 2020; Wang et al., 2020a, 2020b, 2020c; Zheng et al., 2020). In the second half of January 2020, the first-level response to a public health emergency was established across China, and strong epidemic prevention measures were implemented. Some mandatory quarantine measures, such as the implementation of a travel ban (Bao et al., 2020; Wang et al., 2020a, 2020b, 2020c), were adopted throughout China (Xiao, 2020), and other policy measures were proposed successively, such as extending national holidays, closing schools, and postponing classes (Cao et al., 2020).

Y. Ao and H. Bahmani, *Navigating Complexity: Understanding Human Responses to Multifaceted Disasters*, https://doi.org/10.1007/978-981-99-8207-3_6

Public psychology is always affected by epidemics. Patients recovering from the SARS (severe acute respiratory syndrome) outbreak of 2003 had a higher than normal likelihood of emotional distress, including but not limited to anxiety, depression, and fear (Chowell et al., 2015). The 2012 Middle East respiratory syndrome (MERS) outbreak also caused varying degrees of psychological trauma and severely affected quality of life (Al-Rabiaah et al., 2020; Batawi et al., 2019; Breban et al., 2013). The explosion of Ebola in West Africa in 2014 took a heavy toll on people and had a debilitating impact on public psychology (Dembek et al., 2017; Shultz et al., 2016). Not only do people who experience traumatic events directly develop psychological symptoms, but also those who have been exposed to or associated with trauma may experience varying degrees of psychological trauma, including post-traumatic stress disorder (Barbano et al., 2019), melancholia, and depression (Indah, 2018; Zandi et al., 2020). Ample evidence has shown that psychological disorders (e.g., depression and anxiety) affect quality of life and physical and mental health (Apel & Coenen, 2020; Liu et al., 2020a, 2020b; Xiao, 2020). Since the number of confirmed cases of COVID-19 have rapidly gone up and daily life and social activities have been strictly restricted for an indeterminate period, people will inevitably suffer from anxiety and even become suicidal, potentially leading to longer-term damage to the mental health of individuals (Al-Rabiaah et al., 2020; Shang et al., 2017). Thus, the psychological state of the public during COVID-19 should be paid sufficient attention by decision-makers, researchers, and social media (Xiang et al., 2020). During such a crisis, the psychological health of society is arguably a major concern (Wang et al., 2020a, 2020b, 2020c; Zabini et al., 2020), and the implementation of mental health evaluation and support are priority goals for dealing with mental health consequences of COVID-19 (Xiang et al., 2020).

The International Classification of Diseases, eleventh edition (ICD-11), states that anxiety disorder is manifested as excessive anxiety and related behavioral disorders, and the symptoms are severe enough to cause obvious clinical pain or social function impairment (Barbano et al., 2019). Anxiety can be divided into state anxiety and trait anxiety (Spielberger et al., 1983). State anxiety is a short-term emotional state generated by the perception of threatening stimuli, and it is used to assess people's fear, tension, anxiety, and neurotic experience in immediate or recent occurrences. By contrast, trait anxiety refers to a relatively stable behavioral tendency of an individual to respond to various threatening stimuli, and it is used to assess people's personality traits (Spielberger et al., 1983).

Social support often means the care and support received from others (Segrin & Passalacqua, 2010). It can reduce the anxiety level of people (e.g., patients) in stressful life events (Ratajska et al., 2020) and is positively correlated with mental and physical health (Mulvaney-Day et al., 2007; Sakisaka et al., 2017). Studies have revealed that social support can reduce depression and anxiety (Tham et al., 2020). Social support can effectively affect people's psychological condition and alleviate anxiety, depression, and stress, contributing to post-disaster psychological recovery (Apel & Coenen, 2020; Kim et al., 2020). In order to improve public mental health in the context of the COVID-19 outbreak, the Chinese government has taken the following social support measures in terms of social support: (1) information support (Rubin & Wessely, 2020)—government and relevant departments are providing real-time disclosure of COVID-19-related knowledge and updates; (2) economic support—timely treatment is provided to confirmed patients, reducing financial stress and increasing public confidence; and (3) emotional support—the media reports on front-line health caregivers and their work to prevent and control the epidemic, uplifting public confidence in the fight against the virus. Moreover, most previous COVID-19 studies have focused on determining the epidemiological and clinical characteristics of infected patients (Chen et al., 2020a, 2020b; Huang et al., 2020), genomic characteristics of the virus (Lu et al., 2020), global health governance challenges (De Matos et al., 2020; Rubin & Wessely, 2020), and mental health studies of front-line health care providers (Chen et al., 2020a, 2020b; Kang et al., 2020; Xiao et al., 2020). However, the existing literature on the psychological effects of COVID-19 on the general population in China is still scarce (Wang et al., 2020a, 2020b, 2020c), and there are only a handful of studies devoted to this issue. Liu et al. (X. Liu et al., 2020a, 2020b) compared the differences in the mental state of the public among different genders, ages, occupations, education levels, and places of residence in the context of COVID-19, finding that gender and age have a significant impact on psychological anxiety. Shevlin et al. (2020) concluded that income is significantly related to the degree of psychological anxiety caused by COVID-19.

In light of this, this study analyzed the public's psychological anxiety at the early stage of the COVID-19 outbreak in China, describing the anxiety characteristics of the respondents, comparing the anxiety levels of multiple groups, and scrutinizing the relationship between social support and mental anxiety status of the respondents. This China-based study is of paramount importance for the accumulation of scientific evidence since China is where the virus was first reported and the Chinese population have suffered isolation and lockdowns longer than any other country in the world. This study can provide support to government and social institutions seeking to protect people's mental health in mitigating the effects of the COVID-19 outbreak.

The remainder of this paper is structured as follows. Section 6.2 presents materials and methods. Section 6.3 shows and discusses the empirical results. Section 6.4 presents a discussion. Section 6.5 concludes the paper, summarizes research limitations, and lists avenues for future research.

6.2 Materials and Methods

6.2.1 Data Collection

This study was jointly completed by researchers from several universities (primarily located in Southwest China) after Wuhan was closed down during the early COVID-19 outbreak. To explore the psychological health of the Chinese public in the context of the epidemic and the impact of social support on the mental state of the public, the authors initially designed the online structured questionnaire after examining existing literature and psychological evaluation scale manuals. It is worth noting that the online questionnaire survey is the only method available for the collection of psychological anxiety and social support data at the early stage of the pandemic. Notably, the State–Trait Anxiety Inventory (STAI) is used to depict the psychological anxiety profile of the public, and the Social Support Rating Scale (SSRS) is used to obtain the social support profile of the public.

The questionnaire was first filled out by all the team members to check whether or not the content is clear and unambiguous enough. Many university professors were then asked to check the validity of the questionnaires and point out some minor problems to make the language of the questionnaire acceptable to the public. A pilot survey was then conducted by the team members' relatives and friends between January 31st and February 9th, 2020.

The main survey was distributed through the Internet on February 10th. The questionnaire was distributed to various WeChat (a ubiquitous instant messaging/social media mobile phone application in China) groups. The respondents covered a wide range of places in China, including 33 provincial-level administrative regions. By the end of February 18th, 2020, a total of 956 questionnaires were collected, 736 of which were valid.

To facilitate the detection of a valid questionnaire and ensure the cleanness of data for further statistical analyses (data screening), we set 2 test questions in the survey: "This is a test question. Please select'$3'$ to show that you have read the question carefully" and "This is a test question. Please choose'$2'$ to show that you have read the question carefully". If one or both of the questions were erroneous, the questionnaire was considered invalid.

6.2.2 Ethical Context

The ethical contexts of this study are as follows. First, potential participants were well informed of the purpose of the research at the beginning of the questionnaire, and it was promised that the data would be used only for scientific research and policy recommendations. After obtaining consent from respondents to participate in the research, a new page for questionnaire items would automatically open. Second, to preserve the anonymity of the respondents and ensure the confidentiality of the participant information, we decided not to include any personal identification items in the questionnaire.

6.2.3 Participants

Table 6.1 shows the socio-demographic characteristics of the participants. There were 429 (58.3%) females and 307 (41.7%) males. The respondents were mainly aged 18–25 years (30.3%), 26–30 years (19.2%), 31–40 years (30.3%), and 41–50 years (15.4%). A total of 64.9% of the respondents were recent migrants to urban areas, 18.5% were long-term urban residents, and 16.6% were from rural areas.

Table 6.1 Socio-demographic characteristics of the participants

Characteristic	N	%	Characteristic	N	%
Gender			Education attainment		
Male	307	41.7	Junior secondary and below	30	4.1
Female	429	58.3	High school/technical secondary school/vocational high school	39	5.3
Age			Junior college	104	14.1
Below 18 years	1	0.1	Undergraduate	299	40.6
18–25 years	223	30.3	Master's degree	170	23.1
26–30 years	141	19.2	Doctorate	94	12.8
31–40 years	223	30.3	Annual household income (RMB)		
41–50 years	113	15.4	Less than 30,000 (< USD 4560)	61	8.3
51–60 years	31	4.2	30,000–50,000 (= USD 4560–7600)	92	12.5
Above 60 years	4	0.5	60,000–100,000 (= USD 9120–15,200)	164	22.3
Marital status			110,000–150,000 (= USD 16,720–22,800)	122	16.6
Unmarried	315	42.8	160,000–200,000 (= USD 24,320–30,400)	100	13.6
Married	406	55.2	210,000–250,000 (= USD 31,920–38,000)	59	8
Divorced	12	1.6	260,000–300,000 (= USD 39,520–45,600)	52	7.1
Widowed	3	0.4	Over 300,000 (> USD 45,600)	86	11.7
Occupation			Residential location		
Ordinary employee/enterprise employee	238	32.3	Rural	122	16.6
Government/public institution personnel	215	29.2	Town	136	18.5
Farmer	9	1.2	City	478	64.9
Student	190	25.8			
Other	84	11.4			

6.2.4 *Measure*

6.2.4.1 STAI-C

In this study, the STAI (Spielberger et al., 1983), which has been extensively used in research and clinical settings diagnosing anxiety and differentiating it from depressive syndromes, was used to measure the state anxiety and trait anxiety of the public. The STAI has recently been used in COVID-19 settings. A study used the STAI to investigate the state anxiety level of nurses offering care for COVID-19 patients in Turkey (Bahadir-Yilmaz & Yüksel, 2020). Another study used it to explore differences in anxiety among adults in the United Kingdom in the context of COVID-19 (Maaravi & Heller, 2020).

This study used the STAI with the COVID-19 background (STAI-COVID-19, hereinafter referred to as STAI-C) to depict the psychological anxiety profile of the public. All the questions in the STAI-C were attached to the background of the epidemic, such as "before the occurrence of COVID-19, I felt happy" and "during the occurrence of COVID-19, I felt calm". The STAI-C consists of the state anxiety inventory (SAI-C) and the trait anxiety inventory (TAI-C).

The STAI-C has 40 items, 20 for evaluating state anxiety and 20 for evaluating trait anxiety. For state anxiety, 10 items describe negative emotions, and the other 10 capture positive emotions. Likewise, for trait anxiety, 11 items describe negative emotions, and the other 9 characterize positive emotions.

The STAI-C uses a four-point scoring method. For the SAI-C, 1 = not at all, 2 = a little, 3 = somewhat, and 4 = very much so. For the TAI-C, 1 = hardly, 2 = some, 3 = often, 4 = almost always. All positive emotions are reversely scored.

The score range of state anxiety or trait anxiety is 20–80. According to the score, people can be divided into 4 groups: no anxiety ($\leq$ 20), mild anxiety (21–39), moderate anxiety (40–59), and severe anxiety (60–80) (Liu et al., 2020a, 2020b).

6.2.4.2 SSRS

Social support is often defined as spiritual and material support from social networks that makes people feel cared for, loved, respected, and valued, and it has positively impacted the mental health of disaster victims (Sakisaka et al., 2017). It is generally believed that social support can be divided into 2 categories in nature: one is objective, visible, or actual support, including direct material assistance and social networks, the existence of group relationships, and the degree of individual participation; the other is subjective and experienced emotional support, which refers to the emotional experience and satisfaction that an individual is respected, supported, and understood in society. In addition, the individual's use of support should also be assessed. Individuals differ in their use of social support. Moreover, the support of people is an interactive process. While a person supports others, he/she may gain support from others.

This study adopted a 10-item SSRS designed by Xiao (1994), which has been used in many prior studies (Dong & Li, 2020; Sun et al., 2020a, 2020b; Xiao et al., 2020). The SSRS includes the following 3 dimensions: (1) subjective support dimensions (4 items), which reflect the subjective emotional experience and satisfaction of the respondents' feeling that they feel respected, understood, and supported; (2) objective support dimensions (3 items), which reflects the actual support the subjects thought they received, including direct assistance and social relations; and (3) support utilization dimensions (3 items), which reflect respondents' utilization degree of social support. The higher the score, the higher the social support. No COVID-19-related information is included in the scale.

6.2.4.3 Reliability and Validity Tests

In this study, Cronbach's α and McDonald's ω were adopted to test the reliability of the questionnaire data. The results showed that Cronbach's α of the SAI-C was 0.919, Cronbach's α of the TAI-C was 0.862, Cronbach's α of the SSRS was 0.691, McDonald's ω of the SAI-C was 0.920, McDonald's ω of the TAI-C was 0.869, and McDonald's ω of the SSRS was 0.734. Therefore, we can conclude that the reliability of the STAI-C data was at a high level, while the SSRS data had an acceptable level of reliability.

This study adopted factor analysis to test the construct validity of the questionnaire data. The results revealed that the KMO (Kaiser–Meyer–Olkin) values of the state anxiety scale and trait anxiety scale were 0.933 and 0.921, respectively. Two factors were extracted from the SAI-C. The total variance explanation rate was 58.89%. Additionally, two factors were extracted from the TAI-C, and the total variance explanation rate was 50.71%. This result was compared with that of the original scale (Spielberger et al., 1983), and the entries in the two factors of state anxiety and trait anxiety in this study were consistent. The KMO of the SSRS was 0.713, and 3 factors were extracted. The total variance explanation rate was 55.16%. The items in the factor were consistent with the original scale (Xiao, 1994). Therefore, the questionnaire data had good construct validity.

The common method variance test adopted Harman's single factor test. The result showed that the variance explanation percentage of the first common factor was 22.46%, and thus the common method variance problem was not serious (Yang et al., 2020a, 2020b).

6.2.5 Research Methods

In this study, SPSS 24.0 statistical software was used for statistical analysis. Firstly, an exploratory analysis was then used to test the distribution and variance homogeneity of the scale data. If the data were normally distributed or homogenized, the *t*-test or analysis of variance (ANOVA) was used to check the score differences among different groups. Otherwise, the Mann–Whitney U test and the Kruskal–Wallis H test were used to check the score differences among different groups. Thirdly, a pair-wise correlation analysis was used to analyze the correlations among the SAI-C, the TAI-C, and the SSRS. Finally, the mediation effect of variables was analyzed with a bootstrap set of 5000. $p < 0.05$ was considered "statistically significant" (Yang et al., 2020a, 2020b).

6.3 Results

6.3.1 *Anxiety Levels of Different Socio-demographic Groups*

The average score of TAI-C was 38.0 ± 8.2, and that of SAI-C was 48.0 ± 10.4. The detection rate of no anxiety symptoms was 0.5% (4 people), that of mild anxiety was 19.8% (146 people), that of moderate anxiety was 68.5% (504 people), and that of severe anxiety was 11.1% (82 people).

Table 6.2 reveals the mean and standard deviation of anxiety scores among socio-demographic groups. The average TAI-C score for males was 38.0 ± 7.7, and that for females was 38.5 ± 8.5. The average SAI-C score for males was 48.6 ± 10.6, and that for females was 47.5 ± 10.2. There were significant differences in state anxiety scores and positive emotion sub-score of state anxiety between family income groups ($\chi^2 = 18.963, p < 0.01$; $\chi^2 = 22.295, p < 0.01$). Except for the very poor and the very rich, the state anxiety level increased with advancing income. Another interesting observation was that the very rich had the lowest anxiety level. A possible explanation is that they have more resources and thus are more confident in combating the disease (e.g., can pay more to find a cure). However, most characteristics seemingly did not affect anxiety levels. A possible explanation is that the state anxiety was very high for all the respondents facing COVID-19. Future studies are needed to seek out more persuasive explanations.

6.3.2 *Anxiety Differences Among Social Support Groups*

On the basis of the social support score, we divided the respondents into three groups: high, medium, and low. Respondents with a score less than or equal to 35 points were in the low group, and those with a score greater than or equal to 44 points were in the high group. Others (with a score between 35 and 44 points) were set as the medium group.

The results (Table 6.3) show that there were significant differences between the positive and negative emotions in the trait anxiety scale ($\chi^2 = 48.500, p < 0.01$; $\chi^2 = 51.258, p < 0.01$), and trait anxiety showed significant differences in different levels

Table 6.2 Anxiety levels of different socio-demographic groups ($\overline{x} \pm s$)

Characteristic	Category	Positive emotion of TAI-C	Negative emotion of TAI-C	TAI-C	Positive emotion of SAI-C	Negative emotion of SAI-C	SAI-C
Gender	Male (n = 307)	21.2 ± 5.9	16.8 ± 3.6	38.0 ± 7.7	29.8 ± 6.1	18.8 ± 6.6	48.6 ± 10.6
	Female (n = 429)	20.9 ± 6.4	17.1 ± 4.2	38.0 ± 8.5	29.1 ± 6.4	18.5 ± 6.0	47.5 ± 10.2
	z	− 1.056	− 0.615	− 0.219	− 1.341	− 0.252	− 1.152
	p	0.291	0.538	0.826	0.180	0.801	0.249
Age (years)	< 18 (n = 1)	19	17	36	35	19	54
	18–25 (n = 223)	21.1 ± 6.6	16.7 ± 3.5	37.9 ± 8.1	28.6 ± 6.8	18.6 ± 6.7	47.3 ± 11.2
	26–30 (n = 141)	21.2 ± 6.3	16.3 ± 3.7	37.5 ± 8.1	30.0 ± 6.3	18.9 ± 6.4	48.9 ± 10.5
	31–40 (n = 223)	20.6 ± 5.7	17.2 ± 4.3	37.8 ± 8.1	29.7 ± 5.6	18.5 ± 6.0	48.2 ± 9.6
	41–50 (n = 113)	21.3 ± 5.8	17.4 ± 4.4	38.7 ± 8.5	29.2 ± 6.2	18.4 ± 5.7	47.6 ± 9.6
	51–60 (n = 31)	22.2 ± 7.2	17.1 ± 4.6	39.3 ± 9.1	30.5 ± 7.2	19.1 ± 5.9	49.6 ± 11.0
	> 60 (n = 4)	25.3 ± 6.2	17.3 ± 1.7	42.5 ± 7.0	32.5 ± 3.7	17.5 ± 8.8	50.0 ± 9.5
	H	3.571	7.349	3.338	7.739	1.838	5.848
	p	0.734	0.290	0.765	0.258	0.934	0.440
Marital status	Unmarried (n = 315)	21.2 ± 6.4	16.8 ± 4.0	38.1 ± 8.4	29.1 ± 6.7	18.9 ± 6.6	48.0 ± 11.1
	Married/divorced/ widowed (n = 421)	20.9 ± 6.0	17.0 ± 4.0	38.0 ± 8.1	29.7 ± 6.0	18.3 ± 5.9	48.0 ± 9.8
	z	− 0.521	− 0.537	− 0.325	− 1.051	− 0.863	− 0.402
	p	0.603	0.591	0.745	0.293	0.388	0.688
Current location	Rural (n = 122)	20.8 ± 5.7	16.5 ± 3.7	37.3 ± 7.4	28.9 ± 6.1	18.2 ± 5.7	47.1 ± 9.8
	Town (n = 136)	21.4 ± 7.0	17.0 ± 4.4	38.3 ± 9.1	29.7 ± 6.7	18.8 ± 6.5	48.5 ± 10.9

(continued)

Table 6.2 (continued)

Characteristic	Category	Positive emotion of TAI-C	Negative emotion of TAI-C	TAI-C	Positive emotion of SAI-C	Negative emotion of SAI-C	SAI-C
	City (n = 478)	21.1 ± 6.1	17.0 ± 3.9	38.1 ± 8.1	29.5 ± 6.2	18.7 ± 6.3	48.1 ± 10.4
	H	0.788	2.493	1.084	1.541	0.529	1.622
	p	0.674	0.288	0.581	0.463	0.767	0.444
Education attainment	Junior secondary and below (n = 30)	19.6 ± 6.4	16.6 ± 4.1	36.2 ± 8.6	29.5 ± 6.3	18.3 ± 6.1	47.8 ± 10.0
	High school (n = 39)	21.2 ± 5.4	17.2 ± 4.5	38.5 ± 7.5	30.9 ± 4.9	18.6 ± 6.7	49.4 ± 10.5
	Junior college (n = 104)	21.0 ± 6.0	16.3 ± 3.0	37.2 ± 7.2	29.8 ± 6.4	19.4 ± 7.1	49.2 ± 11.3
	Undergraduate (n = 299)	21.1 ± 6.2	17.1 ± 4.1	38.2 ± 8.3	29.5 ± 6.4	18.4 ± 6.4	47.9 ± 10.6
	Master's degree (n = 170)	21.3 ± 6.1	17.1 ± 4.2	38.4 ± 8.3	28.6 ± 6.4	18.9 ± 5.8	47.4 ± 10.0
	Doctorate (n = 94)	21.0 ± 6.9	16.9 ± 4.0	37.9 ± 9.0	29.7 ± 5.9	18.0 ± 5.3	47.7 ± 9.2
	H	2.442	2.497	2.706	4.991	2.634	1.817
	p	0.785	0.777	0.745	0.417	0.756	0.874
Annual household income (RMB)	Under 30,000 (USD 4560) (n = 61)	21.3 ± 6.6	16.8 ± 3.3	38.0 ± 8.1	29.7 ± 6.3	19.0 ± 6.8	48.7 ± 11.2
	30,000–50,000 (USD 4560–7600) (n = 92)	21.0 ± 5.8	17.3 ± 4.7	38.4 ± 8.7	29.1 ± 6.4	19.0 ± 6.2	48.1 ± 10.4

(continued)

Table 6.2 (continued)

Characteristic	Category	Positive emotion of TAI-C	Negative emotion of TAI-C	TAI-C	Positive emotion of SAI-C	Negative emotion of SAI-C	SAI-C
	60,000–100,000 (USD 9120–15,200) (n = 164)	20.0 ± 6.4	16.7 ± 4.1	36.7 ± 8.5	29.2 ± 6.6	18.9 ± 6.7	48.1 ± 10.9
	110,000–150,000 (USD 16,720–22,800) (n = 122)	21.5 ± 6.0	16.9 ± 4.2	38.4 ± 8.0	29.4 ± 6.0	18.7 ± 6.7	48.1 ± 10.5
	160,000–200,000 (USD 24,320–30,400) (n = 100)	22.1 ± 6.0	16.4 ± 2.9	38.5 ± 7.2	30.8 ± 5.7	18.3 ± 5.9	49.1 ± 9.9
	210,000–250,000 (USD 31,920–38,000) (n = 59)	21.9 ± 7.2	16.5 ± 3.9	38.5 ± 8.8	30.1 ± 6.0	18.5 ± 5.1	48.7 ± 8.0
	260,000–300,000 (USD 39,520–45,600) (n = 52)	20.8 ± 6.5	17.4 ± 4.3	38.2 ± 8.7	31.1 ± 6.1	19.6 ± 6.6	50.8 ± 10.6
	Over 300,000 (USD 45,600) (n = 86)	20.8 ± 5.1	17.7 ± 4.1	38.4 ± 7.6	27.0 ± 6.4	16.9 ± 5.1	43.9 ± 9.5
	H	10.271	7.071	8.353	22.295	8.071	18.963
	p	0.174	0.422	0.303	0.002	0.326	0.008

(continued)

Table 6.2 (continued)

Characteristic	Category	Positive emotion of TAI-C	Negative emotion of TAI-C	TAI-C	Positive emotion of SAI-C	Negative emotion of SAI-C	SAI-C
Occupation	Ordinary employee/ enterprise employee (n = 238)	20.7 ± 6.0	16.6 ± 3.8	37.3 ± 7.7	29.9 ± 5.8	18.4 ± 6.4	48.2 ± 10.2
	Government/public institution personnel (n = 215)	21.7 ± 6.3	17.2 ± 4.2	38.9 ± 8.5	29.6 ± 6.2	18.3 ± 5.4	47.9 ± 9.4
	Farmer (n = 9)	19.2 ± 6.3	16.9 ± 3.5	36.1 ± 9.0	31.4 ± 5.5	20.6 ± 7.7	52.0 ± 12.0
	Student (n = 190)	21.4 ± 6.8	17.0 ± 4.1	38.4 ± 8.7	28.3 ± 7.2	18.8 ± 6.9	47.1 ± 11.8
	Others (n = 84)	19.8 ± 4.9	17.1 ± 3.6	37.0 ± 7.1	30.2 ± 5.6	19.3 ± 6.1	49.5 ± 9.5
	H	6.276	4.414	5.877	6.678	3.003	4.998
	p	0.179	0.353	0.209	0.154	0.557	0.287

Table 6.3 Comparison of anxiety levels between Social Support Rating Scale (SSRS) groups ($\overline{x} \pm s$)

Item	SSRS groups			χ^2	p
	Low social Support group	Medium social support group	High social Support group		
Number of observations	201	308	227		
Percentage (%)	27.3	41.8	30.9		
Positive emotion of TAI-C	22.8 ± 5.8	21.5 ± 6.1	19.0 ± 6.1	48.500	< 0.01
Negative emotion of TAI-C	18.2 ± 4.4	17.0 ± 3.9	15.7 ± 3.2	51.258	< 0.01
TAI-C	41.0 ± 8.2	38.5 ± 7.8	34.7 ± 7.5	64.989	< 0.01
Positive emotion of SAI-C	30.3 ± 6.0	29.9 ± 5.9	28.1 ± 6.8	13.807	< 0.01
Negative emotion of SAI-C	19.2 ± 6.5	19.0 ± 6.1	17.5 ± 6.1	11.976	< 0.01
SAI-C	49.5 ± 10.1	48.8 ± 9.9	45.6 ± 10.8	16.762	< 0.01

of social support ($\chi^2 = 64.989, p < 0.01$). Similarly, state anxiety levels are different among different groups.

6.3.3 *Correlation Analysis Between the STAI-C and the SSRS*

Pearson correlation analysis was used in this study to further explore the correlation between various dimensions of state anxiety, trait anxiety, and social support. The analysis results are shown in Table 6.4. We observe that state anxiety and trait anxiety are positively correlated, while the social support measure is negatively related to the anxiety levels.

6.3.4 *Relationship Between Social Support and State Anxiety: Examining the Mediation Effect of Trait Anxiety*

Trait anxiety may mediate the effect of social support on state anxiety. To examine the mediation effect of trait anxiety, we adopted the Hayes establishment of SPSS Model 4 (Model 4 for the simple mediation model) and used the bootstrap for several sets of 5000(Bolin, 2014). The results (shown in Fig. 6.1 and Table 6.5) revealed that social support had a significant and negative effect on state anxiety ($\beta = -0.212, p < 0.01$); the direct effect of social support on state anxiety was still significant when

Table 6.4 Correlation analysis among scales

	M	SD	PE (TAI-C)	NE (TAI-C)	TAI-C	PE (SAI-C)	NE (SAI-C)	SAI-C	SS	OS	SU	SSRS
PE (TAI-C)	21.07	6.193	1									
NE (TAI-C)	16.93	3.98	0.258 **	1								
TAI-C	37.99	8.181	0.883 **	0.682 **	1							
PE (SAI-C)	29.42	6.294	0.227 **	0.059	0.201 **	1						
NE (SAI-C)	18.60	6.243	− 0.077 *	0.309 **	0.092 *	0.366 **	1					
SAI-C	48.02	10.362	0.092 *	0.222**	0.178 **	0.828 **	0.825 **	1				
SS	21.87	3.918	− 0.184 **	− 0.269 **	− 0.270**	− 0.127 **	− 0.098 **	− 0.136 **	1			
OS	10.11	3.881	− 0.200 **	− 0.143 **	− 0.221 **	− 0.119 **	− 0.072	− 0.116 **	0.380 **	1		
SU	7.64	1.91	− 0.196 **	− 0.192**	− 0.242 **	− 0.07	− 0.071	− 0.085 *	0.387 **	0.312 **	1	
SSRS	39.62	7.486	− 0.250 **	− 0.264 **	− 0.318 **	− 0.146 **	− 0.107 **	− 0.153 **	0.819 **	0.797 **	0.620 **	1

Note ** Significant at the 1% level. * Significant at the 5% level. (Positive emotion of TAI-C: PE (TAI-C). Negative emotion of TAI-C: NE (TAI-C). Positive emotion of SAI-C: PE (SAI-C). Negative emotion of SAI-C: NE (SAI-C). SS: subjective support; OS: objective support; SU: support utilization)

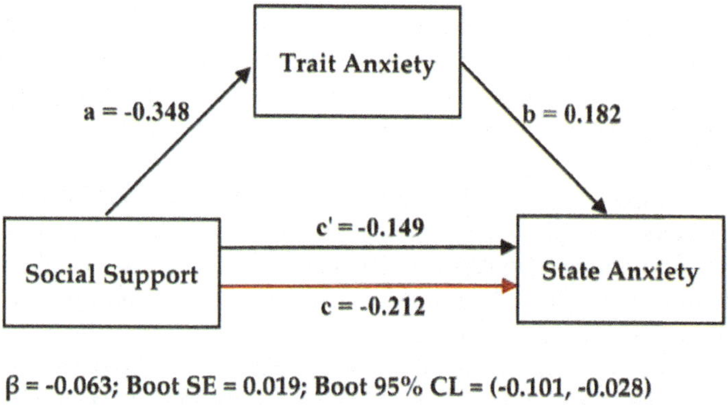

Fig. 6.1 Mediation modeling results

Table 6.5 Mediation effect test of trait anxiety (n = 736)

Effects	Path	β	SE	p	
Effect: Social Support → Trait Anxiety	a	− 0.348	0.038	< 0.01	
Effect: Trait Anxiety → State Anxiety	b	0.182	0.048	< 0.01	
Direct effect: Social Support → State Anxiety	c'	− 0.149	0.053	< 0.01	
Total effect: Social Support → State Anxiety	c	− 0.212	0.051	< 0.01	
State Anxiety total effect model (F = 17.637; p < 0.01; R2 = 0.024)					
Indirect effects		β	Boot SE	Boot 95% CI	
				IL	LL
Total indirect effect		− 0.063	0.019	− 0.101	− 0.028

Note Boot standard error, Boot CI lower limit, and Boot CI upper limit refer to the standard error, lower limit, and upper limit, respectively, of the 95% confidence interval of indirect effects estimated by percentile bootstrap method through deviation correction

the mediating variables were added ($\beta = -0.149$, $p < 0.01$), and social support had a significant negative effect on trait anxiety ($\beta = -0.348$, $p < 0.01$), while trait anxiety had a significant positive effect on state anxiety ($\beta = 0.182$, $p < 0.01$). The direct effect (− 0.149) and mediation effect (− 0.063) of the model accounted for 70.24% and 29.76% of the total effect (− 0.212), respectively.

6.4 Discussion

This study explored the psychological anxiety of the Chinese public during the COVID-19 outbreak and the impact of social support on psychological anxiety. By analyzing the anxiety score of the samples, we found that at the early stage of the pandemic, the respondents were generally anxious, with most presenting moderate anxiety. Our result is consistent with existing research results (Wang et al., 2020a, 2020b, 2020c). From stable trait anxiety to state anxiety during the epidemic, the anxiety level increased significantly. There may be several reasons for the high level of public anxiety: (1) COVID-19 seriously threatens the safety of life; (2) there is currently no specific drug or vaccine against this virus (Xiao et al., 2020); (3) the first-level response adopted by the state was that all localities would be under closed management, and residents would not be allowed to enter or leave at will, thus restricting their mobility (Brooks et al., 2020); and (4) the excessive media exposure makes it hard to find a reliable information source, which makes people confused about what to do (Shigemura et al., 2020).

Therefore, the government should manage the highly credible information flow in an efficient, rapid, and transparent manner to reduce misinformation and speculation (Brooks et al., 2020). Once people get to know the latest development of the epidemic, they can follow the infection and recovery progress (Wang et al., 2020a, 2020b, 2020c), better understand the merits of restricting outside movement, and limit interaction with others. Meanwhile, they can undertake recreational activities and physical exercise at home or in their neighborhood to maintain physical and mental fitness (Bao et al., 2020).

We find that people often have different levels of anxiety. The sudden outbreak of COVID-19 has affected a wide range of people. The government has taken measures to quarantine them at home for some time. People who are isolated from their homes are unable to work (unless they work online), and family income may be severely affected; thus, there are inevitable concerns about the family's economic functioning (Brooks et al., 2020; Cao et al., 2020). Gender, age, marital status, recent residential location, occupation, and education attainment do not affect trait anxiety or state anxiety. This result is different from other empirical results. A possible reason is that life-threatening epidemics are breaking out, and the anxiety of different groups of residents is high. However, some socio-demographic characteristics (e.g., annual household income) truly affect people's state anxiety (Dijkstra-Kersten et al., 2015; Dorahy et al., 2015; Kamberi et al., 2018; Melchior et al., 2010). This observation informs us that different people should be treated and addressed in different ways, and a uniform, one-size-fits-all method is not appropriate.

People with different levels of social support are found to exhibit different levels of state anxiety and trait anxiety. Social support can effectively relieve anxiety, depression, and stress (Xiao et al., 2020). Friends or family members can provide social or emotional support, and social interactions can reduce negative emotions and improve mood (Eray et al., 2017; Segrin & Passalacqua, 2010). As the number of confirmed COVID-19 cases in the country increases, more and more areas are affected, and residents are isolated at home and unable to go out to social events as usual, which aggravates the public's psychological anxiety (Bartoszek et al., 2020; Brooks et al., 2020). If residents have extensive social networks, social support can reduce their psychological loneliness by reducing perceived and physical responses to the threat of stressful events as well as stress-induced inappropriate behavior (Adamczyk & Segrin, 2015).

The trait anxiety score and the state anxiety score showed a positive relationship. This result suggests that improving mental resilience, resilience to stress, and personality traits can make people more resistant to sudden irritant events (Sim et al., 2019; Ying et al., 2016). Trait anxiety and state anxiety are negatively related to social support (Cao et al., 2020), which reveals that the higher the degree of social support, the lower the anxiety level. This finding indicates that social support is an effective intervention approach in people's psychological conditions (Xiao et al., 2020). To reduce the public's psychological anxiety and promote psychological recovery of the public during/after the epidemic, the degree of social support for the public should be increased. For example, online and offline psychological counseling platforms can be set up. On the one hand, residents should have access to accurate information through broadcasting. On the other hand, psychological counseling can be provided to the public through electronic devices to actively and effectively ease the public's psychological anxiety (Duan & Zhu, 2020; Kang et al., 2020).

The results of the mediation model show that social support has a direct effect as well as an indirect effect on state anxiety, and social support influences state anxiety by affecting trait anxiety. The social support scale measures how much social support a person receives on a normal time basis. Trait anxiety measures a relatively stable behavioral tendency of subjects to react to generalized threatening stimuli with anxiety. The formation of this behavioral tendency is influenced by social support. Individuals with higher levels of social support can better resist the negative effects of threat stimuli, and this habitual behavioral tendency is conducive to reducing the level of psychological anxiety under the influence of COVID-19. The public receives high social support at normal times, which is conducive to establishing a healthy and positive attitude and can effectively cope with sudden threatening stimulus events (Cao et al., 2020; Kaniasty, 2019). Therefore, effective psychological interventions and social support networks should be set up even in normal (non-crisis) periods to foster positive attitudes, improve personality traits and mental health, and enhance public resilience in the face of potential societal challenges (Indah, 2018). Such efforts will also effectively help people recover faster in future calamity (Ishibashi et al., 2019).

6.5 Conclusions

This study adopted the STAI and the SSRS to assess the levels of anxiety and social support in the early days of the COVID-19 epidemic. The results are as follows. First of all, under the influence of the COVID-19 epidemic, the public was found to be generally anxious since the virus seriously threatens physical health, with isolation restrictions on freedom and the uncertainty of information adding to this anxiety. Second, psychological anxiety was found to vary significantly among different social support groups. Third, social support was found to be negatively correlated with the degree of anxiety. More specifically, social support adversely affected state anxiety directly and indirectly (through the mediation of trait anxiety). Put in another way, anxiety will be significantly lowered if social support is higher.

This study has some shortcomings. First, it is cross-sectional in nature. Future studies should use the longitudinal research method to compare the changes before and after the public psychology and analyze the influencing factors, which is more conducive to the construction of public psychological recovery after an epidemic. Second, one important dimension that has not been collected is financial resilience. It has been seen in crisis studies is that the level of savings, interfamilial ability and willingness to support, and access to infrastructure (e.g., ability to access funds) contributes to anxiety. Third, at the early stage of the epidemic, Chinese residents were isolated at home, and the only way to obtain the necessary information for researchers was through an online questionnaire survey. In other words, researchers had no other methods available to collect data during such a unique time. Admittedly, the online questionnaire data were not random (e.g., biased towards young undergraduates). This is a common weakness of early COVID-19 social science studies (Liu et al., 2020a, 2020b; Maaravi & Heller, 2020). We agree that the mental health of the Chinese population should be assessed over the long run, and thus more rigorous data collection techniques and more sophisticated research methods are needed to reach a more robust conclusion. Fourth, the explained variance (R^2) in the mediation model was low (0.024). Finally, the SSRS questionnaire does not present high reliability (Cronbach's $\alpha = 0.691$, slightly lower than 0.70).

References

Adamczyk, K., & Segrin, C. (2015). Perceived social support and mental health among single vs. partnered polish young adults. *Current Psychology, 34*(1), 82–96. https://doi.org/10.1007/s12144-014-9242-5

Al-Rabiaah, A., Temsah, M. H., Al-Eyadhy, A. A., Hasan, G. M., Al-Zamil, F., Al-Subaie, S., Somily, A. M., et al. (2020). Middle East Respiratory Syndrome-Corona Virus (MERS-CoV) associated stress among medical students at a university teaching hospital in Saudi Arabia. *Journal of Infection and Public Health*. https://doi.org/10.1016/j.jiph.2020.01.005

Apel, D., & Coenen, M. (2020). Motivational and educational starting points to enhance mental and physical health in volunteer psycho-social support providers after the 2013 flood disaster in Germany. *International Journal of Disaster Risk Reduction, 43*. https://doi.org/10.1016/j.ijdrr.2019.101359

Bahadir-Yilmaz, E., & Yüksel, A. (2020). State anxiety levels of nurses providing care to patients with COVID-19 in Turkey. *Perspectives in Psychiatric Care*. https://doi.org/10.1111/ppc.12661

Bao, Y., Sun, Y., Meng, S., Shi, J., & Lu, L. (2020). 2019-nCoV epidemic: Address mental health care to empower society. *The Lancet, 395*(10224), e37–e38. https://doi.org/10.1016/s0140-6736(20)30309-3

Barbano, A. C., van der Mei, W. F., deRoon-Cassini, T. A., Grauer, E., Lowe, S. R., Matsuoka, Y. J., et al., International Consortium to Prevent. (2019). Differentiating PTSD from anxiety and depression: Lessons from the ICD-11 PTSD diagnostic criteria. *Depress Anxiety, 36*(6), 490–498. https://doi.org/10.1002/da.22881

Bartoszek, A., Walkowiak, D., Bartoszek, A., & Kardas, G. (2020). Mental well-being (depression, loneliness, insomnia, daily life fatigue) during COVID-19 related home-confinement-A study from Poland. *International Journal of Environmental Research and Public Health, 17*(20). https://doi.org/10.3390/ijerph17207417

Batawi, S., Tarazan, N., Al-Raddadi, R., Al Qasim, E., Sindi, A., Al Johni, S., Alraddadi, B. M., et al. (2019). Quality of life reported by survivors after hospitalization for Middle East respiratory syndrome (MERS). *Health and Quality of Life Outcomes, 17*(1), 101. https://doi.org/10.1186/s12955-019-1165-2

Bolin, J. H. (2014). Introduction to mediation, moderation, and conditional process analysis: A regression-based approach. *Journal of Educational Measurement, 51*(3). Hayes, A. F. (2013). The Guilford Press

Breban, R., Riou, J., & Fontanet, A. (2013). Interhuman transmissibility of Middle East respiratory syndrome coronavirus: Estimation of pandemic risk. *Lancet, 382*(9893), 694–699. https://doi.org/10.1016/S0140-6736(13)61492-0

Brooks, S. K., Webster, R. K., Smith, L. E., Woodland, L., Wessely, S., Greenberg, N., & Rubin, G. J. (2020). The psychological impact of quarantine and how to reduce it: Rapid review of the evidence. *The Lancet, 395*(10227), 912–920. https://doi.org/10.1016/s0140-6736(20)30460-8

Cao, W., Fang, Z., Hou, G., Han, M., Xu, X., Dong, J., & Zheng, J. (2020). The psychological impact of the COVID-19 epidemic on college students in China. *Psychiatry Research, 287*, 112934. https://doi.org/10.1016/j.psychres.2020.112934

Chen, N., Zhou, M., Dong, X., Qu, J., Gong, F., Han, Y., Zhang, L., et al. (2020a). Epidemiological and clinical characteristics of 99 cases of 2019 novel coronavirus pneumonia in Wuhan, China: A descriptive study. *Lancet, 395*(10223), 507–513. https://doi.org/10.1016/S0140-6736(20)30211-7

Chen, Q., Liang, M., Li, Y., Guo, J., Fei, D., Wang, L., Zhang, Z., et al. (2020b). Mental health care for medical staff in China during the COVID-19 outbreak. *The Lancet Psychiatry, 7*(4), e15–e16. https://doi.org/10.1016/s2215-0366(20)30078-x

Cheng, V. C. C., Wong, S. C., To, K. K. W., Ho, P. L., & Yuen, K. Y. (2020). Preparedness and proactive infection control measures against the emerging novel coronavirus in China. *Journal of Hospital Infection, 104*(3), 254–255. https://doi.org/10.1016/j.jhin.2020.01.010

Chowell, G., Abdirizak, F., Lee, S., Lee, J., Jung, E., Nishiura, H., & Viboud, C. (2015). Transmission characteristics of MERS and SARS in the healthcare setting: A comparative study. *BMC Medicine, 13*, 210. https://doi.org/10.1186/s12916-015-0450-0

De Matos, D. G., Aidar, F. J., Almeida-Neto, P. F., Moreira, O. C., Souza, R. F., Marçal, A. C., dos Santos, J. L., et al. (2020). The impact of measures recommended by the government to limit the spread of coronavirus (COVID-19) on physical activity levels, quality of life, and mental health of Brazilians. *Sustainability, 12*(21), 9072.

Dembek, Z. F., Mothershead, J. L., Chekol, T., Myers, D. B., Meris, R. G., Meranus, D., & Wu, A. (2017). Operational perspective of lessons learned from the Ebola crisis. *Military Medicine, 182*(1), e1507–e1513. https://doi.org/10.7205/MILMED-D-16-00074

Deng, S. Q., & Peng, H. J. (2020). Characteristics of and public health responses to the coronavirus disease 2019 outbreak in China. *Journal of Clinical Medicine, 9*(2). https://doi.org/10.3390/jcm9020575

Dijkstra-Kersten, S. M., Biesheuvel-Leliefeld, K. E., van der Wouden, J. C., Penninx, B. W., & van Marwijk, H. W. (2015). Associations of financial strain and income with depressive and anxiety disorders. *Journal of Epidemiology and Community Health, 69*(7), 660–665. https://doi.org/10.1136/jech-2014-205088

Dong, Y., & Li, H. (2020). The relationship between social support and depressive symptoms among the college students of Liaoning, China: A moderated mediated analysis. *Psychology, Health & Medicine, 25*(3), 368–378. https://doi.org/10.1080/13548506.2019.1679844

Dorahy, M. J., Rowlands, A., Renouf, C., Hanna, D., Britt, E., & Carter, J. D. (2015). Impact of average household income and damage exposure on post-earthquake distress and functioning: A community study following the February 2011 Christchurch earthquake. *British Journal of Psychology, 106*(3), 526–543. https://doi.org/10.1111/bjop.12097

Duan, L., & Zhu, G. (2020). Psychological interventions for people affected by the COVID-19 epidemic. *The Lancet Psychiatry, 7*(4), 300–302. https://doi.org/10.1016/s2215-0366(20)30073-0

Eray, Ş, Uçar, H. N., & Murat, D. (2017). The effects of relocation and social support on long-term outcomes of adolescents following a major earthquake: A controlled study from Turkey. *International Journal of Disaster Risk Reduction, 24*, 46–51. https://doi.org/10.1016/j.ijdrr.2017.05.026

Huang, C., Wang, Y., Li, X., Ren, L., Zhao, J., Hu, Y., Cao, B., et al. (2020). Clinical features of patients infected with 2019 novel coronavirus in Wuhan, China. *Lancet, 395*(10223), 497–506. https://doi.org/10.1016/S0140-6736(20)30183-5

Indah, R. (2018). Probing problems: Dilemmas of conducting an ethnographic study in a disaster-affected area. *International Journal of Disaster Risk Reduction, 31*, 799–805. https://doi.org/10.1016/j.ijdrr.2018.07.025

Ishibashi, R., Nouchi, R., Honda, A., Abe, T., & Sugiura, M. (2019). A concise psychometric tool to measure personal characteristics for surviving natural disasters: Development of a 16-item power to live questionnaire. *Geosciences, 9*(9). https://doi.org/10.3390/geosciences9090366

Kamberi, M., Hoxha, F., Shala, M., Shahini, M., & Vehapi, S. (2018). Anxiety predictors among college students in Kosovo. *International Journal of Adolescence and Youth, 24*(1), 117–124. https://doi.org/10.1080/02673843.2018.1479277

Kang, L., Li, Y., Hu, S., Chen, M., Yang, C., Yang, B. X., Liu, Z., et al. (2020). The mental health of medical workers in Wuhan, China dealing with the 2019 novel coronavirus. *The Lancet Psychiatry, 7*(3). https://doi.org/10.1016/s2215-0366(20)30047-x

Kaniasty, K. (2019). Social support, interpersonal, and community dynamics following disasters caused by natural hazards. *Current Opinion in Psychology, 32*, 105–109. https://doi.org/10.1016/j.copsyc.2019.07.026

Kim, Y.-J., Lee, S.-Y., & Cho, J.-H. (2020). A study on the job retention intention of nurses based on social support in the COVID-19 situation. *Sustainability, 12*(18), 7276.

Liu, K., Chen, Y., Wu, D., Lin, R., Wang, Z., & Pan, L. (2020a). Effects of progressive muscle relaxation on anxiety and sleep quality in patients with COVID-19. *Complementary Therapies in Clinical Practice, 39*. https://doi.org/10.1016/j.ctcp.2020.101132

Liu, X., Luo, W. T., Li, Y., Li, C. N., Hong, Z. S., Chen, H. L., Xia, J. Y., et al. (2020b). Psychological status and behavior changes of the public during the COVID-19 epidemic in China. *Infectious Diseases of Poverty, 9*(1), 58. https://doi.org/10.1186/s40249-020-00678-3

Lu, R., Zhao, X., Li, J., Niu, P., Yang, B., Wu, H., Tan, W., et al. (2020). Genomic characterisation and epidemiology of 2019 novel coronavirus: Implications for virus origins and receptor binding. *Lancet, 395*(10224), 565–574. https://doi.org/10.1016/S0140-6736(20)30251-8

Maaravi, Y., & Heller, B. (2020). Not all worries were created equal: The case of COVID-19 anxiety. *Public Health, 185*, 243–245. https://doi.org/10.1016/j.puhe.2020.06.032

Melchior, M., Chastang, J. F., Walburg, V., Arseneault, L., Galera, C., & Fombonne, E. (2010). Family income and youths' symptoms of depression and anxiety: A longitudinal study of the French GAZEL Youth cohort. *Depression and Anxiety, 27*(12), 1095–1103. https://doi.org/10.1002/da.20761

Mulvaney-Day, N. E., Alegria, M., & Sribney, W. (2007). Social cohesion, social support, and health among Latinos in the United States. *Social Science and Medicine, 64*(2), 477–495. https://doi.org/10.1016/j.socscimed.2006.08.030

Rahman, M. M., Thill, J.-C., & Paul, K. C. (2020). COVID-19 pandemic severity, lockdown regimes, and people's mobility: Early evidence from 88 countries. *Sustainability, 12*(21). https://doi.org/10.3390/su12219101

Ratajska, A., Glanz, B. I., Chitnis, T., Weiner, H. L., & Healy, B. C. (2020). Social support in multiple sclerosis: Associations with quality of life, depression, and anxiety. *Journal of Psychosomatic Research, 138*, 110252. https://doi.org/10.1016/j.jpsychores.2020.110252

Rubin, G. J., & Wessely, S. (2020). The psychological effects of quarantining a city. *BMJ, 368*, m313. https://doi.org/10.1136/bmj.m313

Sakisaka, K., Yoshida, H., Takahashi, K., Miyashiro, T., Yamamoto, T., Fujiga, M., Nishina, N., et al. (2017). Living environment, health status, and perceived lack of social support among people living in temporary housing in Rikuzentakata City, Iwate, Japan, after the Great East Japan Earthquake and tsunami: A cross-sectional study. *International Journal of Disaster Risk Reduction, 21*, 266–273. https://doi.org/10.1016/j.ijdrr.2016.12.001

Segrin, C., & Passalacqua, S. A. (2010). Functions of loneliness, social support, health behaviors, and stress in association with poor health. *Health Communication, 25*(4), 312–322. https://doi.org/10.1080/10410231003773334

Shang, L. L., Huang, Y. Q., Liu, Z. R., & Chen, H. G. (2017). A cross-sectional survey of disability attributed to mental disorders and service use in China. *Chinese Medical Journal (Engl), 130*(12), 1441–1445. https://doi.org/10.4103/0366-6999.207462

Shevlin, M., Nolan, E., Owczarek, M., McBride, O., Murphy, J., Gibson Miller, J., Bentall, R. P., et al. (2020). COVID-19-related anxiety predicts somatic symptoms in the UK population. *The British Journal of Health Psychology, 25*(4), 875–882. https://doi.org/10.1111/bjhp.12430

Shigemura, J., Ursano, R. J., Morganstein, J. C., Kurosawa, M., & Benedek, D. M. (2020). Public responses to the novel 2019 coronavirus (2019-nCoV) in Japan: Mental health consequences and target populations. *Psychiatry and Clinical Neurosciences, 74*(4), 281–282. https://doi.org/10.1111/pcn.12988

Shultz, J. M., Cooper, J. L., Baingana, F., Oquendo, M. A., Espinel, Z., Althouse, B. M., Rechkemmer, A., et al. (2016). The role of fear-related behaviors in the 2013–2016 West Africa Ebola virus disease outbreak. *Current Psychiatry Reports, 18*(11), 104. https://doi.org/10.1007/s11920-016-0741-y

Sim, T., Lau, J., Cui, K., & Wei, H.-H. (2019). Post-disaster psychosocial capacity building for women in a Chinese rural village. *International Journal of Disaster Risk Science, 10*(2), 193–203. https://doi.org/10.1007/s13753-019-0221-1

Spielberger, C. D., Gorsuch, R. L., Lushene, R., Vagg, P. R., & Jacobs, G. A. (1983). *Manual for the state-trait anxiety inventory*. Consulting Psychologists Press.

Sun, J., Sun, R., Jiang, Y., Chen, X., Li, Z., Ma, Z., Zhang, L., et al. (2020a). The relationship between psychological health and social support: Evidence from physicians in China. *PLoS ONE, 15*(1), e0228152. https://doi.org/10.1371/journal.pone.0228152

Sun, Z., Yang, B., Zhang, R., & Cheng, X. (2020). Influencing factors of understanding COVID-19 risks and coping behaviors among the elderly population. *International Journal of Environmental Research and Public Health, 17*(16). https://doi.org/10.3390/ijerph17165889

Tham, S. M., Ellithorpe, M. E., & Meshi, D. (2020). Real-world social support but not in-game social support is related to reduced depression and anxiety associated with problematic gaming. *Addictive Behaviors, 106*, 106377. https://doi.org/10.1016/j.addbeh.2020.106377

Wang, C., Pan, R., Wan, X., Tan, Y., Xu, L., Ho, C. S., & Ho, R. C. (2020). Immediate psychological responses and associated factors during the initial stage of the 2019 Coronavirus disease (COVID-19) epidemic among the general population in China. *International Journal of Environmental Research and Public Health, 17*(5). https://doi.org/10.3390/ijerph17051729

Wang, H., Wang, Z., Dong, Y., Chang, R., Xu, C., Yu, X., Cai, Y., et al. (2020b). Phase-adjusted estimation of the number of Coronavirus disease 2019 cases in Wuhan, China. *Cell Discovery, 6*, 10. https://doi.org/10.1038/s41421-020-0148-0

Wang, Y., Di, Y., Ye, J., & Wei, W. (2020). Study on the public psychological states and its related factors during the outbreak of coronavirus disease 2019 (COVID-19) in some regions of China. *Psychology Health and Medicine*, 1–10. https://doi.org/10.1080/13548506.2020.1746817

Xiang, Y. T., Yang, Y., Li, W., Zhang, L., Zhang, Q. E., Cheung, T., & Ng, C. H. (2020). Timely mental health care for the 2019 novel coronavirus outbreak is urgently needed. *Lancet Psychiatry, 7*(3), 228–229. https://doi.org/10.1016/S2215-0366(20)30046-8

Xiao, C. (2020). A novel approach of consultation on 2019 novel coronavirus (COVID-19)-related psychological and mental problems: Structured letter therapy. *Psychiatry Investigation, 17*(2), 175–176. https://doi.org/10.30773/pi.2020.0047

Xiao, H., Zhang, Y., Kong, D., Li, S., & Yang, N. (2020). The effects of social support on sleep quality of medical staff treating patients with coronavirus disease 2019 (COVID-19) in January and February 2020 in China. *Medical Science Monitor: International Medical Journal of Experimental and Clinical Research, 26*, e923549. https://doi.org/10.12659/MSM.923549

Xiao, S.-Y. (1994). The theoretical basis and research application of social support rating scale. *Journal of Clinical Psychiatry, 4*(2), 98–100.

Yang, L., Chau, K., Szeto, W., Cui, X., & Wang, X. (2020a). Accessibility to transit, by transit, and property prices: Spatially varying relationships. *Transportation Research Part D: Transport and Environment, 85*, 102387.

Yang, L., Chu, X., Gou, Z., Yang, H., Lu, Y., & Huang, W. (2020b). Accessibility and proximity effects of bus rapid transit on housing prices: Heterogeneity across price quantiles and space. *Journal of Transport Geography, 88*, 102850.

Ying, L., Wang, Y., Lin, C., & Chen, C. (2016). Trait resilience moderated the relationships between PTG and adolescent academic burnout in a post-disaster context. *Personality and Individual Differences, 90*, 108–112. https://doi.org/10.1016/j.paid.2015.10.048

Zabini, F., Albanese, L., Becheri, F. R., Gavazzi, G., Giganti, F., Giovanelli, F., Viggiano, M. P., et al. (2020). Comparative study of the restorative effects of forest and urban videos during COVID-19 lockdown: Intrinsic and benchmark values. *International Journal of Environmental Research and Public Health, 17*(21). https://doi.org/10.3390/ijerph17218011

Zandi, G., Shahzad, I., Farrukh, M., & Kot, S. (2020). Supporting role of society and firms to COVID-19 management among medical practitioners. *International Journal of Environmental Research and Public Health, 17*(21). https://doi.org/10.3390/ijerph17217961

Zheng, C., Huang, W. Y., Sheridan, S., Sit, C. H., Chen, X. K., & Wong, S. H. (2020). COVID-19 pandemic brings a sedentary lifestyle in young adults: A cross-sectional and longitudinal study. *International Journal of Environmental Research and Public Health, 17*(17). https://doi.org/10.3390/ijerph17176035

Zhou, T., Liu, Q., Yang, Z., Liao, J., Yang, K., Bai, W., Zhang, W., et al. (2020). Preliminary prediction of the basic reproduction number of the Wuhan novel coronavirus 2019-nCoV. *Journal of Evidence-Based Medicine, 13*(1), 3–7. https://doi.org/10.1111/jebm.12376

Chapter 7
Influence of COVID-19 Home-Based Learning on Future Online Study Intentions in China

Abstract This chapter applies Flow Experience and Theory of Planned Behavior to construct a theoretical framework for assumption making and the assumptions made are validated by data gained from questionnaires. 6933 students from 54 institutions in China participated in the investigation, with 5456 valid questionnaires returned. This study employs partial least squares (PLS) regression and confirmative factor analysis (CFA) to analyze and estimate the measurement model and the structural model. The results indicate that the experience of home-based learning significantly influenced the attitudes of Chinese university students, which in turn had a positive influence on their intention to continue online learning. The research findings provide a theoretical framework and practical guidelines on building a scientific online learning platform with appropriate online learning environments and tasks for a post-COVID-19 era blended-learning in Chinese universities.

Keywords Online learning · Home-based learning experience · Flow experience · Theory of planned behavior · COVID-19 · Blended learning design

7.1 Introduction

During the COVID-19 pandemic, many countries suspended classes worldwide, affecting over 1.7 billion students, according to United Nations Educational, Scientific and Cultural Organization (UNESCO) statistics. In China, the Ministry of Education issued a Notice on the Postponement of the Spring Semester on January 27th, 2020, and the State Council held a press conference on the joint prevention and control mechanism and issued a directive, "classes suspended but learning continues" on February 12th, 2020, requiring universities and colleges in China to conduct online education. According to statistics, as of May 8th, 1454 colleges and universities across the country were engaged in online education, with 1.03 million teachers

Y. Ao and H. Bahmani, *Navigating Complexity: Understanding Human Responses to Multifaceted Disasters*, https://doi.org/10.1007/978-981-99-8207-3_7

offering 1.07 million courses, a total of 12.26 million course sessions for both theoretical and experimental contents. In total, 17.75 million university students participated in 2.3 billion sessions, making online education the main mode of education in China's colleges and universities during the COVID-19 pandemic (MOE, 2020).

Online learning is a form of learning that requires a high degree of independence and autonomy from students and is blended with traditional face-to-face classroom instruction to form a blended learning model. Online learning not only changes the existing curriculum but also creates a new form of teaching and learning (Bonk & Graham, 2005). The online teaching model, based on modern information technology, is beneficial for expanding teaching scale, increasing teaching efficiency, and improving learning outcomes, but it also creates many new problems (Wang et al., 2020). Research indicates that although online learning models give learners a high degree of autonomy, their motivation to continue learning remains low (Kang et al., 2014). The virtuality of the medium makes learners feel disconnected from the real world, resulting in feelings of isolation and reduced learning engagement (Rovai, 2003; Waugh & Su-Searle, 2014). In addition, online learning lacks face-to-face communication, making it difficult to access timely guidance and attention (Guo et al., 2016).

While past online learning experiences were characterized by high self-selection and large differences in individual learning environments (Cao, 2014), online learning during the COVID-19 pandemic had two new features: on the one hand, it was highly mandatory and prescriptive as an alternative to offline offline learning during the pandemic and was consistent with the planned offline classroom in terms of learning contents. On the other hand, online teaching during the pandemic was the first large-scale experiment on a global scale. According to Yan Wu, Director General of the Department of Higher Education of the Ministry of Education of China, "Higher education can no longer and should not revert to the state of teaching and learning before the pandemic, and online education will become the new normal in the future" (MOE, 2020). Therefore, exploring the impact of the large scale college students' online learning experience during the epidemic on their willingness to continue online learning has important theoretical and practical implications for further design of the blended learning model of higher education.

The paper is structured as follows: Sect. 7.2 presents a detailed literature review. Section 7.3 illustrates the research framework and the research hypothesis. Section 7.4 explains the research design, followed by data analysis results in Sect. 7.5. Conclusions are presented in Sect. 7.6.

7.2 Literature Review

7.2.1 Flow Experience Theory in Online Learning

The concept of Flow Experience was first proposed by psychologist Csikszentmihalyi in the 1970s and 1980s. It refers to the overall feeling people get when they are fully engaged in an activity; i.e., when people are fully absorbed in an activity, they often forget the passage of time and the perception of things around them, or even lose their sense of self-awareness, and feel a constant sense of pleasure (Csikszentmihalyi & LeFevre, 1989). Originally applied to the field of psychology, this theory has been utilized in many different studies, such as those on games, dance, rock climbing, shopping, and sports (Lu et al., 2009). It has now been introduced into the area of online learning (Lee & Choi, 2013), to study the impact of network user experience on usage intention and behavior.

The Flow Experience model, constructed by a process-driven approach, has been widely used in empirical studies of online learning. For instance, Davis and Wong (2007b), who combined the Flow Experience and Technology Acceptance Model theory, discovered that the acquisition of the necessary skills to overcome course-related challenges is critical to students' ability to generate a Flow Experience and to continue learning. Shin (2006) verified that the balance of skills and challenges for online learners could significantly affect their Flow Experience and found that the Flow Experience greatly affected students' satisfaction with online learning. The many factors that contribute to the user's Flow Experience can be grouped into three dimensions, person (P), artifact (A), and task (T), according to the Person–Artifact–Task Model (PAT Model) proposed by Finneran and Zhang in (2003). The artifact restricts the possibility of the user's access to new media, while the user's characteristics and state determine the experience of each use of the product. Only when users have clear task goals and are proficient with the artifact can they generate mental flow and enter a state of ecstasy (Chen et al., 2019).

7.2.2 Behavioral Science in the Online Learning Experience

Existing research shows that learners' experiences in online environments are influenced by a variety of factors (Liu & Zhang, 2017), and online learning experiences largely influence learners' intentions and behaviors to continue online learning (Davis & Wong, 2007; Lee, 2010). The online learning environment, online learning resources, teacher-student and peer interactions (Paechter et al., 2010), collaborative learning (Sakulwichitsintu et al., 2015), course tasks, and the design of course activities (Jiang et al., 2019) are considered to be important factors influencing online learning experiences. To predict learning behaviors and explore directions and paths for improvement, researchers have applied influence factor modeling to screen and identify influences on learning behaviors (Alraimi et al., 2015). Behavioral science theory is often used to construct online learning influence factor models and to study human behavior, mainly from the perspective of human needs, desires, motivations, purposes, and other psychological aspects (Liu & Sun, 2005).

Behavioral science is mainly based on theories such as the Theory of Rational Behavior (TRA), Theory of Planned Behavior (TPB), Technology Acceptance Model (TAM), and Expectation Confirmation Model (ECM-ISC), among others. Different behavioral science theories carry different assumptions about the environment in which learning behavior occurs. For example, TAM and ECM-ISC are only applicable regarding the willingness of adopting information systems or online learning systems, while TRA and TPB are applicable to a wider range of online and blended learning environments (Kong et al., 2017).

The Theory of Rational Behavior was originally proposed by American scholars (Fishbein & Ajzen, 1975). In the Theory of Rational Behavior, behavioral intention is the key factor that causes the individual to perform a certain behavior, which can be understood as jointly determined by the subjective norm and the individual's attitude towards the behavior to be performed. Attitude is a predictor of behavioral intentions and is an individual's overall evaluation of the behavior (Conner & Armitage, 1998). Specifically, in the study of online learning behavior, "attitude" can be understood as the predisposition, positive or negative, that learners hold towards online learning behavior, which is determined by the individual's beliefs about the outcome of the behavior and their assessment of its importance (Kong et al., 2017). The basic assumption of the Theory of Rational Behavior is that most human behavior is rational and self-controllable and therefore is weak in explaining factors influencing uncontrollable behavior.

Accordingly, Ajzen (1991) extended the Theory of Rational Behavior by adding a third determinant, namely perceived behavioral control, and proposed the Theory of Planned Behavior (TPB). Perceived behavioral control is defined as behavior in which beliefs are related to whether a person can successfully obtain the necessary resources and opportunities to perform the completed behavior, weighted by the perceived ability of each factor to facilitate or inhibit the behavior. Perceptions of factors that may facilitate or inhibit behavioral performance are referred to as control beliefs. These factors include internal control factors (information, skills, emotions, etc.) and external control factors (opportunity, dependence on others, barriers, etc.). The new theory constructed by introducing the concept of perceived behavioral control goes beyond the main hypothesis proposed by TRA regarding the complete control of behavior by personal willpower (Sheppard et al., 1988). TPB has been found in different empirical studies to significantly improve the explanatory power of behavior (Ajzen, 1991). In this model, attitudes, subjective norms, and perceived behavioral control are the three main determinants of behavioral intentions. The theory assumes that the more positive the attitude, the greater the subjective norm, and the stronger the perceived behavioral control, the stronger the behavioral willingness and the more likely the actual behavior will occur.

7.2.3 *Combination of Theories*

Researchers tend to draw on other disciplines and well-established theories to guide their research so that they not only develop and test more comprehensive and integrated models to fill in missing causal relationships, but also use other well-established research to supplement their arguments and theoretical foundations (Karahanna & Straub, 1999). Recent research has shown that either extending the original rational choice model theory or combining it with other social psychological theories can yield better results and explanatory power than simply using the original rational choice model (Han, 2015). A major limitation of TPB theory is considered to be its lack of attention to the drivers of belief-based behavioral intentions (Hagger & Chatzisarantis, 2009; Hagger et al., 2002), so researchers often test and develop more complex models to improve the explanatory power of the theory. Lee (2010), for example, has comprehensively integrated and extended the Expectation Confirmation Model (ECM-ISC), Technology Acceptance Model (TAM), Theory of Planned Behavior (TPB), and theory of Flow Experience in his research explaining learners' willingness to continue using online learning. Several past studies have also validated the rationale for integrating the Theory of Planned Behavior and the theory of Flow Experience (Alzahrani et al., 2017).

7.3 Research Framework and Hypothesis

As mentioned in Sect. 7.2.1, online learning experiences largely influence learners' intentions and behaviors to continue online learning and Flow Experience is an extremely enjoyable psychological state and temporary subjective experience which is deemed to be an intrinsic motivator in explaining learner's willingness. Davis and Wong (2007, Lee (2010). Therefore, an important prerequisite to predicting future learners' intention to continue engaging in online learning is the flow experience of large-scale online learning during the pandemic.

As explained in Sect. 7.2.2, the Theory of Planned Behavior (TPB) has been found in different empirical studies to significantly improve the explanatory power of behavior (Ajzen, 1991) and is considered to be one of the most dominant theories in the field of behavior science (Armitage & Conner, 2001). Many studies have also extended the theory, adapting the theoretical framework to enhance its validity and adequacy. Therefore, the present study starts from the theory of Flow Experience which identifies the experience of large-scale online learning during pandemic and combines it with the Theory of Planned Behavior to predict learners' intention to continue to engage in online learning after the pandemic in the Chinese higher education context.

Furthermore, due to the isolation of learners' home-based learning, the influence of other participants in the teaching session on learners is reduced. Before the formal investigation, we found that the interaction effect between students was really

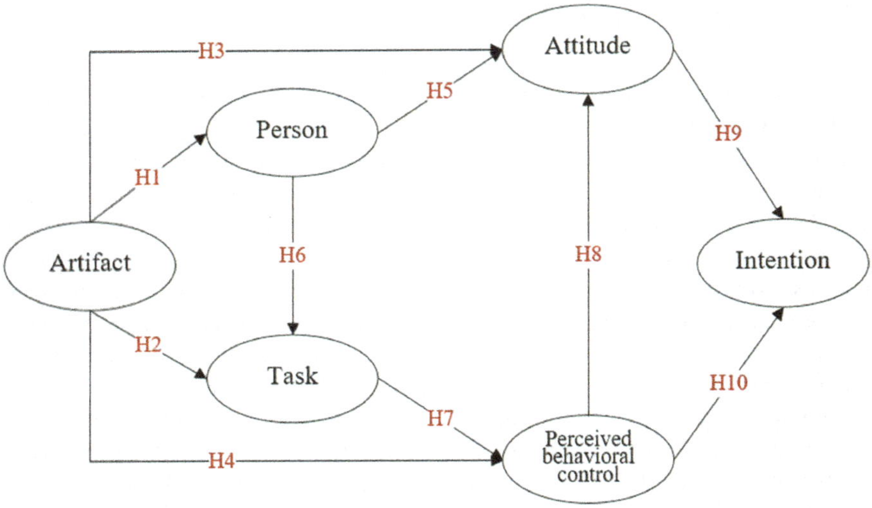

Fig. 7.1 Research framework

small through network interview. As such, this study assumes that subjective external pressure (perceptions) has no effect on online learners during the pandemic. With this in mind, this paper builds a research framework as shown in Fig. 7.1, taking the perspective of endogenous psychological factors concerning the three dimensions of the PAT (Person–Artifact–Task) Model of mental flow (Finneran & Zhang, 2003) and considering primarily the influence on the two elements in the Theory of Planned Behavior, namely attitudes and perceived behavioral control, on the students' intention to further online learning after the pandemic.

During the COVID-19 pandemic, Chinese universities have made a large-scale transformation from offline learning to online learning, which placed high demands on the teaching platforms and shaped the characteristics of the learners' experience of using the platforms during this period. The experience of this artifact is the most intuitive in the shifting process of learning mode. In the field of computer-mediated environments, some studies indicate that users can experience mental flow when using artifacts such as spreadsheets and word processors (Ghani, 1995; Webster et al., 1993). According to Csikszentmihalyi (1990), the experience of mental flow is more typical and pronounced when the user interacts with a toy. For online learning, mental flow is facilitated when the learner's experience of artifacts and platforms is as good and pleasant as that of toys. In the antecedent conditions that produce flow experiences, artifacts and tasks are closely related. Norman (1988) describes the correlation between them in this way: "When I use a direct manipulation system—whether for text editing, drawing pictures, or creating and playing video games—I do think of myself not as using a computer but at doing the particular task. The computer is, in effect, invisible." This means that when a good artifact experience is matched with a task challenge, it allows the learner to concentrate on the task when facing the challenge. Therefore, this paper proposes the following hypotheses:

H1. During the pandemic, learners' experience with artifacts used in online learning has a significant positive impact on individual learners' states regarding online learning.

H2. During the pandemic, learners' experience with artifacts used in online learning has a significant positive effect on task challenges undertaken.

According to the PAT model, artifact factors include system quality (e.g., interface usability, perceived ease of use, controllability, and interactivity), information quality, and service quality (Li & Xiao, 2019). For online learning, the artifact experience mainly includes learners' perceptions of artifact usability and ease of use when using the platform, while ease of use of the system is influenced by the resource form and task characteristics, and system usefulness is influenced by resource content and source (Ji, 2013). Therefore, the artifact experience also includes learners' perceptions of the richness and quality of online learning platform resources. Among them, perceived usefulness and perceived ease of use are a pair of concepts developed by Davis (1989) based on the Theory of Rational Behavior to construct a Technology Acceptance Model.

The model assumes that perceived usefulness and perceived ease of use have a direct effect on behavioral attitudes and, consequently, indirectly affect behavioral intentions. The Technology Acceptance Model has been proposed for the use of information systems, and existing research has found that usefulness and ease of use act as antecedent variables in the generation of mental flow, which in turn affects attitudes, intentions, and usage behavior (Sánchez-Franco, 2007). This theory is also applicable to the study of online learning systems: if users perceive that online learning platforms have better usefulness and ease of use as well as powerful features, they will advance their learning process to a certain extent, making them more willing to embrace online learning (Yang & Yan, 2011). In addition, it has been found that when users enjoy using artifacts to look for information, they also regard them as easy to use and are willing to use them in the future (Liaw & Huang, 2003). On this basis, this paper proposes the following hypotheses:

H3. During the pandemic, learners' experience with artifacts used in online learning has a significant positive impact on their attitude towards continuing online learning after the pandemic.

H4. During the pandemic, learners' experience with artifacts used in online learning has a significant positive effect on their perceived behavioral control.

In the PAT model (Finneran & Zhang, 2003), human factors include skills, motivation (Hoffman & Novak, 1996), curiosity, intrinsic interest (Webster et al., 1993), and other aspects. Characteristics of an individual and psychological states influence the generation of the mental flow experience. Wang et al. (2017) categorized highly focused attention, loss of self-awareness, and time distortion as experiential factors based on the nine dimensions of mental flow experience proposed by Csikszentmihalyi (1997). All of these factors point to the individual state of a person during the mental flow experience. Presence is also a very important individual state in online learning, which is the feeling of being immersed in an environment shaped by a medium and having a sense of reality, or "being personally on the scene" (Kim & Biocca, 1997). The sense of presence in online learning is a feeling like taking a class in a real space, with a strong sense of the presence of the teacher and other students, which reflects the feeling of "being with others". Social presence helps reduce the sense of loneliness that may appear in the online learning environment, which makes up for this deficiency in online learning (Zhao et al., 2018). It has been shown that presence is an important condition for generating a mental flow experience (Finneran & Zhang, 2005; Nah et al., 2011) and that if learners can enjoy the experience of online learning activities, they will not only engage in online learning but will also be intrinsically motivated to do so by consistently engaging in online learning to acquire these positive feelings (Davis & Wong, 2007). Research shows that increased learning, exploratory, and participatory behaviors, and positive subjective experiences are important outcomes of mental flow (Hoffman & Novak, 1996) and that with the mental flow, learners will devote more time and energy to the learning task, have a more adequate initiative to deal with the challenge, and show significantly enhanced attitudes (Richard & Chandra, 2005; Sánchez-Franco, 2007). Therefore, this paper proposes the following hypotheses:

H5. During the pandemic, individual learners' states of online learning have a significant positive effect on their attitude towards continuing online learning after the pandemic.

H6. During the pandemic, individual learners' states of online learning have a significant positive effect on online task challenges undertaken after the pandemic.

For the study of Flow Experience, researchers generally agree on the use of three flow stages as a framework: flow conditions, flow experience, and flow outcomes (Ghani, 1995; Ghani & Deshpande, 1994; Trevino & Webster, 1992). Among the condition factors, Novak (1999) place clear goals, immediate feedback, and the match of challenges with skills as antecedents to the occurrence of mental flow experiences, while Chen et al. (2000) sees clear goals, immediate feedback, potential control, and a combination of action and awareness as antecedents to flow. Although the division of antecedents varies, it can be inferred that goals, skills, and challenges are important antecedents, and the achievement of learning goals in the online learning domain must experience the completion of tasks and challenges in the learning process. According to Csikszentmihalyi's (1990) definition, the mental flow experience is optimal when challenges and skills are balanced, and this state will promote perceived control. The more motivated the learner is, as well as the better their self-monitoring of learning time management and task completion, the better the learning process will be to ensure a good experience and thus a high level of satisfaction (Jiang et al., 2017), thereby giving learners a more positive attitude. On this basis, this paper proposes the following hypotheses:

H7. During the pandemic, learners' online task challenges have a significant positive effect on their perceived behavioral control of online learning after the pandemic.

H8. During the pandemic, learners' perceived behavioral control of online learning have a significant positive effect on their attitude to continuing online learning after the epidemic.

According to the Theory of Planned Behavior, attitudes, subjective norms, and perceptual behavioral control all have a significant influence on willingness. This theory is also widely used in the field of online learning. During the pandemic, university students' home-based learning changed from the traditional offline learning style, and others' opinions could hardly influence students' online learning norms. Therefore, from the perspective of psychological, internal factors of university students, this paper mainly predicts learners' willingness to engage in future online learning based on their attitude according to planned behavior theory and perceived behavioral control. As is confirmed by many studies, when learners have a positive attitude towards online learning, their willingness to participate in online learning is higher (Chu & Chen, 2016; Liu et al., 2009; Park, 2009; Sawang et al., 2014; Valtonen et al., 2015; Yi & Hwang, 2003). Perceived behavioral control refers to how learners evaluate their ability and resources and is also a key factor in predicting behavioral intention. Ndubisi (2004) found that perceived behavioral control has a significant positive impact on online learning intention. Zhou (2016) found that learning attitude and perceived behavioral control have a significant impact on students' willingness to use Massive Open Online Course (MOOC) in a study of Chinese university students' acceptance of MOOC. Therefore, this paper takes attitude and perceived behavioral control as the intermediary variables of the three elements of flow experience theory to predict the influence that learners have on their willingness to participate in future online learning. The following hypotheses are proposed:

H9. Learners' attitude towards online learning has a significant positive impact on their willingness to use online learning after the pandemic.

H10. Learners' perceived behavior control of online learning has a significant positive impact on their willingness to use online learning after the pandemic.

7.4 Research Methods

This section is divided into two parts: questionnaire design (Sect. 7.4.1) and sample selection and preliminary data statistics description (Sect. 7.4.2).

7.4.1 Questionnaire Design

The questionnaire designed for this study mainly includes three parts, namely research background, respondents' basic information, and measurement scale of the theoretical model. The scale of the theoretical model refers to the existing research literature and the online learning environment during the pandemic. It adopts a five-point Likert Scale, where 1 means that the question in the questionnaire is totally inconsistent with the respondent's situation (feeling), while 5 means it is completely consistent. According to Csikszentmihalyi's (1989) description of mental flow, this study adopts the questions about concentration raised by Ghania and Deshpande (1994), those about pleasure raised by Moon and Kim (2001), and another question about sense of immediacy. Artifact experience refers to learners' feelings about the usability and ease of use of artifacts when using the platform, as well as their feelings about the richness and quality of resources of the online learning platform (Ji, 2013; Li & Xiao, 2019), for which three questions were developed. According to the framework of three flow stages of mental flow (Ghani, 1995; Ghani & Deshpande, 1994; Trevino & Webster, 1992), goals, skills, and challenges are important conditions for the generation of mental flow, for which four questions of task challenges were developed (Guo et al., 2016; Jackson & Marsh, 1996). In the theory of planned behavior, the attitude was covered by the three questions raised by Thompson et al. (1991), perceived behavior control covered by three questions based on Taylor and Todd (1995) and Zhou et al. (2016), and the willingness to participate in online learning answered by two questions raised by Davis (1989) and Niu (Lung-Guang, 2019). Appendix 1 lists the complete questions used in this study. To ensure the quality of questionnaire collection, the research group sets a question in the questionnaire to test whether the respondents fill in the questionnaire carefully, which asks them to select option D among the options presented. If the respondents choose any option other than D, the questionnaire was judged as invalid and excluded.

7.4.2 Sample Selection and Data Statistics

This study took the undergraduate students from 1243 undergraduate schools and colleges who participated in online learning during the pandemic as the research objects. According to Yamane's (1967) formula, the sample size of universities is $n = 42.91$, and the calculated sampling rate is $r = 3.45\%$ (the calculation formula is shown in Formula (7.1)).

$$n = N/(1 + Ne_2) \qquad (7.1)$$

Here, "n" represents the sample size of universities, "N" is the number of all Chinese universities, and e is the accuracy, set to 15%.

A sample survey was conducted according to the geographical division of Chinese universities (seven regions, namely Northeast China, North China, East China, South China, Central China, Northwest China, and Southwest China). The research group carried out random sampling for each region whose number was not less than 150%r and conducted a questionnaire survey in universities selected by random sampling. The research group obtained permission from teachers in sampled universities, mainly through telephone or network, and then collected data through a questionnaire that was spread by these teachers. From June 8th to 30th, 2020, the research group picked the sampled universities and collected data. Ultimately, the group received effective responses from 54 universities. Information on the number of universities in each region and their sampling quantity is listed in Table 7.1. The effective sampling rate of universities in all regions was higher than the calculated sampling rate except that in East China, which is slightly lower than the calculated sampling rate. The distribution of effective sampled universities is listed in Fig. 7.2.

Table 7.1 Number of universities in different regions and their sampling numbers

Region	Number of universities	The sample size of 150%r in each region	Number of universities by random sampling	Number of universities with effective response	Effective sampling rate	Comparison of effective sampling rate with nationally calculated sampling rate (r)
Northeast China	140	7.25	9	6	4.29%	> 3.45%
North China	202	10.46	12	12	5.94%	> 3.45%
East China	373	19.31	20	12	3.22%	< 3.45%
South China	108	5.59	6	5	4.63%	> 3.45%
Central China	173	8.96	9	6	3.47%	> 3.45%
Northwest China	107	5.54	8	5	4.67%	> 3.45%
Southwest China	140	7.25	8	8	5.71%	> 3.45%

Thanks to the active cooperation of teachers in randomly sampled universities, the research group finally collected 6933 questionnaires and excluded 1477 questionnaires that failed the validity testing question. In the end, 5456 valid questionnaires were collected, with an effective rate of 78.64%. Among the respondents, 2341 were male students, accounting for 57.1%, and 3115 were female students, accounting for 42.9%. A total of 3550 respondents majored in science and engineering, accounting for 65.1%, and 1906 respondents majored in culture and management, accounting for 34.9%. The vast majority were freshmen, sophomores, and juniors, accounting for 93.71%. Among the respondents, 45.1% attended 6–8 online learning courses during the pandemic, and 40.1% spent an average of 5–6 h on online learning every day during the pandemic. Details of respondents are listed in Table 7.2.

7.5 Data Analysis Results

This study applied the partial-least-squares regression analysis tool in evaluating the validity and reliability of measurement model and quality in testing the hypothesis between structural models and in assessing the significance of load and path coefficient.

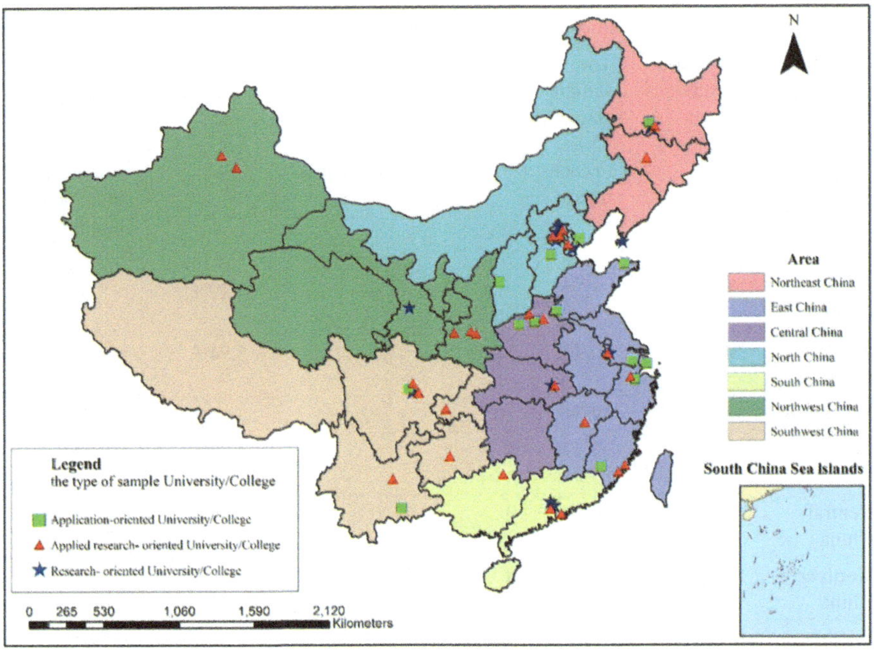

Fig. 7.2 Distribution of questionnaires. *Note* Research-oriented University/College is a University/ College that offers a comprehensive bachelor's degree program, puts Research first, and is committed to high-level talent development and scientific Research and development. Applied research-oriented University/College is a new type of research University/College which emphasizes more on practical application ability, practical ability, innovation ability and creativity than academic research University/College. Application-oriented University/College refers to undergraduate institutions of higher learning with application-oriented orientation rather than scientific research orientation, focusing on undergraduate education, as opposed to the concept of academic University/College

7.5.1 Measurement Model

The study first calculated the value of Cronbach's Alpha, which was used to test the overall reliability of the questionnaire. Generally, when the Cronbach's Alpha value is greater than 0.9, the result is highly credible; when it is between 0.7 and 0.9, it is very credible (Wang et al., 2017; Zhao et al., 2018). In this study, the Cronbach's Alpha value of the total scale is 0. 984, that of each sub-scale is greater than 0.7, and the compound reliability (CR) of each variable is greater than 0.7 (Hair et al., 2009). This implies that the scale has high reliability, good stability, and internal consistency. In terms of convergence validity, the factor loading of each variable is greater than 0.7, and the average variance of each variable is also greater than 0.5 (Hair Jr et al., 2014), indicating that the scale has a good convergence effect (Table 7.3). From the perspective of discrimination validity (Table 7.4), the correlation coefficients among

Table 7.2 Profile of respondents

	Frequency	Percentage
Gender		
Male	2341	42.91
Female	3115	57.09
Grade		
Freshman	1930	35.37
Sophomore	1554	28.48
Junior	1629	29.86
Senior	343	6.29
Professional category		
Science and engineering	3550	65.07
Culture and management	1906	34.93
Number of online courses during the pandemic		
2 and less	173	3.17
3–5 courses	593	10.87
6–8 courses	2462	45.12
9–10 courses	1431	26.23
11 courses or more	797	14.61
Average daily online learning hours duration during the pandemic		
1 h or less	131	2.40
1–2 h (including 2 h)	483	8.85
3–4 h (including 4 h)	1804	33.06
5–6 h (including 6 h)	2190	40.14
7 h and above	848	15.54

all variables are less than the square root of AVE, indicating that the scale has good discrimination validity (Chin, 1998; Fornell & Larcker, 1981).

7.5.2 *Structural Model*

This study repeatedly sampled 5456 cases using the bootstrapping function of SmartPLS3. 0 software to verify the relationships between model variables. Figure 7.3 shows all the path coefficients of the model and the variance of interpretation. The data shows that the highest R2 value of task is 0.645. This means that the artifact and person have the greatest influence on the task of Chinese undergraduates' online learning during the pandemic, as their interpretation variance reached 64.5%. In descending order of interpretation variance, the influence of attitude and perceived behavioral control on intention ranks second at $R^2 = 0.62$. For the influence

Table 7.3 Factor loading, Cronbach's α, CR, AVE of constructs

Variable	Item	Loadings	Cronbach's Alpha	AVE	CR
Person (Online learning state of respondents)	P1	0.846	0.789	0.703	0.877
	P2	0.856			
	P3	0.813			
Artifact (Online learning platforms and resources)	A1	0.883	0.803	0.717	0.884
	A2	0.817			
	A3	0.839			
Task (Online learning task)	T1	0.847	0.864	0.711	0.908
	T2	0.877			
	T3	0.813			
	T4	0.834			
Perceived behavioral control	PBC1	0.864	0.792	0.705	0.878
	PBC2	0.828			
	PBC3	0.826			
Attitude	ATT1	0.800	0.842	0.761	0.905
	ATT2	0.902			
	ATT3	0.911			
Intention	INT1	0.958	0.909	0.917	0.917
	INT2	0.957			

Table 7.4 Simple correlation matrix and discriminatory validity

Construct	Task	Artifact	Person	Attitude	Intention	Perceived behavioral control
Task	0.843					
Artifact	0.692	0.847				
Person	0.765	0.665	0.839			
Attitude	0.581	0.588	0.619	0.872		
Intention	0.572	0.59	0.596	0.776	0.958	
Perceived Behavioral Control	0.659	0.714	0.595	0.584	0.575	0.840

of artifact and task on perceived behavioral control, $R^2 = 0.563$; for the influence of artifact and perceived behavioral control on attitude, $R^2 = 0.467$; finally, for the influence of artifact on person, $R^2 = 0.442$. The Standardized Root Mean Square Residual (SRMR) of the model is 0.065 (less than 0.08), which indicates that the model fits well (Hu & Bentler, 1998). The test results of the research hypothesis are

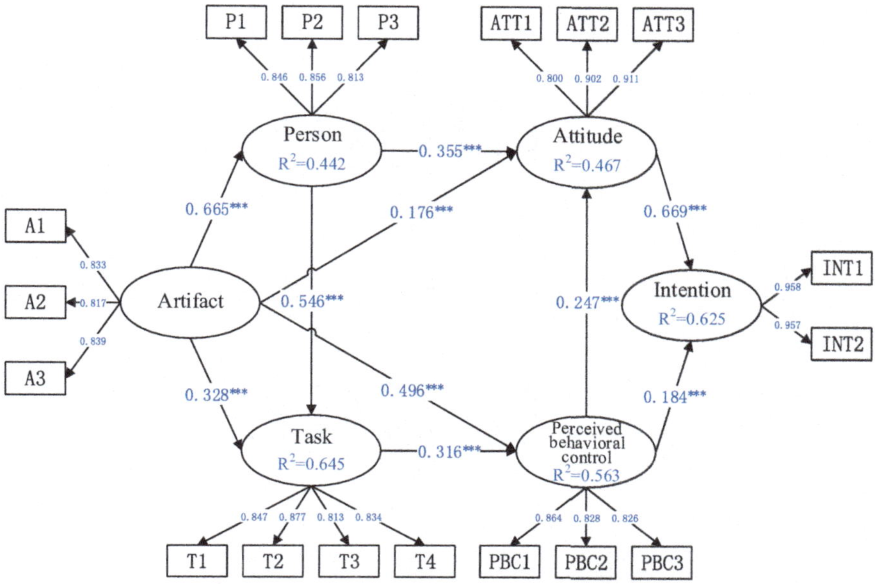

Fig. 7.3 Structural model and paths coefficient

shown in Table 7.5, and the significance of each path reached the level of $P < 0.001$, indicating that the research hypothesis is well supported.

7.5.3 Discussion and Implications

During the pandemic, students in Chinese universities have had intensive access to various online learning platforms through online learning. As learning delivery has shifted from onsite to online, university students have experienced massive online learning activities. Online learning, like other forms of learning, is an experience wrought with emotions (Sharpe & Benfield, 2005). Learners may feel depressed, desperate, happy, warm, lonely, or a sense of belonging (Xiao, 2020). This study reveals the mental flow experience (artifact experience, respondent's online learning state, online learning tasks and challenges) of online learning of Chinese undergraduates during the pandemic. The research results are generally consistent with those of Wang et al. (2017), Davis and Wong (2007), and Lee (2010). Therefore, even in the case of the sudden outbreak, flow experience and theory of planned behavior still have their applicability.

According to the PAT model (Finneran & Zhang, 2003), the artifact is one of the conditions for creating mental flow. In the context of this pandemic, both large-scale online learning and artifact experience (learning platform and resources) have had a positive impact on the process and results of online learning. This study shows that

Table 7.5 The t-value of research hypotheses and path coefficients

Hypothesis	Relationship	Path coefficient	t-value	*p*-value	Conclusion
H1	Artifact–person	0.665	74.022	***	Support
H2	Artifact–task	0.328	24.032	***	Support
H3	Artifact–attitude	0.176	9.658	***	Support
H4	Artifact–perceived behavioral control	0.496	33.538	***	Support
H5	Person–attitude	0.355	22.342	***	Support
H6	Person–task	0.546	42.325	***	Support
H7	Task–perceived behavioral control	0.316	20.672	***	Support
H8	Perceived behavioral control–attitude	0.247	14.471	***	Support
H9	Attitude–intention	0.669	62.682	***	Support
H10	Perceived behavioral control–intention	0.184	15.242	***	Support

the online learning artifact experience has had significant and positive impacts on the individual state of Chinese undergraduates during the pandemic (R = 0.665, P = 0. 000, H1) and also actively promoted their online learning tasks (R = 0.328, P = 0. 000, H2). According to Norman (1988), a good artifact experience enables learners to ignore the artifact itself and focus on the task, thus promoting learners' task attainment. At the same time, the artifact experience of online learning has also significantly and positively influenced the perceived behavioral control of Chinese undergraduates during the pandemic (R = 0.496, P = 0. 000, H4). Through learning platform artifacts, learners become familiar with the functions of the online learning platform, which promotes their control ability of these platforms and is conducive to improving their ability to control the perceived behavior of online learning in future blending learning. According to the empirical results, the positive artifact experience of online learning during the pandemic has significantly and positively influenced students' attitudes toward further online learning (R = 0.176, P = 0. 000, H3). The better the experience of the online learning platform and learning resources, the more students found online learning interesting and the more inclined they were to choose online learning in future learning. This study describes the artifact experience according to the aspects of interactivity (Chen & Jia, 2020), usefulness, resource quality, and richness (Lee, 2010), which constitute important links for learners in using the platform. On the whole, with a good artifact experience, students can significantly improve their learning state (better concentration and happier learning), and be actively driven to complete tasks and challenges. Besides, a good artifact experience also has a positive role in promoting the further use of perceived behavior control and learning attitude in online learning.

For Chinese undergraduates, the individual's psychological state has a significant positive impact on completing tasks and challenges in online learning (R = 0.546, P = 0. 000, H6) and on the attitude to further adopting online learning (R = 0.355, P = 0. 000, H5). This is consistent with the research of Richard (2005) and Sanchez-Franco (2007). The research of Gong et al. (2016) finds that when students experience positive emotions during engagement with blending learning, they accomplish online learning tasks with better performance and higher satisfaction. These results show that learners' individual learning state affects their perception of task and challenges. Similarly, a positive state during online learning can both improve their ability to cope with tasks and challenges in learning and also effectively facilitate further adoption of online learning. During the pandemic, students' positive experiences of tasks and challenges in online learning significantly improved their perceived behavior control in online learning (R = 0.316, P = 0. 000, H7). When students realize that they have clear goals, they can control learning progress and learning outcomes, satisfaction with online learning is effectively enhanced (Jiang et al., 2017), subjective initiative in using the online learning platform and learning resources can be improved, ability to solve problems in the face of difficulties can be strengthened, and they will have greater confidence and greater control of their perception and behavior in further adopting online learning platforms and resources (Jiang et al., 2017).

Consistent with most research results, this study concludes that Chinese undergraduates' perceived behavior control in further adopting online learning has a significant positive impact on their attitude (R = 0.247, $P = 0.000$, H8) and intention to further adopt online learning (R = 0.184, $P = 0.000$, H10) (Moon & Kim, 2001; Zhou, 2016). Through the online learning experience during the pandemic, learners' positive attitude (interest and likeness) to further adopt online learning can significantly enhance their intention to further adopt online learning (R = 0.669, $P =$ 0. 000), which is consistent with the research results of Park (2009), Sawang et al. (2014), and Valtonen et al. (2015).

In the context of the COVID-19 pandemic, this study provides a new perspective for a wide range of online teaching and learning. It also proves the wide applicability of Flow Experience and Theory of Planned Behavior. However, this study only considered the core elements of flow experience and the theory of planned behaviour, in fact, there may be more key influencing factors, such as the impact of COVID-19 itself was not considered. This lays a foundation for further research.

7.6 Conclusion

Due to the long and large scale online education during the pandemic, the Chinese higher education is experiencing dramatic changes and this would result in blended learning in the post-pandemic era. This study aims to provide theoretical and practical support for a better design of a blended learning model in Chinese higher education institutions in the post-pandemic era. The influence of the online learning experience of university undergraduates on their willingness to continue with online learning in the context of the COVID-19 pandemic is explored to help shape post-pandemic education. This is the first study on the recent reforms in teaching and learning in Chinese higher education in the post-pandemic era from a large scale.

To understand the past experience on future intentions for Chinese students to continue online learning, this paper integrates the theory of mental flow experience and the theory of planned behavior, to measure the learning experience of online learners during the COVID-19 pandemic and predict the impact on learners' planned future online learning behavior. This research results in a theoretical basis for the reform of undergraduate education in Chinese universities in the post-pandemic era.

The results show that during the pandemic, factors such as learners' personal learning state, online learning artifacts, and online learning task and challenges all have positive impacts on learners' perceived behavior control and attitude in further adopting online learning, thus actively promoting their willingness to engage in future online learning. Therefore, in order to promote educational reform towards a blended learning model in Chinese universities in the post-pandemic era, this study offers the following specific recommendations regarding online learning platform design, curriculum task design, and learners' learning state.

Learners always perceive online learning as a lonely experience (Simpson, 2014). Therefore, it is important to consider how to fully exploit the unique advantages of online learning artifacts, improve the emotional presence of online learning, enhance the individual learning state of learners, and formulate scientific and reasonable online learning tasks and schedules for the good experience of online learning (Cleveland-Innes & Campbell, 2012). In this way, students could gain pleasure, eliminate their loneliness, and improve the learning effect during online learning. First, as for artifacts, online learning platforms should provide functions able to fully meet the learning needs of learners and provide reliable quality and convenient access to learning resources. In terms of interactivity, there are three types of interaction among online participants, namely teacher–learner, learner–content, and learner–learner (Moore, 1989). Effective interaction in online learning significantly improves online learners' learning attitude, learning satisfaction, and willingness to continue to adopt online learning (Hrastinski, 2009), so online courses should effectively make use of four-dimensional interactions between learner, teacher, resource, and learning. Secondly, as is indicated by the significant positive influence of individual learning state on willingness to engage in online learning, the proposals and developers of online learning should beware of making online learning a burden and solely a form of learning. Instead, they should enable learners to fully feel the pleasure of being immersed in it. For example, some researchers find that colorful and personalized learning content in a multimedia environment could successfully trigger students' positive emotions and improve their academic performance (Plass et al., 2014). Meanwhile, online learners also need to actively organize and adjust their learning state in order to improve the learning effect. Finally, as for the design of curriculum tasks, it is necessary to give learners clear goals and to consider students' level so that curriculum challenges accurately match students' abilities. The task requirements should conform to students' cognitive ability, be able to facilitate students to accomplish tasks, and stimulate their internal motivation for online learning (Joo et al., 2013). For example, by associating tasks with students' real lives and future careers, students can realize the importance and value of learning tasks (Gong et al., 2016), after all, only a good mental flow experience can provide strong support for the normalized development of online learning.

Appendix 1

Survey items

Constructs	No	Items	References
Artifact	A1	The rich online learning platforms and their functions during the pandemic can meet the needs of my online learning	Hoffman and Novak (1996), Finneran and Zhang (2003)

(continued)

(continued)

Constructs	No	Items	References
	A2	In the online study during the pandemic, I can interact effectively with teachers and classmates through the platform	
	A3	There are rich resources (audio, graphics, and text) in online learning during the pandemic, which can satisfy my demands for learning resources	
Task	T1	When I start online learning during the pandemic, I have clear learning goals	Ghani (1995), Finneran and Zhang (2003), Guo et al. (2016)
	T2	When pursuing online during the pandemic, I have strong time management ability and I'm able to control study progress	
	T3	When pursuing online during the pandemic, I am able to complete the course tasks of online learning	
	T4	During the pandemic, I learn online and achieve the same effect as the previous offline offline learning	
Person	P1	When learning online during the pandemic, I can concentrate well	Csikszentmihalyi and LeFevre (1989), Ghani and Deshpande (1994), Moon and Kim (2001)
	P2	I feel very happy in online learning during the pandemic	
	P3	When pursuing online during the pandemic, I have the same feeling as offline offline learning in the past	
Perceived behavioral control	PBC1	I am equipped with the skills and knowledge necessary for online learning	Davis and Wong (2007), Ajzen (1991), Taylor and Todd (1995)
	PBC2	I am familiar with the functions of various online learning platforms and can operate them skillfully	
	PBC3	With the online learning experience during the pandemic, I find it easy to use the online learning platform	
Attitude	ATT1	I use online learning during the pandemic, and it will difficult for me to stop using it now and in the future	Zhou (2016), Thompson et al. (1991)
	ATT2	I think online learning is a more interesting way of learning than offline learning	
	ATT3	I like learning online	

(continued)

(continued)

Constructs	No	Items	References
Intention	INT1	After the pandemic, I am willing to continue to use online learning	Davis (1989), Lung-Guang (2019)
	INT2	I am willing to stick to online learning in the future	

References

Ajzen, I. (1991). The theory of planned behavior. *Organizational Behavior and Human Decision Processes, 50*(2), 179–211. https://doi.org/10.1016/0749-5978(91)90020-t

Alraimi, K. M., Zo, H., & Ciganek, A. P. (2015). Understanding the MOOCs continuance: The role of openness and reputation. *Computers & Education, 80*, 28–38. https://doi.org/10.1016/j.compedu.2014.08.006

Alzahrani, A. I., Mahmud, I., Ramayah, T., Alfarraj, O., & Alalwan, N. (2017). Extending the theory of planned behavior (TPB) to explain online game playing among Malaysian undergraduate students. *Telematics and Informatics, 34*(4), 239–251. https://doi.org/10.1016/j.tele.2016.07.001

Armitage, C. J., & Conner, M. (2001). EYcacy of the theory of planned behaviour: A meta-analytic review. *British Journal of Social Psychology, 40*(4), 471–499.

Bonk, C. J., & Graham, C. R. (2005). *The handbook of blended learning: Global perspectives*. Wiley.

Cao, L. L. (2014). Study on the analysis of learning path and the characteristics of learning behavior in online learning. *Distance Education in China, 4*, 25–30.

Chen, H., Wigand, R. T., & Nilan, M. (2000). Exploring Web users' optimal flow experiences. *Information Technology & People, 13*(4), 263–281. https://doi.org/10.1108/09593840010359473

Chen, P. L., Yin, X. C., & Liu, J. (2019). Research on mobile reading application design based on PAT model. *Design, 9*, 100–103.

Chen, W. Y., & Jia, W. J. (2020). Research on contributing factors of university students' online learning experience. *Journal of East China Normal University(Educational Sciences), 7*, 42–53. https://doi.org/10.16382/j.cnki.1000-5560.2020.07.003

Chin, W. W. (1998). *The partial least squares approach to structural equation modeling*. Psychology Press.

Chu, T.-H., & Chen, Y.-Y. (2016). With good we become good: Understanding e-learning adoption by theory of planned behavior and group influences. *Computers & Education, 92–93*, 37–52. https://doi.org/10.1016/j.compedu.2015.09.013

Cleveland-Innes, M., & Campbell, P. (2012). Emotional presence, learning, and the online learning environment. *International Review of Research in Open and Distance Learning, 13*(4), 269–292. https://doi.org/10.19173/irrodl.v13i4.1234

Conner, M., & Armitage, C. J. (1998). Extending the theory of planned behavior: A review and avenues for further research. *Journal of Applied Social Psychology, 28*(15), 1429–1464.

Csikszentmihalyi, M. (1990). *Flow: The psychology of optimal experience*. Harpers Perennial.

Csikszentmihalyi, M. (1997). *Finding flow: The psychology of engagement with everyday life*. Basic Books.

Csikszentmihalyi, M., & LeFevre, J. (1989). Optimal experience in work and leisure. *Journal of Personality and Social Psychology, 56*(5), 815–822.

Davis, F. D. (1989). Perceived usefulness, perceived ease of use, and user acceptance of information technology. *MIS Quarterly, 13*(3), 319–339. https://doi.org/10.2307/249008

Davis, R., & Wong, D. (2007). Conceptualizing and measuring the optimal experience of the eLearning environment. *Decision Sciences Journal of Innovative Education, 5*(1), 97–126.

Hair Jr, J. F., Sarstedt, M., Hopkins, L., & Kuppelwieser, V. G. (2014). Partial least squares structural equation modeling (PLS-SEM). *European Business Review, 26*(2), 106–121. https://doi.org/10.1108/ebr-10-2013-0128

Finneran, C. M., & Zhang, P. (2003). A person–artefact–task (PAT) model of flow antecedents in computer-mediated environments. *International Journal of Human-Computer Studies, 59*(4), 475–496. https://doi.org/10.1016/s1071-5819(03)00112-5

Finneran, C. M., & Zhang, P. (2005). Flow in computer-mediated environments: Promises and challenges. *Communications of the Association for Information Systems, 15*(4), 82–101.

Fishbein, M., & Ajzen, I. (1975). *Belief, attitude, intention and behaviour: An introduction to theory and research.* Addison-Wesley.

Fornell, C., & Larcker, D. F. (1981). Evaluating structural equation models with unobservable variables and measurement error. *Journal of Marketing Research (JMR), 18*(1), 39–50. https://doi.org/10.2307/3151312

Ghani, J. (1995). *Flow in human computer interactions: Test of a model.* Ablex Publishing Corp.

Ghani, J. A., & Deshpande, S. P. (1994). Task characteristics and the experience of optimal flow in human—Computer interaction. *The Journal of Psychology, 128*(4), 381–391.

Gong, S. Y., Han, Y. S., Wang, L. X., Gao, L., & Xiong, J. M. (2016). The relationships among task value, academic emotions and online learning satisfaction. *e-Education Research, 3*, 72–77. https://doi.org/10.13811/j.cnki.eer.2016.03.010

Guo, Z., Xiao, L., Van Toorn, C., Lai, Y., & Seo, C. (2016). Promoting online learners' continuance intention: An integrated flow framework. *Information & Management, 53*(2), 279–295. https://doi.org/10.1016/j.im.2015.10.010

Hagger, M. S., & Chatzisarantis, N. L. (2009). Integrating the theory of planned behaviour and self-determination theory in health behaviour: A meta-analysis. *The British Journal of Health Psychology, 14*(Pt 2), 275–302. https://doi.org/10.1348/135910708X373959

Hagger, M. S., Chatzisarantis, N. L. D., & Biddle, S. J. H. (2002). A meta-analytic review of the theories of reasoned action and planned behavior in physical activity: Predictive validity and the contribution of additional variables. *Journal of Sport and Exercise Psychology, 24*(1), 3–32. https://doi.org/10.1123/jsep.24.1.3

Hair, J. F., Black, W. C., Babin, B. J., & Anderson, R. E. (2009). *Multivariate data analysis* (7th ed.). Prentice Hall.

Han, H. (2015). Travelers' pro-environmental behavior in a green lodging context: Converging value-belief-norm theory and the theory of planned behavior. *Tourism Management, 47*, 164–177. https://doi.org/10.1016/j.tourman.2014.09.014

Hoffman, D. L., & Novak, T. P. (1996). Marketing in hypermedia computer-mediated environments: Conceptual foundations. *Journal of Marketing, 60*(3), 50–68. https://doi.org/10.2307/1251841

Hrastinski, S. (2009). A theory of online learning as online participation. *Computers & Education, 52*(1), 78–82. https://doi.org/10.1016/j.compedu.2008.06.009

Hu, L.-T., & Bentler, P. M. (1998). Fit indices in covariance structure modeling: Sensitivity to underparameterized model misspecification. *Psychological Methods, 3*(4), 424–453.

Jackson, S. A., & Marsh, H. W. (1996). Development and validation of a scale to measure optimal experience: The flow state scale. *Journal of Sport & Exercise Psychology, 18*(1), 17–35.

Ji, Z. (2013). Online learning behaviour of college students: An empirical study. *Distance Education in China, 5*, 59–63. https://doi.org/10.13541/j.cnki.chinade.2013.05.003

Jiang, Y. J., Bai, X. M., Wu, W. C., & Luo, X. J. (2019). Research on the structural relationship of influencing factors of online learning experience. *Modern Distance Education, 181*(1), 27–36.

Jiang, Z. H., Zhao, C. L., Li, H. X., Liang, Y. Z., & Huang, Y. (2017). A study on learner satisfaction in online open courses: Development, influencing factors and direction. *Modern Distance Education, 171*(3), 34–43. https://doi.org/10.13927/j.cnki.yuan.2017.0026

Joo, Y. J., Lim, K. Y., & Kim, J. (2013). Locus of control, self-efficacy, and task value as predictors of learning outcome in an online university context. *Computers & Education, 62*, 149–158. https://doi.org/10.1016/j.compedu.2012.10.027

Kang, M., Liew, B. T., Kim, J., & Park, Y. (2014). Learning presence as a predictor of achievement and satisfaction in online learning environments. *International Journal on E-Learning, 13*(2), 193–208.

Karahanna, E., & Straub, D. W. (1999). The psychological origins of perceived usefulness and perceived ease-of-use. *Information & Management, 35*(4), 237–250.

Kim, T., & Biocca, F. (1997). Telepresence via television: Two dimensions of telepresence may have different connections to memory and persuasion. *Journal of Computer-Mediated Communication, 3*(2).

Kong, L. L., Ma, Z. Q., Yi, Y. H., & Yang, H. (2017). Influence factors on online learning behavior: Based on the theory of behavioral science. *Journal of Open Learning, 22*(5), 46–53. https://doi.org/10.19605/j.cnki.kfxxyj.2017.05.007

Lee, M.-C. (2010). Explaining and predicting users' continuance intention toward e-learning: An extension of the expectation–confirmation model. *Computers & Education, 54*(2), 506–516.

Lee, Y., & Choi, J. (2013). A structural equation model of predictors of online learning retention. *The Internet and Higher Education, 16*, 36–42. https://doi.org/10.1016/j.iheduc.2012.01.005

Li, Y., & Xiao, S. (2019). A study on the influence of the flow experience on the continuous knowledge sharing willingness of users in knowledge question and answer community—From the perspective of PAT model. *Journal of Modern Information, 39*(2), 111–120.

Liaw, S.-S., & Huang, H.-M. (2003). An investigation of user attitudes toward search engines as an information retrieval tool. *Computers in Human Behavior, 19*(6), 751–765. https://doi.org/10.1016/s0747-5632(03)00009-8

Liu, B., & Zhang, W. L. (2017). Research on the influence factors and structure of the learning experience of online course. *Modern Educational Technology, 27*(9), 107–113. https://doi.org/10.3969/j.issn.1009-8097.2017.09.016

Liu, J. Y., & Sun, S. R. (2005). A summary of theoretical research on behavioral science. *Exploration of Financial Theory, 5*, 36–37. https://doi.org/10.16620/j.cnki.jrjy.2005.05.012

Liu, S.-H., Liao, H.-L., & Pratt, J. A. (2009). Impact of media richness and flow on e-learning technology acceptance. *Computers & Education, 52*(3), 599–607. https://doi.org/10.1016/j.compedu.2008.11.002

Lu, Y., Zhou, T., & Wang, B. (2009). Exploring Chinese users' acceptance of instant messaging using the theory of planned behavior, the technology acceptance model, and the flow theory. *Computers in Human Behavior, 25*(1), 29–39. https://doi.org/10.1016/j.chb.2008.06.002

Lung-Guang, N. (2019). Decision-making determinants of students participating in MOOCs: Merging the theory of planned behavior and self-regulated learning model. *Computers & Education, 134*, 50–62. https://doi.org/10.1016/j.compedu.2019.02.004

MOE. (2020). *The relevant situation of online education in colleges and universities and considerations for the next step.*

Moon, J.-W., & Kim, Y.-G. (2001). Extending the TAM for a world-wide-web context. *Information & Management, 38*(4), 217–230.

Moore, M. G. (1989). Three types of interaction. *American Journal of Distance Education, 3*(2), 1–7. https://doi.org/10.1080/08923648909526659

Nah, Eschenbrenner, & DeWester. (2011). Enhancing brand equity through flow and telepresence: A comparison of 2D and 3D virtual worlds. *MIS Quarterly, 35*(3). https://doi.org/10.2307/23042806

Ndubisi, N. O. (2004). *Factors influencing e-learning adoption intention: Examining the determinant structure of the decomposed theory of planned behaviour constructs* In HERDSA 2004 Conference

Norman, D. A. (1988). *The design of everyday things.* Doubleday.

Novak, T. P. (1999). *Internet marketing: influences, themes and directions* The AMA/Sheth Foundation Doctoral Consortium, University of Southern California.

Paechter, M., Maier, B., & Macher, D. (2010). Students' expectations of, and experiences in e-learning: Their relation to learning achievements and course satisfaction. *Computers & Education, 54*(1), 222–229. https://doi.org/10.1016/j.compedu.2009.08.005

Park, S. Y. (2009). An analysis of the technology acceptance model in understanding university students' behavioral intention to use e-learning. *Educational Technology & Society, 12*(3), 150–162.

Plass, J. L., Heidig, S., Hayward, E. O., Homer, B. D., & Um, E. (2014). Emotional design in multimedia learning: Effects of shape and color on affect and learning. *Learning and Instruction, 29*, 128–140. https://doi.org/10.1016/j.learninstruc.2013.02.006

Richard, M.-O., & Chandra, R. (2005). A model of consumer web navigational behavior: Conceptual development and application. *Journal of Business Research, 58*(8), 1019–1029. https://doi.org/10.1016/j.jbusres.2004.04.001

Rovai, A. P. (2003). In search of higher persistence rates in distance education online programs. *Internet and Higher Education, 6*(1), 1–16.

Sakulwichitsintu, S., Colbeck, D., Ellis, L., & Turner, P. (2015). *Online peer learning: What influences the students' learning experience* 2015 IEEE 15th International Conference on Advanced Learning Technologies.

Sánchez-Franco, M. J. (2007). Exploring the influence of gender on the web usage via partial least squares. *Behaviour & Information Technology, 25*(1), 19–36. https://doi.org/10.1080/01449290500124536

Sawang, S., Sun, Y., & Salim, S. A. (2014). It's not only what I think but what they think! The moderating effect of social norms. *Computers & Education, 76*, 182–189. https://doi.org/10.1016/j.compedu.2014.03.017

Sharpe, R. J., & Benfield, G. (2005). The student experience of e-learning in higher education: A review of the literature. *Brookes eJournal of Learning and Teaching, 1*(3), 1–9.

Sheppard, B. H., Hartwick, J., & Warshaw, P. R. (1988). The theory of reasoned action: A meta-analysis of past research with recommendations for modifications and future research. *Journal of Consumer Research, 15*(3), 325–343. https://doi.org/10.1086/209170

Shin, N. (2006). Online learner's 'flow' experience: An empirical study. *British Journal of Educational Technology, 37*(5), 705–720. https://doi.org/10.1111/j.1467-8535.2006.00641.x

Simpson, O. (2014). *Supporting students for success in online and distance education* (3rd ed.). Routledge.

Taylor, S., & Todd, P. A. (1995). Understanding information technology usage: A test of competing models. *Information Systems Research, 6*(2), 144–176.

Thompson, R. L., Higgins, C. A., & Howell, J. M. (1991). Personal computing: Toward a conceptual model of utilization. *MIS Quarterly, 15*(1), 124–143.

Trevino, L. K., & Webster, J. (1992). Flow in computer-mediated communication: Electronic mail and voice mail evaluation and impacts. *Communication Research, 19*(5), 539–573.

Valtonen, T., Kukkonen, J., Kontkanen, S., Sormunen, K., Dillon, P., & Sointu, E. (2015). The impact of authentic learning experiences with ICT on pre-service teachers' intentions to use ICT for teaching and learning. *Computers & Education, 81*, 49–58. https://doi.org/10.1016/j.compedu.2014.09.008

Wang, L. Y., He, Y. F., & Tian, J. H. (2020). Constructing and verifying a model of integrating multimodal data from online learning behaviors. *Distance Education in China, 6*, 22–30. https://doi.org/10.13541/j.cnki.chinade.2020.06.002

Wang, W., Shi, R. H., & Li, X. N. (2017). Factors influencing the continuance intention of online learning from the perspective of flow theory: The case of MOOC platforms. *Distance Education in China, 5*, 17–23. https://doi.org/10.13541/j.cnki.chinade.20170517.004

Waugh, M., & Su-Searle, J. (2014). Student persistence and attrition in an online M. S. Program: Implications for program design. *International Journal on E-Learning, 13*(1), 101–121.

Webster, J., Trevino, L. K., & Ryan, L. (1993). The dimensionality and correlates of flow in human-computer interactions. *Computers in Human Behavior, 9*(4), 411–426.

Xiao, J. H. (2020). Aspects of online learning-a review of online learning: The student experience. *Distance Education in China, 8*, 62–68.

Yang, L. N., & Yan, Z. J. (2011). An empirical study of online learning behavior from the perspective of information technology adoption. *Distance Education in China, 4*, 36–40. https://doi.org/10.13541/j.cnki.chinade.2011.04.015

Yemane, T. (1967). *Statistics: An introductory analysis* (2nd ed). Harper and Row.

Yi, M. Y., & Hwang, Y. (2003). Predicting the use of web-based information systems: Self-efficacy, enjoyment, learning goal orientation, and the technology acceptance model. *International Journal of Human-Computer Studies, 59*(4), 431–449. https://doi.org/10.1016/s1071-5819(03)00114-9

Zhao, C. L., Wang, X., & Ma, C. X. 2018. Impact of Perceived Interactivity on the online learners' continuance intention: Based on S-O-R Perspective. *Modern Distance Education, 177*(3), 12–20. https://doi.org/10.13927/j.cnki.yuan.2018.0024

Zhou, M. (2016). Chinese university students' acceptance of MOOCs: A self-determination perspective. *Computers & Education, 92–93*, 194–203. https://doi.org/10.1016/j.compedu.2015.10.012

Chapter 8
Seismic Evacuation Decision-Making During COVID-19 Lockdown-Lunding Earthquake Case Study

Abstract This chapter contributes to evacuation and emergency research by examining seismic evacuation decision-making by surveying the areas affected by the Luding earthquake on 5 September 2022, when most parts of Sichuan province were experiencing strict pandemic restrictions. Using these data and per the emergency evacuation decision-making mechanism, we developed six hierarchical series of logistic regression models. Our major results suggest that (1) Socio-demographic parameters have shown different correlations with the study's dependent variables in each stage of those hierarchical models, (2) Respondents at home at the time of the earthquake were more likely to identify the earthquake risk than those who stayed outdoors; the former group showed less willingness to evacuate, (3) Rural residents have perceived higher earthquake risks than urban residents, and (4) Loss of job, change in income due to COVID-19 restrictions, and difficulty accessing daily supplies during the lockdown affected the residents' risk assessment and evacuation decision-making negatively. Insights into these aspects are expected to contribute to a better understanding of evacuation behavior during double disasters by modifying emergency response regulations and providing the residents with information about emergencies during pandemic restrictions.

Keywords Evacuation behavior · COVID-19 · Lockdown · Luding earthquake · Emergency

8.1 Introduction

In locations where emergencies happen, evacuation of people is considered an integral aspect of protecting public safety. Apart from the crucial role of evacuation management (Beverly & Bothwell, 2011; Yongjun et al., 2022) and emergency management (Homa & Wei, 2022), evacuation behavior and decision-making have become important research topic in the field of emergency management and public safety. Raphael mentioned that human behavior during a disaster depends on "preexisting social grouping and, level of perceived physical threat or damage,

Y. Ao and H. Bahmani, *Navigating Complexity: Understanding Human Responses to Multifaceted Disasters*, https://doi.org/10.1007/978-981-99-8207-3_8

attachment to the family" (Raphael, 2005). Experience and the results of pivotal studies revealed that human behavior during an emergency varies based on risk perception (Ştefan et al., 2021; Yiannis & Georgia, 2022). People's risk perception can be affected by many factors, including individual characteristics and environmental factors (Bahmani et al., 2023; Verdière et al., 2015). In recent research, scholars evaluated "the effects of demographic, socio-economic features of people, and disaster experience/knowledge" on evacuation behavior. (Dongkwan et al., 2018) The results illustrate that age, income, housing type, drill experience, and knowledge of the shelter location affect people's evacuation behavior. Another similar research also explores the effects of "human physical, environmental, and psychological/sociological features" on human evacuation behavior during a disaster (Rodrigueza & Estuar, 2021). Wu et al. explored the impact of public knowledge on disaster avoidance before, during, and after the disaster, (Jiaojiao et al., 2022) showing that the residents' disaster knowledge can positively affect their evacuation choice. Studies in the field of earthquake risk perception and preparedness manifested socio-demographic characteristics of residents such as age, education, borning place, and disaster experience and loss have significant impacts on earthquake preparedness (Ao et al., 2021; Ao et al., 2020a, 2020b, 2020c) Considering the environmental impact on people's behavior during an emergency, the built environment features considerably affect the occupants' evacuation behavior (Ao et al., 2020b; Ştefan et al., 2021).

On the other hand, scholars have also studied evacuation decision-making under different scenarios. Koshiba et al. explored earthquake post-evacuation choices of students considering thirteen scenarios and analyzed under which situation the respondents were more likely to return to the teaching building after the evacuation (Koshiba & Suzuki, 2018). The subject of research conducted by Chen et al. has been self-evacuation and guided evacuation of children under non-emergency and emergency situations (2020). Similarly, public evacuation behavior considering emergency and non-emergency egress revealed differences in exit choice and directional flow (Haghani & Sarvi, 2016). The matter of low visibility during an evacuation and its effect on the evacuees' evacuation route choice was another crucial topic studied by Ding and Sun and Xie et al. (2020), (2020). In spite of the extensive research on human evacuation behavior, to the best of our knowledge, research on how the occurrence of two disasters simultaneously affects individuals' evacuation behavior is limited.

In early January 2020, the World Health Organization (WHO) announced that the coronavirus disease 2019 (COVID-19) is a pandemic. Later, WHO suggested that every country take public health actions to suppress the spread of COVID-19 (Organization, 2020); however, COVID-19 has soon become a global health crisis, causing unpredictable health and social challenges. Frequent outbreaks of COVID-19 across the globe resulted in lockdown strategies limiting residents' mobility and affecting daily life. The existing studies pointed out that lockdowns have had more adverse side effects than positive consequences (Schippers, 2020). From 2020–2021, 90% of countries experienced a decline in economic activity, more severe than two World Wars, the Great Depression of the 1930s, the emerging economy debt crises of the 1980s, and the 2007–09 global financial crisis (Bank, 2022). Apart from the negative impact of lockdowns on economic growth (Xuan et al., 2020), there are several studies demonstrating lowered socio-economic status, family violence and conflict (Malika et al., 2021; Obja & Sarmistha, 2021), food insecurity (Anjani et al., 2022), and increased stress and anxiety (Hassan et al., 2020; Malika et al., 2021) imposed by lockdown strategies. A study by Whytlaw et al. has identified that COVID-19 and lockdowns exacerbated health, economic, and social vulnerabilities associated with the increased concern for hurricane evacuation and sheltering during a pandemic (Whytlaw et al., 2021). That research broadened the perspective about the increased disaster vulnerabilities due to the COVID-19 pandemic and has also stated the increasing need to study policy adjustments to meet the vulnerable groups' needs. Given the rising prevalence of lockdowns due to pandemics, studying human evacuation behavior facing emergencies during lockdowns is infrequently discussed in contrast to other topics.

This study aims to understand seismic evacuation behavior during a pandemic lockdown when people's choices for taking proactive measures might be limited. We hypothesized that a significant correlation exists between the restrictions of COVID-19 lockdowns and individuals' evacuation decision-making. Moreover, considering the strict pandemic restrictions, socio-demographic and built environment features will lead to different results in modeling evacuation behavior. In order to prove this hypothesis, the study has chosen Sichuan province as a case study, which experienced a wide lockdown in September 2022. With restricted control, the Luding earthquake happened on 5 September when many residents were not allowed to leave their living compounds. The remainder of the paper is structured as follows. Section 8.2 presents the research materials and methodology by explaining sample and data collection methods, variables selection, and model construction. The study's detailed results are given in Sect. 8.3, and Sect. 8.4 is allocated to the discussion. Finally, the paper will finish by stating the conclusions and policy implementations in Sect. 8.5, while Sect. 8.6 suggests future research directions.

8.2 Materials and Methods

8.2.1 Sample and Data Collection

Considering Sichuan province is one of the most quake-prone provinces in China with a history of some of the deadliest earthquakes in the last decades, such as the Wenchuan earthquake, 2008 (Huang et al., 2011), our study focuses on one of the last earthquakes that happened in Sichuan province, Luding earthquake, 5 September 2022. The Luding earthquake occurred in Luding county, Ganzi prefecture, with an epicenter of 29.59° north latitude and 102.08° east longitude and a focal depth of 16 km. The earthquake magnitude was 6.8 Richter, affecting a wide area (Administration, 2022).

Besides the high magnitude, during the Luding earthquake, most parts of Sichuan province were in lockdown due to the COVID-19 pandemic. Chengdu, the capital of the Sichuan province, has reported 1332 local COVID-19 cases from 12 August to 4 September. The majority of shops (except for those providing daily public necessities) and all public places were closed due to the emergence of COVID-19. During the pandemic restrictions, the residents in Sichuan province were required to undergo nucleic acid tests once a day. Each resident has been allocated by health code to show their health status, nucleic acid test results, and traveling history. Those who held the red health code, including the confirmed and suspicious Civid-19 patients, their close contacts who have not passed isolations, and people from high risks areas, were prohibited from entering public places. Those who have shown respiratory symptoms and close contact with COVID-19 patients who passed isolation held a yellow code, and their traveling was limited. Other residents not included in the last two categories were supposed to have a green health code without traveling restrictions upon providing nucleic acid tests. Except for special vehicles carrying out emergency tasks, other vehicles were prohibited from passing on the road. Public transportation maintained regular operation, but providing a green health code and negative nucleic acid tests for 24 h was necessary to enter all public places. Gatherings were strictly prohibited, and the residents were asked to return home quickly after the daily nucleic acid tests. The high-risk areas with confirmed COVID-19 cases were under restricted lockdown, and the residents were not allowed to leave their apartments, and medium-risk areas strictly implemented the principle of "people are not allowed to leave the communities". Outside the high and middle-risk areas, residents were advised to stay home, and traveling was strictly forbidden, letting each household only arrange for one person to go out for two hours a day to buy living materials from nearby markets. Therefore, this study has chosen the affected cities/counties by the Luding earthquake, as shown in Fig. 8.1, to conduct the survey.

8.2.2 Variables Selection and Hypothesis Construction

Several models and frameworks have been applied to understand human evacuation behavior and decision-making during an emergency, among which the Belief- Desire-Intention (BDI) model (Bratman et al., 1988), Protection Motivation Theory (PMT) (Rogers, 1975), social force model (Helbing & Molnár, 1995), Physical, Emotional, Cognitive, and Social Factors (PECS) (Urban, 1997), and fast and frugal decision-making tree (Gigerenzer, 2007) are the most popular ones. However, considering the unpredictable nature of most emergencies, not only is studying individuals' responses necessary to understand the individuals' evacuation behavior but also gaining a clear perception of the reasons behind that response is critical to investigating the evacuation decision-making process. Furthermore, this study has applied the Emergency Decision-Making Mechanism (EDMM) proposed by Perry and Greene (1982). As shown in Fig. 8.2, human emergency decision-making should undergo "risk identification, risk assessment, and risk mitigation/reduction". The first step involves

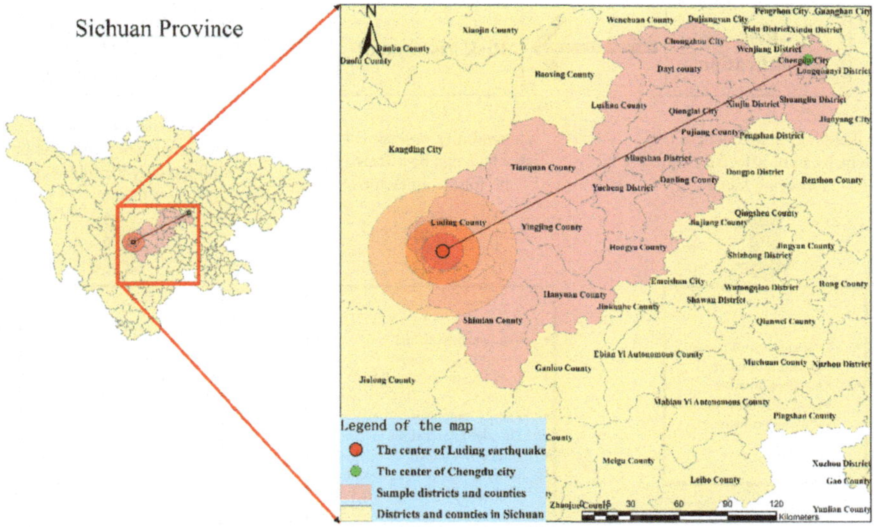

Fig. 8.1 Location map of the sample villages, counties, and cities

identifying whether the emergency risk really exists or not, which can be affected by the warning notice and its content and the authority of the warning notice sources. Once the threat's existence is confirmed, the individuals will assess their vulnerability against the risk by evaluating the likelihood and severity of the risk. Finally, if the person ensures the threat's existence and their vulnerability against it, the next step would be evaluating the effectiveness of risk mitigation/reduction measures and taking proactive measures (Gantt and Gantt, 2012; Perry & Green, 1982).

In order to provide firm logic, an extensive literature review has been conducted, and a wide range of factors influencing human behavior during emergencies has been identified. Subsequently, the researchers construct the study's hypothesis referring to the EDMM. The influential factors on each stage of the EDMM have been extracted from the literature, as shown in Table 8.1. Note that because the study tended to understand human seismic evacuation behavior during the COVID-19 lockdown, the researchers have hypothesized that the relevant potential factors such as areas' COVID-19 risk level, personal health code of individuals, access to daily supplies during the lockdown, and vaccination statues of the individuals may have correlations with the three phases of the EDMM. Those factors have been identified through brainstorming techniques with the research team members who have also experienced the Luding earthquake. During the online brainstorming meeting, the research question and the applied theory were clearly explained, and all the suggestions were recorded unbiasedly. Later, we reviewed those suggestions based on the study by Saunders et al. (2007). The study then revised the research hypothesis and constructed the questionnaire. Table 8.1 demonstrates the study's hypothesis while each independent variable's expected impact on the dependent variables has been shown by " + " if the literature agreed on the positive effect of the predictor variable

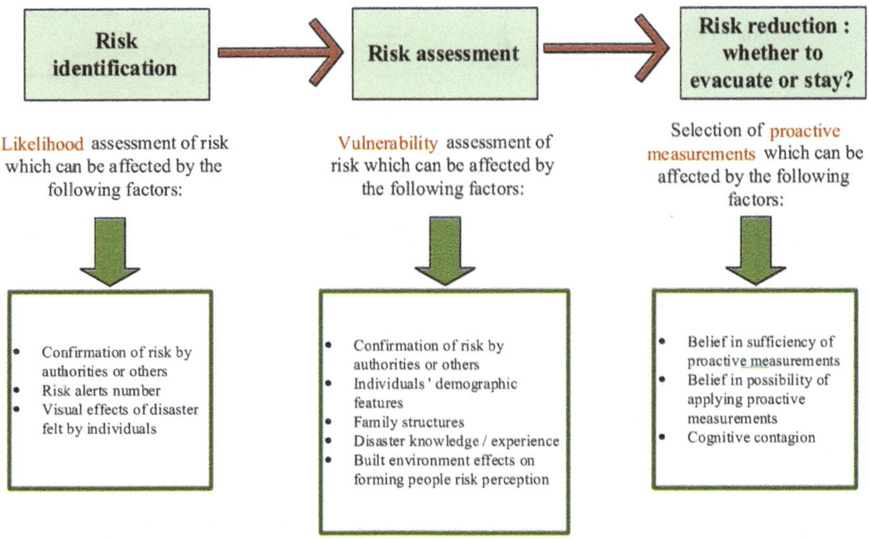

Fig. 8.2 Emergency decision-making mechanism (EDMM), based on research by Perry and Greene

on the dependent one, while " – " indicates the negative association between dependent and independent variable. In case of no clear association between the variables, the study constructs a neutral hypothesis, and the association between the variables has been examined throughout the case-by-case analysis. Those factors marked as "not included" have not been assumed in modeling the corresponding dependent variables.

The dependent variables in this study are earthquake risk identification, risk assessment, and risk mitigation/reduction strategies. Among them, earthquake risk identification and mitigation/reduction are a 0–1 type dichotomous variable, and earthquake risk assessment is an ordered multi-categorical variable consisting of four sub-factors (more details can be found in Sect. 8.2.3). Based on the characteristics of dependent variables, this study applied binary logistic regression to assess relationships between the selected independent factors and earthquake risk identification and mitigation/reduction. Moreover, ordinal logistic regression has been used to explore associations between the predictor variables and earthquake risk assessment. The models can be described through equations such as follows:

$$Y_1 = LogitP = \alpha_0 + \alpha_1 \times age + \alpha_2 \times gender + \cdots + \alpha_i \times i_{independentfactor} + \varepsilon_n$$

$$Y_2 = OlogitP = \beta_0 + \beta_1 \times age + \beta_2 \times gender + \cdots + \beta_j \times j_{independentfactor} + \delta_n$$

$$Y_3 = LogitP = \gamma_0 + \gamma_1 \times age + \gamma_2 \times gender + \cdots + \gamma_k \times k_{independentfactor} + \theta_n$$

Table 8.1 Hypothesis structure; list of dependent and independent variables

Independent variables		Dependent variables		
Factors	Sources	Risk identification	Risk assessment	Risk mitigation/ reduction
Age	[9, 42, 43]	(+)	(+)	(+)
Gender	[9, 43]	(+)	(+)	(+)
Community type	Brainstorming	(case-by-case basis)	(case-by-case basis)	(case-by-case basis)
Job	[9]	(case-by-case basis)	(case-by-case basis)	(case-by-case basis)
Education	[9]	(+)	(+)	(+)
Marital statues	[9]	(+)	(+)	(−)
Homeownership	[9]	(case-by-case basis)	(+)	(+)
Household size	[9]	(+)	(+)	(−)
Presence of vulnerable groups in family	[9]	(not included)	(+)	(-)
Social ties between family members	[19, 44]	(not included)	(+)	(+)
Earthquake experience	[9, 11, 45], [46]	(+)	(+)	(+)
Family income	[9]	(not included)	(+)	(+)
Housing type	[9]	(case-by-case-basis)	(+)	(−)
Housing structure	[14]	(not included)	(case-by-case-basis)	(case-by-case-basis)
House floor	[14]	(+)	(+)	(−)
Indoor/outdoor during the earthquake	Brainstorming	(+)	(+)	(−)
Residential area's COVID-19 risk level	Brainstorming	(−)	(−)	(−)
Health code statues	Brainstorming	(case-by-case-basis)	(−)	(−)
COVID-19 vaccination statues	Brainstorming	(not included)	(case-by-case-basis)	(−)
Access to daily necessities	Brainstorming	(not included)	(−)	(−)
Epidemic control policies	Brainstorming	(−)	(−)	(−)
Income change (decrease)	[9, 28]	(not included)	(+)	(−)

(continued)

Table 8.1 (continued)

Independent variables		Dependent variables		
Factors	Sources	Risk identification	Risk assessment	Risk mitigation/ reduction
Lost of job	Brainstorming	(not included)	(+)	(−)
Personal safety attitude	Brainstorming	(not included)	(case-by-case-basis)	(+)
Intention to follow others	[44]	(+)	(+)	(+)
Sensitivity to ground shaking	[14]	(+)	(not included)	(not included)
Access to earthquake information sources	[47]	(+)	(not included)	(not included)
Trust in information sources	[48, 49]	(+)	(not included)	(not included)
Risk identification	[48, 49]		(+)	(+)
Risk assessment	[48, 50]	(not included)	(not included)	(+)

*** Risk identification will impact risk assessment. Risk identification and assessment will impact risk reduction

where dependent variables of the models are Y1, Y2, and Y3, referring to earthquake risk identification, assessment, and mitigation/reduction strategies, respectively. The independent variable of each model is shown in Table 8.1, and $\alpha 0, \ldots, \alpha i, \beta 0, \ldots, \beta j, \gamma 1, \ldots, \gamma 1$ are the coefficient values of the independent variables. The models' residuals are εn, δn, θn, and SPSS 26 has been employed to analyze the data. To construct the models, the independent variables have been entered within three steps, the hierarchical method, to explore the contribution of each type of variable to the proposed models.

8.2.3 *Questionnaire Design and Survey Data Collection*

The questionnaire generated after brainstorming based on the research's hypothesis has undergone another round of brainstorming and modification procedure. The final questionnaire consisted of four parts focusing on (1) socio-demographic information, (2) housing and built environment information, (3) COVID-19 and lockdown information, (4) risk identification, assessment, and mitigation strategies, and (5) disaster information resources and individuals' attitude during the earthquake. In addition, the validity detection test was performed by asking the respondents to choose "agree" if they read the questionnaire carefully.

The first section asks respondents basic information, including age, gender, location during the earthquake, education, marital status, household size and social ties, presence of vulnerable groups, income, homeownership, community type, and earthquake experience. On the other hand, the participants were asked about their housing type, structure, and floor. The third section attempts to collect information about the respondents' community pandemic strategies and situations during the earthquake. COVID-19 risk levels, personal health code statutes, vaccination statutes, lockdown policies, access to daily necessities, and change in income and job due to the pandemic regulations have been presented in this section. Three dependent variables have been asked by proposing six questions, among which four were to measure people's earthquake risk assessment. Concern about loss of life or injuries, loss of properties or financial loss, suffered by further disasters such as being trapped inside the building or experiencing aftershocks, and anxiety were the measurements of earthquake risk assessment (Lazo et al., 2015). To calculate the final score of this dependent variable, if each respondent's answer to all the four sub-factors is "strongly agree", then the final score of risk assessment is "strongly agree". If the answers include three "strongly agree", we categorized the total score as "agree" and so on, as shown in Table 8.2.

Although previous studies in China applied face-to-face surveying methods (Hongtai et al., 2022), because of the large size of the study's regions, mobility limitations caused by the pandemic control regulations, and in order to represent a sample that is a qualified representative of the study's population by giving all the population an equal chance to participate in the survey (Sharma, 2017), from 13 September through 21 September 2022, we randomly surveyed the regions' households shown in Fig. 8.1 by contacting local colleagues. Later by the snowball sampling method, other people were familiar with the research and filled out the questionnaire. The study has also illustrated that every household chooses one representative to participate to avoid possible correlations among the study's variables. Additionally, to ensure no possible impact of household membership on the regression models, the study carefully monitored the respondents' geographical location during the earthquake, which was obtained by asking the participants. During the data acquisition process, the research team kept a daily update of the filled questionnaires in each district/county and re-shared the online survey through different platforms if the filled questionnaires in some districts were insufficient. Finally, by running a public survey in the affected areas, the database included 1512 questionnaires has been collected. The validity detection test, performed by asking the respondents to choose "agree" if they read the questionnaire carefully, has identified 254 unacceptable filled

Table 8.2 Risk assessment score

Strongly agree	Agree	Neutral	Disagree	Strongly disagree
If all answers are "strongly agree"	If three of the answers are "strongly agree"	If two of the answers are "strongly agree"	If one of the answers is "strongly agree"	If none of the answers are "strongly agree"

Table 8.3 The study's sample distribution across the region

City/district/county	Num. of participants	City/district/county	Num. of participants
Chengdu city	679	Tianquan county	72
Shuanliu district	93	Yingjing county	32
Chongzhou city	103	Hanyuan county	120
Dayi county	54	Shimian county	54
Qionglai county	66	Luding county	64
Pujiang county	53	Yingjing County	32
Mingshan county	7	Leshan city	14
Danling county	51	Dongpo District	13
Lushan county	36	Renshou County	16
Yucheng district	58	Other areas	52
Hongya county	57	Total number	1257

questionnaires. Moreover, the first round of data screening omitted one more filled questionnaire due to its geographical location outside our study location. Therefore, the sample number is 1257 valid questionnaires, and the geographical location of the respondents during the earthquake is given in Table 8.3.

8.3 Results

8.3.1 Descriptive Analysis

Women formed 61% of the study's population. The age of the respondents ranged from 11 to 73, while the mean age of the population was nearly 40 years old. The study has categorized the jobs into a detailed categorization system, among which "civil servant/public institution worker/government worker" and "employee of state-owned enterprises" account for almost 40% of the sample population. In addition, most of the sample population comes from non-agricultural communities in Sichuan province.[1] Regarding education and marriage status, most respondents hold undergraduate/ junior college degrees (n = 842) and are married (n = 792). Seventy percent of the sample population live in self-purchased houses, and 89.7% of the participants have responded that the number of their family members present at the time of the earthquake was four or fewer, among which children were the most populated vulnerable group[2] at the time of the quake. Family social ties referring to the average

[1] Agricultural community, as a unique component in the process of urbanization transition, refers to the successive marginal community in the process of urbanization.

[2] Vulnerable groups in this study has been divided into children under 12 years old, elderlies above 65 years old, and people with disabilities.

time spent with the family within the last 15 days before the earthquake, show that 44.6% of the participants have always stayed with their families. The study also asked the participants to choose their family's monthly income, and 81.5% of the respondents have experience dealing with earthquakes which were expectable due to the geographical location of Sichuan province and the occurrence of several earthquakes. Respondents' socio-demographic information is presented in Table 8.4.

Figure 8.3 compares the residents' mitigation/reduction choices with their earthquake risk identification. 93% of the residents have successfully gained information about the earthquake occurrence, while only 65% of the respondents have evacuated. On the other hand, based on Fig. 8.4, the residents were worried more about the loss of life or injuries caused by the earthquake. On the other hand, the respondents were less concerned about financial loss since this factor has gained the lowest mean.

8.3.2 Models Results

8.3.2.1 Test of Logistic Regression Assumptions

The Variance Inflation Factor (VIF) was used to test multicollinearity among the dependent variables. All VIF values of the variables used in this study were less than two, and the tolerance value is higher than 0.1, indicating no multicollinearity problem exists between all explanatory variables considering three different dependent variables (Yunong et al., 2015). Moreover, by running covariate analysis, no correlation between independent variables has been found since the Spearman rho scores for correlation coefficient are less than 0.7 (Saunders et al., 2007). Thirdly, the researchers analyzed the data to identify outliers by calculating the cook's distance. The results indicate that all the scores were below 0.015, much less than one (Anderson et al., 2016). Moreover, no extreme outliers have been detected based on the scatter plot of the cook's distance.

8.3.2.2 Regression Results of the Risk Identification Stage

A stepwise binary logistic regression was used to examine whether the study's independent variables were associated with the likelihood of people's earthquake risk identification. The variables have been entered within three steps (hierarchical method) to explore the contribution of each type of variable to the proposed models. Results are presented in Table 8.5 with 95% confidence intervals (CIs) and *P*-values.

The first model, including demographic and built environment information, was statistically significant (X2 (28, N = 1257) = 48.630, p-value = 0.009 < 0.05), suggesting that this model could distinguish between those who have and have not identified the earthquake risk. The model explained between 3.08% (Cox and Snell R-square) and 9.07% (Nagelkerke R-square) of variances in dependent variables and

Table 8.4 Respondents' socio-demographic information

	Maximum	Minimum	Mean	Mode	Std. Deviation
Age	11	73	34.04	24	10.968
	Options			Frequency	%
Gender	Male = 0			489	38.9
	Female = 1			768	61.1
Community type	Rural areas = 1			359	28.6
	Agricultural communities = 2			152	12.1
	Urban areas (not agriculture-related) = 3			746	59.3
Job	Unemployed = 1			71	5.6
	Farmer = 2			25	2.0
	Professional (such as teacher/doctor/lawyer, and others) = 3			104	8.3
	Civil servant/Public institution worker/Government worker = 4			267	21.2
	Employee of state-owned enterprises = 5			249	19.8
	Employee of private enterprises = 6			128	10.2
	Part-time worker = 7			46	3.7
	Self-employed individual = 8			49	3.9
	Manufacturing employee = 9			45	3.6
	Student = 10			217	17.3
	Retired = 11			13	1.0
Education level	Primary school and below = 1			24	1.9
	Middle school/Secondary vocational school = 2			106	8.4
	High school/Technical secondary school/Higher vocational school = 3			147	11.7
	Undergraduate/Junior college = 4			842	67.0
	Postgraduate = 5			127	10.1
	Doctor degree = 6			11	0.9
Marriage statues	Unmarried = 0			465	37.0
	Married = 1			792	63.0
Homeownership	Self-purchased house = 1			882	70.2
	Renting house = 2			234	18.6
	Dormitories and rooms provided by companies = 3			43	3.4
	Living with families/friends = 4			6	0.5
	Self-built house = 5			65	5.2
	Others = 6			27	2.1

(continued)

Table 8.4 (continued)

	Maximum	Minimum	Mean	Mode	Std. Deviation
Household size	Single person = 1			285	22.7
	1 people = 2			263	20.9
	2 people = 3			264	21.0
	3–4 people = 4			315	25.1
	5–6 people = 5			88	7.0
	More than 7 people = 6			42	3.3
Presence of children	No = 0			979	77.9
	Yes = 1			278	22.1
Presence of elderlies	No = 0			1058	84.2
	Yes = 1			199	15.8
Presence of disabled people	No = 0			1213	96.5
	Yes = 1			44	3.5
Family social ties	Never = 1			169	13.4
	Rarely = 2			87	6.9
	Sometimes = 3			112	8.9
	Often = 4			328	26.1
	Always = 5			561	44.6
Income	Less than 3000 RMB = 1			164	13.0
	3000–5000 RMB = 2			337	26.8
	5000–7000 RMB = 3			255	20.3
	7000–9000 RMB = 4			142	11.3
	More than 9000 RMB = 5			359	28.6
Earthquake experience	No = 0			1024	81.5
	Yes = 1			233	18.5

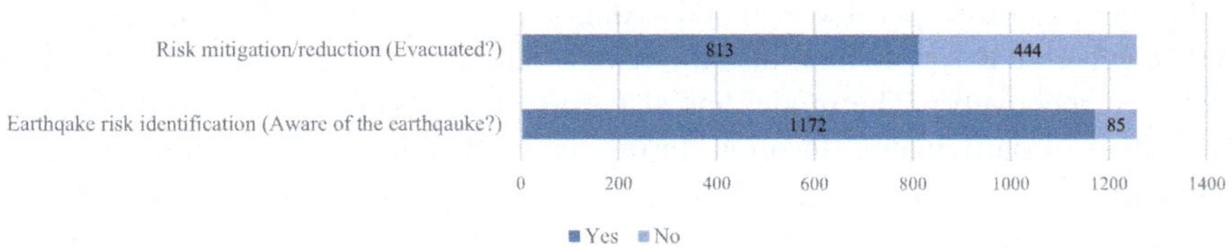

Fig. 8.3 Earthquake risk identification and mitigation/reduction among the residents

correctly classified 93.2% of the cases. People's location during the earthquake significantly affected the risk identification (X2 (2, N = 1257) = 9.722, *p*-value = 0.008 < 0.05). Individuals at home were 2.316 times more likely to identify earthquake risk than participants outdoors during the earthquake.

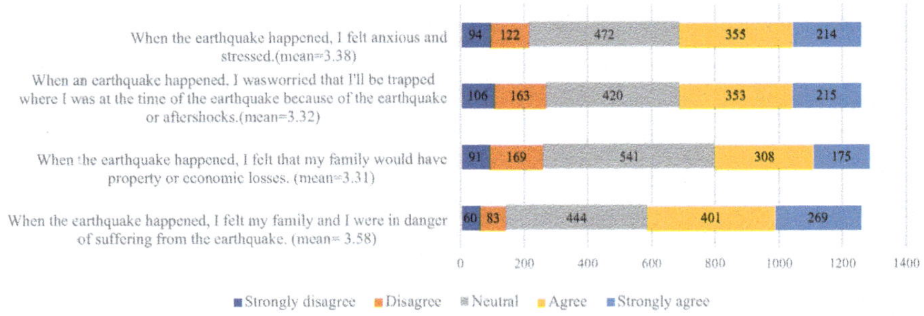

Fig. 8.4 The mean value of the residents' earthquake risk assessment factors

The second model, including COVID-19 variables, assessed their impacts on risk identification and did not show significant improvement compared to the first mode (X2 (9, N = 1257) = 13.924, *p*-value = 0.124 > 0.05). However, the second model has significantly improved compared to the null model (X2 (40, N = 1257) = 65.993, *p*-value = 0.006 < 0.05). In terms of Cox & Snell R-square and Nagelkerke R-square, the second model has successfully explained 4.9% and 12.4% of the variances in dependent variables, respectively. Location during the earthquake significantly impacted risk identification as people who stayed at home were 2.140 times more likely to identify the earthquake risk than those in outside areas. Having prior experience dealing with earthquakes is associated with earthquake risk recognition. Keeping other factors consonant, people with earthquake experience are 2.384 times more likely to identify the earthquake risk than those without experience. Moreover, the results confirmed that residents living in single houses show a reverse likelihood of understanding earthquake risk (OR = 0.585).

Lastly, the study entered the information regarding disaster information resources into the third model resulting in a significant improvement in coefficient score compared to the second model (X2 (15, N = 1257) = 69.126, *p*-value = 0.000 < 0.05), and overall improved significantly comparing to the null mode (X2 (52, N = 1257) = 131.699, *p*-value = 0.000 < 0.05). The Cox and Snell R-square scores and Nagelkerke R-square have also improved compared to the last model (9.9% and 25.5%, respectively). The model has also correctly classified 93.5% of the cases. The effect of participants' location during the earthquake is still significant, indicating that people who stayed at home were 2.106 times more likely to identify the earthquake than those who reported their location in outdoor areas such as streets. Interestingly, there was no association between household size and the dependent variable in the third model; however, the odds ratio of people with earthquake experience has increased compared to the second model (OR = 2.751). Statistical tests illustrated a significant difference between the people with different sensitivity to ground shaking and their earthquake risk identification (X2 (2, N = 1257) = 63.205, *p*-value = 0.000 < 0.05). Logistic regression reveals that both groups of people who are and are not sensitive to ground shaking are less likely to identify earthquake risk than those with unclear perspectives regarding their sensitivity to ground shaking.

Table 8.5 Binary logistic regression results of risk identification

Variables		Model 1		Model 2		Model 3	
		Coefficient	Odds ratio	Coefficient	Odds ratio	Coefficient	Odds ratio
Age		0.003	1.003	0.006	1.006	-0.001	0.999
Gender-female		– 0.168	0.845	– 0.187	0.830	-0.092	0.912
Community type (ref = urban areas) X2 (2, N = 1257) = 2.656, *p*-value = 0.256	Agricultural communities	– 0.181	0.835	– 0.100	0.905	– 0.182	0.834
	Rural areas	(– 0.582*)	0.559	(– 0.561*)	0.571	– 0.428	0.652
Job (ref = others) X2 (11, N = 1257) = 13.751 *p*-value = 0.274	Unemployed	– 0.717	0.488	– 0.532	0.587	– 0.825	0.438
	Farmer	18.601	########	18.776	########	18.113	########
	Professionals	– 0.580	0.560	– 0.504	0.604	– 0.943	0.389
	Civil servants etc	– 0.252	0.777	– 0.193	0.824	– 0.379	0.685
	Employee of state-owned enterprises	– 0.834	0.434	– 0.730	0.482	– 0.990	0.372
	Employee of private enterprises	– 0.757	0.469	– 0.479	0.620	– 0.706	0.494
	Part-time worker	18.350	########	18.466	########	18.371	########
	Self-employed individual	0.880	2.412	1.090	2.974	0.846	2.330
	Manufacturing employee	0.982	2.670	0.952	2.591	0.666	1.946
	Student	– 0.486	0.615	– 0.366	0.694	– 0.812	0.444
	Retired	– 0.814	0.443	– 0.658	0.518	– 0.259	0.772
Education		0.134	1.144	0.179	1.196	0.177	1.193
Houseownership (ref = others) X2 (5, N = 1257) = 4.310, *p*-value = 0.506	Self-purchased house	0.292	1.339	0.191	1.210	0.604	1.830
	Renting house	0.579	1.784	0.417	1.518	0.681	1.976
	Dormitories and rooms provided by companies	0.593	1.810	0.364	1.440	0.700	2.013

(continued)

Table 8.5 (continued)

Variables		Model 1		Model 2		Model 3	
		Coefficient	Odds ratio	Coefficient	Odds ratio	Coefficient	Odds ratio
	Living with families/friends	19.343	########	19.143	########	19.563	########
	Self-built house	2.364*	10.629	2.278*	9.759	2.534*	12.605
Household size		– 0.017	0.983	– 0.029	0.971	– 0.096	0.908
Earthquake experience? Yes		0.760	2.138	0.869*	2.384	1.012*	2.751
Housing type- single building		– 0.509	0.601	(– 0.536*)	0.585	– 0.510	0.601
House floor		0.156	1.169	0.172	1.188	0.066	1.068
Location during the earthquake (ref = outdoor areas) X2 (2, N = 1257) = 9.722, *p*-value = 0.008	Indoor area (home)	0.840**	2.316	0.761**	2.140	0.745**	2.106
	Indoor area (other indoor spaces besides home)	0.317	1.372	0.334	1.396	0.243	1.275
Marriage status-married		0.224	1.251	0.208	1.231	0.194	1.214
Area's COVID-19 risk				0.000	0.000	0.000	0.000
Personal health code (ref = no code) X2 (3, N = 1257) = 11.428, *p*-value = 0.01	Green			– 19.181	0.000	– 19.936	0.000
	Yellow			-20.698	0.000	– 21.552	0.000
	Red			– 0.134	0.874	– 0.643	0.526
Epidemic control policies (ref = others) X2 (5, N = 1257) = 7.988, *p*-value = 0.157	Isolation at special isolation points			1.095	2.988	2.208	9.095
	Strict home isolation			0.200	1.221	– 0.043	0.958
	Home isolation, you rarely can go out			0.585	1.796	0.456	1.578
	Home isolation is not strictly implemented			0.161	1.175	– 0.063	0.939

(continued)

Table 8.5 (continued)

Variables		Model 1		Model 2		Model 3	
		Coefficient	Odds ratio	Coefficient	Odds ratio	Coefficient	Odds ratio
	No clear epidemic control measures			– 0.152	0.859	– 0.516	0.597
Sensitivity to ground shaking (ref = unclear) X2 (2, N = 1257) = 63.205, *p*-value = 0.000	Yes					(– 1.811***)	0.163
	No					(– 1.536***)	0.215
Access early warning notice (ref = unclear) X2 (2, N = 1257) = 18.982, *p*-value = 0.000	Yes					(– 1.072**)	0.342
	No					(– 1.114***)	0.328
Access to official websites (ref = unclear) X2 (2, N = 1257) = 2.054, *p*-value = 0.358	Yes					-0.349	0.706
	No					0.324	1.382
Access to social media (ref = unclear) X2 (2, N = 1257) = 1.090, *p*-value = 0.580	Yes					0.164	1.178
	No					0.348	1.417
Blind follow behavior (ref = unclear) X2 (2, N = 1257) = 5.060, *p*-value = 0.08	Yes					– 0.580	0.560
	No					– 0.108	0.898
Trust in personal feeling						0.285*	1.331
Trust in early warning notice						– 0.304	0.738
Trust in social media						0.095	1.099
Trust in official websites						0.4755**	1.609
Trust in strangers (blind follow)						– 0.175	0.839

(continued)

Table 8.5 (continued)

Variables	Model 1		Model 2		Model 3	
	Coefficient	Odds ratio	Coefficient	Odds ratio	Coefficient	Odds ratio
Constant	0.675	1.964	18.926	########	20.395	########
Number of observations	1257		1257		1257	
Nagelkerke R-square	0.97		0.124		0.255	
LR chi2(χ2)	48.63		62.572		131.698	
Prob > chi2(χ2)	**0.009**		**0.005**		**0.000**	

* $p < 0.10$, ** $p < 0.05$ and *** $p < 0.01$

On the other hand, the more the person trusts his personal feeling, the more likely he perceives the occurrence of the earthquake (OR = 1.331). Keeping other factors consonant, people with access to early warning notice showed a lower likelihood of identifying earthquake risk. Finally, people who trusted more on the earthquake information provided by the official websites have a higher odds ratio to identifying earthquake risk (OR = 1.609).

8.3.2.3 Regression Results of the Risk Assessment Stage

Ordinal logistic regression was used to examine whether the sets of predictor variables were associated with people's risk assessment likelihood. Firstly, the researchers have input socio-demographic and built environment information to construct the model. The first model has shown significant improvement in correctly predicting the cases compared to the null model (X2 (38, N = 1257) = 67.69, *p*-value = 0.002 < 0.05). The model explained between 5.2% (Cox and Snell R-square, 5.7% (Nagelkerke R-square), and 2.2% (McFadden R-square) of variances of dependent variables. On the other hand, the study has hierarchically added factors related to COVID-19 policies and risk identification and built second and third models, both statistically significant, as shown in Table 8.6. During the nesting modeling, the amount of the explained variance in dependent variables has also increased (Cox and Snell R-square model-2 = 0.089, Nagelkerke R-square model-2 = 0.098, McFadden R-square model-2 = 0.038, Cox and Snell R-square model-3 = 0.112, Nagelkerke R-square model-3 = 0.122, McFadden R-square model-3 = 0.048).

The first model indicates that residents who live in rural areas were 1.450 times more likely to perceive higher earthquake risk. However, the second and third models failed to prove an association between community type and risk assessment. On the other hand, gender's effect on the risk assessment is significant based on the first and second models. The findings reveal that males evaluated the earthquake's potential risks lower than women (ORmodel-1 = 0.827, ORmodel-2 = 0.895). Different categories of jobs have also been proven to impact individuals' risk assessment significantly; however, only "part-time workers" and "farmers" were significantly associated with the likelihood of peoples' earthquake risk perception based on all three models. Keeping other factors consonant except for job, farmers are the likeliest group to assess the earthquake risk higher (ORmodel-3 = 3.166). Interestingly, per all three models, the individuals without children at the time of the earthquake are less likely to perceive a high earthquake risk than people with children around. Moreover, the regression analysis of all the models could prove that age positively affected the risk assessment.

Change in family incomes has also had a significant negative association with the risk assessment level as people with a higher loss of income have an odds ratio less than one (ORmodel-2 = 0.918, ORmodel-3 = 0.869), illustrating that these group of people were less likely to perceive the earthquake risk highly. In addition, people who lost their jobs due to the pandemic have shown a higher tendency to evaluate the earthquake risk higher (the odds ratio for people who have not experienced

Table 8.6 Ordinal logistic regression results of risk assessment

Variables		Model 1		Model 2		Model 3	
		Coefficient	Odds ratio	Coefficient	Odds ratio	Coefficient	Odds ratio
Gender-male		− 0.190	0.827	(− 0.111*)	0.895	− 0.148	0.863
Community type (ref = urban areas) X2 (8, N = 1257) = 21.79 *p*-value = 0.006	Rural areas	0.372***	1.450	0.118	1.125	0.212	1.236
Job (ref = others) X2 (44, N = 1257) = 72.818, *p*-value = 0.004	Agricultural communities	0.188	1.207	0.044	1.045	0.137	1.147
	Unemployed	0.868**	2.382	0.3724*	1.451	0.482	1.620
	Farmer	1.519**	4.566	0.703**	2.020	1.152**	3.166
	Professionals	0.480	1.615	0.169	1.184	0.089	1.093
	Civil servants etc	0.492	1.635	0.177	1.194	0.127	1.136
	Employee of state-owned enterprises	0.604*	1.829	0.272	1.313	0.252	1.286
	Employee of private enterprises	0.895***	2.448	0.440**	1.554	0.543	1.721
	Part-time worker	1.091***	2.977	0.489**	1.630	0.714*	2.043
	Self-employed individual	0.997**	2.709	0.45*	1.568	0.459	1.582
	Manufacturing employee	0.764*	2.147	0.306	1.358	0.265	1.303
	Student	0.363	1.437	0.108	1.114	0.019	1.019
	Retired	0.237	1.268	0.029	1.029	− 0.187	0.830
Marriage status- unmarried		0.187	1.206	0.113	1.120	0.223	1.249
Houseownership (ref = others) X2 (20, N = 1257) = 21.119, *p*-value = 0.197	Self-purchased house	− 0.484	0.616	− 0.321	0.726	− 0.444	0.641
	Renting house	− 0.333	0.717	− 0.273	0.761	− 0.306	0.736

(continued)

Table 8.6 (continued)

Variables		Model 1		Model 2		Model 3	
		Coefficient	Odds ratio	Coefficient	Odds ratio	Coefficient	Odds ratio
	Companies dormitories	– 0.237	0.789	– 0.109	0.897	– 0.124	0.883
	Living with families/ friends	– 0.088	0.916	– 0.247	0.781	– 0.187	0.829
	Self-built house	– 0.655	0.519	– 0.363	0.695	– 0.556	0.573
Children?- no		(– 0.329**)	0.720	(– 0.225***)	0.799	(– 0.362**)	0.696
Elderlies? - no		– 0.086	0.917	– 0.017	0.983	– 0.014	0.986
Disabled people?- no		– 0.291	0.747	– 0.146	0.864	– 0.124	0.883
BE housing type- no-single building		0.093	1.097	0.047	1.048	0.091	1.095
BE housing structure (ref = others) X2 (20, N = 1257) = 23.02, *p*-value = 0.288	Stone and wood structure	0.813	2.254	0.628	1.874	1.113	3.042
	Brick and wood structure	0.545	1.725	0.372	1.450	0.498	1.646
	Brick and concerete structure	0.150	1.162	0.217	1.243	0.289	1.335
	Steel and concerete structure	0.053	1.054	0.143	1.153	0.171	1.187
	I do not know	0.932	2.540	0.629	1.876	1.085	2.961
Earthquake experience?- no		0.414	1.514	0.140	1.150	0.327	1.387
Location during the earthquake (ref = outdoor areas) X2 (39, N = 1257) = 8.464, *p*-value = 0.39	Indoor area (home)	0.078	1.081	0.090	1.094	0.235	1.265
	Indoor area (other indoor spaces besides home)	– 0.031	0.969	– 0.020	0.981	– 0.064	0.938
Age		0.017***	1.017	0.008***	1.008	0.017***	1.017
Education		0.115	1.122	0.065	1.067	0.106	1.112

(continued)

Table 8.6 (continued)

Variables		Model 1		Model 2		Model 3	
		Coefficient	Odds ratio	Coefficient	Odds ratio	Coefficient	Odds ratio
Household size		– 0.053	0.948	– 0.045	0.956	– 0.060	0.942
Family social ties		0.019	1.019	0.026	1.026	0.044	1.045
Income		– 0.038	0.963	– 0.027	0.973	– 0.010	0.990
BE house floor		– 0.009	0.991	0.011	1.011	0.004	1.004
Area's COVID-19 risk				– 0.011	0.989	– 0.020	0.980
Income change				(– 0.085***)	0.918	(– 0.140**)	0.869
Personal health code (ref = no code) X2 (12, N = 1257) = 16.143, *p*-value = 0.185	Green			– 1.325	0.266	(– 2.527*)	0.080
	Yellow			(– 1.564*)	0.209	(– 3.044**)	0.048
	Red			– 0.908	0.403	– 1.561	0.210
Vaccination status (ref = no vaccinated) X2 (8, N = 1257) = 5.222, *p*-value = 0.734	Basic vaccinations			0.070	1.072	0.079	1.082
	Booster shot			0.069	1.072	0.094	1.098
Epidemic control policies (ref = others) X2 (20, N = 1257) = 36.321, *p*-value = 0.014	Isolation at special isolation points			0.219	1.244	0.561	1.752
	Strict home isolation			– 0.223	0.800	– 0.321	0.725
	Home isolation, you rarely can go out			– 0.257	0.773	– 0.455	0.634
	Home isolation is not strictly implemented			– 0.219	0.803	– 0.442	0.643
	No clear epidemic control measures			– 0.061	0.941	– 0.068	0.934
Lost of job? No				(– 0.17**)	0.844	(– 0.331**)	0.718

(continued)

Table 8.6 (continued)

Variables		Model 1		Model 2		Model 3	
		Coefficient	Odds ratio	Coefficient	Odds ratio	Coefficient	Odds ratio
Access to daily necessities (ref = special distribution channels are available at any time.) X2 (16, N = 1257) = 51.932, *p*-value = 0.000	Very difficult			(− 0.528*)	0.590	− 0.736	0.479
	Difficult			0.031	1.032	− 0.025	0.975
	As usual			− 0.216	0.806	(− 0.462*)	0.630
	More convenient than usual			− 0.255	0.775	− 0.498	0.608
Feeling safe going outside? No						0.151	1.163
Blind follow behavior (ref = unclear) X2 (8, N = 1257) = 43.186, *p*-value = 0.000	Yes					(− 0.560***)	0.571
	No					(− 0.674***)	0.510
Risk identification = yes						0.0311	1.031
Number of observations		1257		1257		1257	
Scaled deviance		3036.462 (df = 4822)		2999.566 (df = 4869)		2957.397 (df = 4865)	
LR chi2(χ2)		67.69		107.358		149.527	
Prob > chi2(χ2)		0.002		0.000		0.000	

* $p < 0.10$, ** $p < 0.05$ and *** $p < 0.01$

the loss of job is 0.844 and 0.718 in the second and third models, respectively.) Although the regression analysis showed a significant association between people with yellow health code and risk assessment, the risk assessment did not significantly differ across the sub-groups of personal health code (X2 (12, N = 1257) = 16.143, *p*-value = 0.185 > 0.05). Therefore, the study did not conclude whether having a yellow personal health code is associated with higher earthquake risk assessment. Access to daily necessities during lockdown has also been recognized to influence the individuals' earthquake risk assessment. However, the second and third models identified different sub-groups associated with the dependent variable. Based on the second model, individuals with difficulty getting daily necessities compared to people with special distribution channels during lockdown were less likely to perceive high earthquake risk (ORmodel-2 = 0.59). In contrast, people who have been provided with regular access to daily necessities during lockdown have shown less likelihood to evaluate the earthquake risk as high per the third model (ORmodel-2 = 0.63). The third model has proved the significant association between blind follow behavior[3] and earthquake risk assessment; however, giving the people who had an unclear perspective on whether they would follow strangers in case of an earthquake as the reference category, there is not that much difference between odds ratio of individuals who will or will not follow others.

8.3.2.4 Regression Results of the Risk Mitigation/reduction Stage

Binary logistic regression was used to evaluate whether the selected independent variables were associated with the likelihood of people's earthquake risk reduction/mitigation decision-making. The predictor variables have been entered within three steps to explore the contribution of each type of variable to the proposed models (Table 8.7). The first model, including demographic and built environment information, showed acceptable goodness of fit (X2 (38, N = 1257) = 143.468, *p*-value = 0.000 < 0.05), demonstrating the improvement of model fit compared to the null model. The model has also correctly classified 66.3% of the cases (specificity = 30%, sensitivity = 86.2%). 10.8% (Cox and Snell R-square) and 14.8% (Nagelkerke R-square) of the variances in dependent variables have been explained by the model. The second and third models hierarchically added pandemic regulation and their subsequent impacts, risk identification, and assessment. The overall fitness of the second and third models has significantly increased compared to their last model (X2model-2 (17, N = 1257) = 38.77, *p*-value = 0.000 < 0.05), X2model-3 (5, N = 1257) = 115.798, *p*-value = 0.000 < 0.05), while both models show significant improvement compared to the null mode. While the second model has successfully explained 13.5% (Cox and Snell R-square) and 18.6% (Nagelkerke R-square) of variances in dependent variables, Nagelkerke R-square shows great improvement in the third model (0.29). However, Cox and Snell R-square has decreased (0.211). The models' ability to correctly predict the cases has increased to 69.7% and 73.2% of the cases were correctly predicted by the second and third models, respectively, among which those models could considerably predict people who chose to evacuate.

The logistic regression illustrated the significant impact of community type on the residents' risk mitigation decisions. Based on all models, residents who live in agricultural communities were more likely to evacuate than residents in urban areas. Meanwhile, rural residents have shown a higher likelihood of evacuating than agricultural communities based on the first model. Keeping other factors consonant, the job categories have also been proven to influence earthquake mitigation decision-making and demonstrated a significant correlation with the dependent variable, among which self-employed individuals were the likeliest group to evacuate based on the first and third model (ORmodel-1 = 5.601, ORmodel-3 = 3.426). The statistical test of house ownership did not reveal a significant difference between this predictor's sub-categories and risk mitigation strategies (X2 (5, N = 1257) =

[3] Blind follow behavior indicates the evacuees' intention to follow the crowd's direction during evacuation.

Table 8.7 Binary logistic regression results of risk mitigation/reduction

Variables		Model 1		Model 2		Model 3	
		Coefficient	Odds ratio	Coefficient	Odds ratio	Coefficient	Odds ratio
Age		0.009	1.009	0.004	1.004	− 0.001	0.999
Gender-female		0.328	1.389	0.354*	1.425	0.23***	1.350
Community type (ref = urban areas) X2 (2, N = 1257) = 39.799, *p*-value = 0.000	Agricultural communities	0.92***	2.509	0.820***	2.269	0.798***	2.220
	Rural areas	0.206***	1.229	0.115	1.122	0.193	1.212
Job (ref = others) X2 (11, N = 1257) = 30.117, *p*-value = 0.002	Unemployed	0.852*	2.344	1.017**	2.765	0.758	2.133
	Farmer	1.275**	3.580	1.53**	4.637	1.005	2.732
	Professionals	1.094**	2.986	1.081**	2.949	0.688	1.989
	Civil servants etc	1.226***	3.406	1.20**	3.320	0.884**	2.421
	Employee of state-owned enterprises	0.917**	2.502	0.902**	2.467	0.550	1.734
	Employee of private enterprises	0.864**	2.373	0.859**	2.359	0.461	1.586
	Part-time worker	0.703	2.020	0.655	1.926	0.207	1.231
	Self-employed individual	1.723***	5.601	1.781***	5.938	1.150**	3.159
	Manufacturing employee	1.67***	5.314	1.797***	6.012	1.231**	3.426
	Student	0.529	1.698	0.448	1.566	0.162	1.176
	Retired	1.509**	4.520	1.372*	3.943	1.220	3.388
Education		− 0.012	0.988	0.020	1.020	− 0.023	0.977
Houseownership (ref = others) X2 (5, N = 1257) = 10.807, *p*-value = 0.055	Self-purchased house	(-0.97*)	0.379	(− 0.935*)	0.393	− 0.950	0.387
	Renting house	(− 0.96*)	0.377	− 0.893	0.410	− 0.944	0.389
	Dormitories and rooms provided by companies	(− 1.234*)	0.291	− 0.985	0.374	(− 1.308*)	0.270

(continued)

Table 8.7 (continued)

Variables		Model 1		Model 2		Model 3	
		Coefficient	Odds ratio	Coefficient	Odds ratio	Coefficient	Odds ratio
	Living with families/ friends	− 2.471	0.084	(− 2.641**)	0.071	(-3.039***)	0.048
	Self-built house	(− 1.507**)	0.222	(− 1.539**)	0.215	(− 1.717**)	0.180
Household size		(− 0.103*)	0.902	− 0.074	0.929	− 0.048	0.953
Earthquake experience? yes		0.541*	1.718	0.398	1.489	0.296	1.345
Housing type- single house		− 0.218	0.804	− 0.214	0.807	− 0.151	0.860
House floor		− 0.060	0.941	− 0.054	0.947	− 0.066	0.936
Location during the earthquake (ref = outdoor areas) X2 (2. N = 1257) = 40.588, *p*-value = 0.000	Indoor area (home)	(− 0.585**)	0.557	(− 0.463**)	0.629	(− 0.427*)	0.653
	Indoor area (other indoor spaces besides home)	0.198	1.219	0.222	1.248	0.261	1.299
Marriage status- married		(− 0.482**)	0.618	(− 0.553**)	0.575	(− 0.5***)	0.589
Children? Yes		0.063	1.065	0.003	1.003	− 0.104	0.902
Elderlies? Yes		0.242	1.274	0.202	1.224	0.142	1.153
Disabled people? Yes		0.055	1.057	0.164	1.178	0.070	1.072
Family social ties		0.045	1.046	0.054	1.055	0.022	1.022
Income		− 0.08	0.923	− 0.080	0.923	− 0.043	0.958
BE housing structure (ref = others) X2 (5, N = 1257) = 8.502, *p*-value = 0.131	Stone and wood structure	− 1.113	0.328	− 1.246	0.288	− 1.507	0.222
	Brick and wood structure	− 0.801	0.449	− 0.815	0.443	− 1.068	0.344
	Brick and concerete structure	− 1.209	0.299	− 1.200	0.301	(− 1.435*)	0.238
	Steel and concerete structure	(− 1.323*)	0.266	(− 1.289*)	0.276	(− 1.486*)	0.226
	I do not know	− 1.273	0.280	− 1.154	0.315	− 1.485	0.226

(continued)

Table 8.7 (continued)

Variables		Model 1		Model 2		Model 3	
		Coefficient	Odds ratio	Coefficient	Odds ratio	Coefficient	Odds ratio
Area's COVID-19 risk				0.143	1.154	0.147	1.158
Personal health code (ref = no code) X2 (3, N = 1257) = 0.881, *p*-value = 0.83	Green			2.431	11.369	2.216	9.170
	Yellow			2.349	10.475	2.053	7.792
	Red			2.141	8.504	1.593	4.917
Epidemic control policies (ref = others) X2 (5, N = 1257) = 44.430 value = 0.000	Isolation at special isolation points			– 0.326	0.722	– 0.267	0.766
	Strict home isolation			– 0.477	0.621	– 0.599	0.549
	Home isolation, you rarely can go out			– 0.412	0.662	– 0.495	0.610
	Home isolation is not strictly implemented			0.355	1.426	0.221	1.247
	No clear epidemic control measures			0.051	1.052	-0.070	0.933
Vaccination status (ref = no vaccinated) X2 (2, N = 1257) = 3.421, *p*-value = 0.183	Basic vaccinations			0.649	1.914	0.638	1.894
	Booster shot			0.770	2.161	0.776	2.173
Income change				(– 0.207**)	0.813	(– 0.195**)	0.822
Lost of job? Yes				0.094	1.098	0.059	1.061

(continued)

Table 8.7 (continued)

Variables		Model 1		Model 2		Model 3	
		Coefficient	Odds ratio	Coefficient	Odds ratio	Coefficient	Odds ratio
Access to daily necessities (ref = very difficult) X2 (4, N = 1257) = 5.373, *p*-value = 0.251	Difficult			0.630	1.877	0.192	1.212
	As usual			0.836	2.307	0.535	1.708
	More convenient than usual			0.321	1.378	– 0.023	0.977
	Special distribution channels are available at any time			1.322**	3.754	0.897	2.453
Feeling safe going outside? Yes						– 0.097	0.907
Blind follow behavior (ref = unclear) X2 (2, N = 1257) = 126.679, *p*-value = 0.000	Yes					(– 1.222***)	0.295
	No					(– 1.166***)	0.311
Risk identification = yes						0.774***	2.169
Risk assessment						0.362	1.437
Constant		1.836	0.280	– 1.938	0.144	– 1.684	0.186
Number of observations		1257		1257		1257	
Nagelkerke R-square		0.148		0.186		0.29	
LR chi2(χ2)		143.468		182.238		298.036	
Prob > chi2(χ2)		0.000		0.000		0.000	

* $p < 0.10$, ** $p < 0.05$ and *** $p < 0.01$

10.807, *p*-value = 0.055 > 0.05); therefore, the study cannot make a conclusion about association between homeownership and evacuation decision. The impact of gender on the dependent variable was significant in the third model, pointing out that women were 1.350 times more likely to evacuate during the earthquake. Individuals' location during an earthquake was associated with risk mitigation decision-making based on all the models. Also, people at home during the earthquake are less likely than individuals staying outdoors to evacuate. The odd ratio for this statement increased from the first model to the third one. The study has also validated the negative correlation between marriage and evacuation decision-making since married people were less likely to evacuate. Although the results of the first model indicate the negative association between household size and evacuation decision-making (OR = 0.902), the presence of vulnerable groups and family social ties did not significantly impact the individuals' risk mitigation decision-making. The study could not conclude whether housing structure affected earthquake risk mitigation strategies since statistical tests failed to detect any significant difference between different housing structures considering the residents' evacuation choice (X2 (5, N = 1257) = 8.502, *p*-value = 0.131 > 0.05).

The second and third models agreed upon the negative association between the loss of income caused by the pandemic restrictions and earthquake risk mitigation strategies. The higher loss of income the individuals experienced, the lower the likelihood of evacuation (ORmodel-2 = 0.813, ORmodel-3 = 0.822). Finally, the third model has identified the significant correlations between successful risk identification and evacuation decision-making. The residents who were successfully notified of an earthquake were 2.169 times more likely to evacuate. Statistical tests illustrated there is a significant difference between the people with a different attitude toward blind follow behavior and their earthquake risk mitigation decision (X2 (2, N = 1257) = 126.679, *p*-value = 0.000 < 0.05). However, logistic regression reveals that people who would and would not follow a stranger in case of an earthquake are less likely to evacuate than those who have an uncertain attitude toward following others.

8.4 Discussion

This study attempted to examine Emergency Decision-Making Mechanism (EDMM) during a vast epidemic lockdown. The research has constructed sets of hypotheses by considering three dependent variables based on the EDMM and recognizing the predictor variables. Finally, the researchers empirically analyzed the relationships between the dependent and independent variables using binary and ordinal logistic regressions based on the survey data of the residents who experienced the Luding earthquake during the COVID-19 epidemic prevention time. Socio-economic and built environment information, pandemic situation and lockdown policies, and individuals' attitudes and earthquake information have been input to construct hierarchical models to illustrate the improvement of models' predicting ability in each step. Compared to the existing studies examining evacuation decision-making in a single step, this research has applied EDMM to identify stages leading to seismic evacuation decision-making. Moreover, this study has pioneered in analyzing evacuation decision-making during the COVID-19 prevention strategies to examine whether the special conditions imposed by the pandemic restrictions affected the residents' earthquake risk identification, assessment, and mitigation/reduction strategy.

The findings of this study demonstrated some similarities and differences with the existing literature. Several scholars have evaluated the associations between age and gender and evacuation decision-making. The empirical results of this study illustrated that while age is significantly correlated with the residents' risk assessment, risk identification and evacuation decision-making are not associated with the age of the participants, which did not provide enough evidence to accept the study's hypothesis in this regard. On the other hand, risk assessment and risk mitigation/reduction strategies are highly dependent on gender since females were more likely to highly assess the earthquake risks and decide to evacuate, possibly due to gender differences in social roles and evacuation motivations. Although some studies did not find a significant direct relationship between age and gender and disaster risk perception (Lazo et al., 2010; Wim et al., 2012), this research demonstrated that the impact of gender and age on the EDMM is consistent with the findings of the study by some studies (Fatemeh & Babak, 2021; Lazo et al., 2015). Marital status has only negatively affected the residents' evacuation behavior, as married people were less willing to evacuate and could partially confirm the research's hypothesis; however, this finding is contrary to the outcomes of a study by Mohajeri et al. (Fatemeh & Babak, 2021). Although we expected that individuals being accompanied by vulnerable groups at the time of the earthquake might positively affect earthquake risk assessment while having a negative correlation with the residents' evacuation choice, a different pattern of association was observed in the presence of children at the time of the earthquake and the residents' risk perception which is consistent with findings by Murray-Tuite et al. (2012).

No association has been found between having earthquake experience and risk assessment, contrary to the findings by Zheng et al. (Lingyu et al., 2020); however, at this point, we must recognize that there is a tradeoff between experience in dealing with earthquakes and earthquake risk identification and evacuation decision-making. The study has also hypothesized that education and homeownership would positively affect EDMM; however, the findings could not confirm the existence of any association between the mentioned variables. In addition, a significant negative association was observed between household size and evacuation decision-making, which contradicts the findings of some studies (Fatemeh & Babak, 2021); however, no association has been detected between household size and risk assessment and identification. The study has also found that job categories correlate with the residents' risk assessment and evacuation choice. Although framers have perceived higher vulnerability against the earthquake, manufacturing employees were the most likely group to evacuate. Additionally, community type and the residents' risk assessment and risk mitigation/reduction strategies indicated that they are very closely linked. The rural residents perceived higher earthquake risks than urban residents; however, participants residing in agricultural communities were more willing to evacuate.

Although the respondents who were at home at the time of the earthquake were more likely to identify the earthquake risk than those who stayed outdoors, the former group showed less willingness to evacuate. Therefore, the study could also accept the null hypothesis regarding a correlation between the residents' location at the time of the earthquake and risk identification and evacuation choice. Among the variables rendering built environment impact on EDMM, only housing type has shown a negative association with earthquake risk identification, as the residents in single houses were less likely to recognize the occurrence of the earthquake, which might be due to less crowdedness in their environment. Interestingly, the housing floor and structure did not correlate with any stage of EDMM.

The research could also conclude that the residents' risk assessment and evacuation decision-making are associated with the broader aspects of epidemic conditions and lockdown regulations. In contrast, no correlation between this group's factor and earthquake risk identification has been found. Those who have experienced a higher loss of income due to the pandemic and lockdown have perceived lower earthquake risk and were less willing to evacuate, which can be potentially related to their tendency to protect their properties. Loss of job due to the epidemic and lockdown regulations has also been positively correlated with the residents' earthquake risk assessment. Moreover, difficulty accessing daily necessities during the lockdown is associated with the earthquake risk assessment.

The correlation between the risk identification and following predicting variables has also been proven by the empirical analysis, which the previous studies have confirmed; access to early warning notification (Lingyu et al., 2020), sensitivity and trust in personal feeling for ground-shaking (Ao et al., 2020a, 2020b, 2020c), and trust in official websites announcing earthquakes (Lingyu et al., 2020; Rainear & Lin, 2021). In addition, the intention to blindly follow strangers during emergencies has shown significant association with risk assessment and evacuation choice. The analysis results indicate that those who do not intend to follow others during emergencies have perceived higher earthquake risk than the respondents who would follow others. However, the first group was less likely to evacuate, confirming previous studies' findings (Lingyu et al., 2020). Finally, the study proved the significant positive correlation between risk identification and evacuation decision-making, as indicated by Ge et al. (2021) Gu et al. (2016). Earthquake risk assessment did not show a significant association with risk mitigation/reduction strategies, which is in line with the research findings by Rainear et al. (2021). Similarly, the research failed to accept the hypothesis regarding the association between risk identification and assessment.

8.5 Conclusion and Policy Implementations

8.5.1 Conclusion

Based on the results of the empirical results and as discussed in the discussion section, the following conclusions were drawn:

- 93% of respondents successfully identified the earthquake risk, although fewer people (65%) have evacuated. Moreover, the study has identified that successful earthquake risk identification will positively affect evacuation decision-making; however, considering other predicting variables affecting the evacuation choice of the residents, the evacuation rate has considerably decreased. In addition, the residents were mainly worried about the loss of life/injuries rather than the loss of properties.
- Contrary to most existing studies, this research put forward a seismic evacuation decision-making framework based on the EDMM for seismic evacuation decision-making. The analysis of the survey data demonstrated that every stage

of the proposed framework, including earthquake risk identification, risk assessment, and risk mitigation/reduction strategies, might be differently affected by predicting variables. While epidemic prevention regulations, specifically the loss of jobs and property and difficulty in having daily necessities during the lockdown, have significantly affected the residents' earthquake risk perception and evacuation decision-making, risk identification has been majorly impacted by the residents' access to earthquake information and trust in those sources.

- The hierarchical modeling of each EDMM framework has shown significant improvement in the models' perfection, indicating that not only socio-economic and the built environment variables have been associated with the participants' seismic risk detection, assessment, and evacuation behavior but also the epidemic control regulations at the time of the earthquake has impacted the mentioned outcomes.
- Living in the high-rise buildings did not significantly correlate with the evacuation choice of the residents, which is contrary to the prevalent belief that the higher floor the residents live on, the lower their tendency to evacuate. However, the results indicate that those staying home during the earthquake were less willing to take a mitigation decision, although they were more likely to be notified about the quake.
- The study has illustrated that rural residents felt more vulnerable to the earthquake than urban residents, which might imply less sufficient facilities in rural areas to answer emergencies. Moreover, urban residents were less likely to evacuate during the quake.
- Family structure (household size and marriage status) has been proven to affect the evacuation choice of individuals negatively. Additionally, those who had children around during the earthquake felt more in danger.
- Previous experience in coping with an earthquake which has been widely discussed in the literature, has shown a significant impact on the residents' earthquake risk identification and taking risk mitigation/reduction strategies. In contrast, the participants' earthquake experience has not affected earthquake risk perception.

8.5.2 Policy Implementations

- According to the findings of this study, access to disaster information has played a critical role in risk recognition. Although the majority of respondents of this study have successfully identified the earthquake risks, the empirical results demonstrate that the respondents have trusted disaster information distributed by the official website. The early warning system has been activated in most Sichuan province regions; however, people who trusted government official disaster dissemination websites were likelier to perceive the quake than those who trusted early warning systems, which might imply the inefficiency of the early warning systems to inform people timely. Considering the geological location of Sichuan province, governments should debug the early warning systems, constantly update official disaster information dissemination channels, and facilitate public access to these channels by using more accessible platforms such as having official webpages on social media.

- Epidemic control and lockdown regulations have significantly impacted seismic risk perception and mitigation strategies. Although there was no substantial evidence showing the direct correlation between the EDMM stages and restrictions in daily activities based on COVID-19 risk zones, governments, especially in disaster-prone areas, should actively modify the emergencies regulations during the wide lockdown to bring more insights into pandemic control regulations. The disaster information dissemination systems can be linked to real data of pandemic situations and provide critical risk prevention and mitigation suggestions to guide public evacuation decision-making and educate the residents about emergency guidelines during pandemic restrictions.
- Considering the impact of COVID-19's economic decline on the household level, which has been proven to influence evacuation decision-making negatively, authorities should provide disaster financial support for low-income and economically suffering households to mitigate the outcomes of inappropriate evacuation-decision making in case of severe disasters. In addition, the study's results indicated that vulnerable special groups such as females, families with children, and elderlies felt more endangered by the quake's imminent outcomes; however, their concern did not necessarily lead to an evacuation decision. Hence, special attention should be given to those groups whose perceived physical vulnerability possibly obstacled their evacuation decision. Therefore, it is recommended that local authorities in each residential compound link the households' information database to the emergency strategic plans to provide special assistance for the vulnerable groups mentioned above.
- Since at the time of the Luding earthquake, public quarantine facilities accommodated infected people or those with a high risk of infection, disaster dissemination systems and evacuation protocols in such areas should be planned to prevent the occurrence of a catastrophe facing tremendous disaster.

8.6 Limitations and Recommendations for Future Study

Due to the sudden emergence of the earthquake during the lockdown in Sichuan province, the researchers designed the research and structured the survey quickly after the Luding earthquake to prevent loss of memories and better access to accurate information. With the frequent emergence of the COVID-19 pandemic and lockdowns and the necessity of evaluating human evacuation behavior in case of emergencies, it is recommended that scholars provide fundamental theories and assumptions beforehand. In addition, future studies can focus on the evacuation behavior of vulnerable groups during lockdown to recognize their needs and limitations on the way to a safe emergency evacuation.

References

Administration, S. E. (2022). *The emergency management department issued the intensity map of Sichuan Luding M6.8 earthquake*. Sichuan Earthquake Administration. Retrieved 2022–09–11 from https://www.scdzj.gov.cn/xwzx/fzjzyw/202209/t20220911_53440.html

Anderson, D. R., Sweeney, D. J., Williams, T. A., Camm, J. D., & Cochran, J. J. (2016). *Statistics for business & economics*. Cengage Learning.

Anjani, K., Ashok, K. M., Sunil, S., & Shahidur, R. (2022). Government transfers, COVID-19 shock, and food insecurity: Evidence from rural households in India. *Agribusiness, 38*(3).

Ao, Y., Zhou, X., Ji, F., Wang, Y., Yang, L., Wang, Q., & Martek, I. (2020). Flood disaster preparedness: experience and attitude of rural residents in Sichuan, China. *Natural Hazards, 104*(3).

Ao, Y., Martek, I., Ao, Y., & Martek, I. (2020). Seismic evacuation preparedness behavior of rural residents. *Rural Built Environment of Sichuan Province, China*, 317–340.

Ao, Y., Huang, K., Wang, Y., Wang, Q., & Martek, I. (2020). Influence of built environment and risk perception on seismic evacuation behavior: Evidence from rural areas affected by Wenchuan earthquake. *International Journal of Disaster Risk Reduction, 46*(C).

Ao, Y., Zhang, H., Yang, L., Wang, Y., Martek, I., & Wang, G. 2021. Impacts of earthquake knowledge and risk perception on earthquake preparedness of rural residents. ***Natural Hazards***, 107(2).

Bahmani, H., Ao, Y., Yang, D., & Wang, D. (2023). Students' evacuation behavior during an emergency at schools: A systematic literature review. *International Journal of Disaster Risk Reduction, 103584*.

Bank, W. (2022). *Emerging risks to the recovery*. In.

Beverly, J. L., & Bothwell, P. (2011). Wildfire evacuations in Canada 1980–2007. *Natural Hazards, 59*(1).

Bratman, M. E., Israel, D. J., & Pollack, M. E. (1988). Plans and resource-bounded practical reasoning. *Computational Intelligence, 4*(3).

Chen, L., Tang, T. Q., Song, Z., Guo, R. Y., & Huang, H. J. (2020). Empirical investigation of child evacuation under non-emergency and emergency situations. *Journal of Transportation Safety & Security, 14*(4).

Ding, N., & Sun, C. (2020). Experimental study of leader-and-follower behaviours during emergency evacuation. *Fire Safety Journal*.

Dongkwan, L., Soyeon, Y., Seon, P. E., Yuseung, K., & Yoon, D. K. (2018). Factors contributing to disaster evacuation: The case of South Korea. *Sustainability, 10*(10).

Fatemeh, M., & Babak, M. (2021). Studying the role of personality traits on the evacuation choice behavior pattern in urban road network in different severity scales of natural disaster. *Advances in Civil Engineering, 2021*.

Gantt, P., & Gantt, R. (2012). Disaster psychology: Dispelling the myths of panic. *Professional Safety, 57*(08).

Ge, Y., Yang, G., Wang, X., Dou, W., Lu, X., & Mao, J. (2021). Understanding risk perception from floods: A case study from China. *Natural Hazards, 105*(3).

Gigerenzer, G. (2007). *Gut feelings: The intelligence of the unconscious*. Penguin.

Gu, Z., Liu, Z., Shiwakoti, N., & Yang, M. (2016). Video-based analysis of school students' emergency evacuation behavior in earthquakes. *International Journal of Disaster Risk Reduction, 18*.

Haghani, M., & Sarvi, M. (2016). Human exit choice in crowded built environments: Investigating underlying behavioural differences between normal egress and emergency evacuations. *Fire Safety Journal, 85*, 1–9.

Hassan, R. S., Wajiha, H., & Muhammad, S. (2020). COVID-19: A psychosocial perspective. *Frontiers in Psychology, 11*.

Helbing, D., & Molnár, P. (1995). Social force model for pedestrian dynamics. *Physical Review. E, Statistical Physics, Plasmas, Fluids, and Related Interdisciplinary Topics, 51*(5).

Homa, B., & Wei, Z. (2022). Why do communities recover differently after socio-natural disasters? Pathways to comprehensive success of recovery projects based on Bam's (Iran) neighborhoods' perspective. *International Journal of Environmental Research and Public Health, 19*(2).

Hongtai, Y., Guocong, Z., Linchuan, Y., & Kun, X. (2022). How does the suspension of ride-sourcing affect the transportation system and environment? *Transportation Research Part D, 102.*

Huang, Y., Zhou, L., & Wei, K. 2011. 5.12 Wenchuan earthquake recovery government policies and non-governmental organizations' participation. *Asia Pacific Journal of Social Work and Development, 21*(2), 77–91.

Jiaojiao, W., Xue, Y., Xin, D., & Dingde, X. (2022). Does disaster knowledge affect residents' choice of disaster avoidance behavior in different time periods? Evidence from China's earthquake-hit areas. *International Journal of Disaster Risk Reduction, 67.*

Koshiba, Y., & Suzuki, Y. (2018). Factors affecting post-evacuation behaviors following an earthquake: A questionnaire-based survey. *International Journal of Disaster Risk Reduction, 31.*

Lazo, J. K., Bostrom, A., Morss, R. E., Demuth, J. L., & Lazrus, H. (2015). Factors affecting hurricane evacuation intentions. *Risk analysis: An official publication of the Society for Risk Analysis, 35*(10).

Lazo, J. K., Waldman, D. M., Morrow, B. H., & Thacher, J. A. (2010). Household evacuation decision making and the benefits of improved hurricane forecasting: Developing a framework for assessment. *Weather and Forecasting, 25*(1), 207–219.

Lingyu, Z., Yuntao, G., Srinivas, P., & Bing, W. (2020). Impacts of information from various sources on the evacuation decision-making process during no-notice evacuations in campus environment. *Journal of Transportation Safety & Security, 12*(7).

Malika, S., Sara, R., Barry, P. S., Leanne, M., Enrica, L., Chermaine, T., & Laurie, D. (2021). Growing up under COVID-19: Young people's agency in family dynamics. *Frontiers in Sociology, 6.*

Murray-Tuite, P., Yin, W., Ukkusuri, S. V., & Gladwin, H. (2012). Changes in evacuation decisions between Hurricanes Ivan and Katrina. *Transportation Research Record, 2312*(1).

Obja, B. H., & Sarmistha, D. (2021). Paid and unpaid work during the Covid-19 pandemic: A study of the gendered division of domestic responsibilities during lockdown. *Journal of Gender Studies, 30*(4).

Organization, W. H. (2020). *COVID-19 strategy update—14 April 2020.* World Health Organization. https://www.who.int/publications/i/item/covid-19-strategy-update---14-april-2020

Perry, R. W., & Green, M. R. (1982). The role of ethnicity in the emergency decision-making process*. *Sociological Inquiry, 52*(4).

Rainear, A. M., & Lin, C. A. (2021). Communication factors influencing flood-risk-mitigation motivation and intention among college students. *Weather, Climate, and Society, 13*(1), 125–135.

Raphael, B. (2005). Crowds and other collectives: Complexities of human behaviors in mass emergencies. *Psychiatry: Interpersonal and Biological Processes, 68*(2), 115–120.

Rodrigueza, R. C., & Estuar, M. R. J. E. (2021). A spatial agent-based model for preemptive evacuation decisions during typhoon. In *Proceedings of the 2021 IEEE/ACM international conference on advances in social networks analysis and mining*

Saunders, M., Lewis, P., & Thornhill, A. (2007). Research methods. *Business Students 4th edition Pearson Education Limited, England, 6*(3), 1–268.

Schippers, M. C. (2020). For the greater good? The devastating ripple effects of the Covid-19 crisis. *Frontiers in Psychology, 11.*

Sharma, G. (2017). Pros and cons of different sampling techniques. *International Journal of Applied Research, 3*(7), 749–752.

Ştefan, I., Ionuţ, N., & Nora, C. (2021). Cybernetics approach using agent-based modeling in the process of evacuating educational institutions in case of disasters. *Sustainability, 13*(18).

Urban, C. (1997). PECS: A reference model for the simulation of multi-agent systems. In *Tools and techniques for social science simulation* (pp. 83–114). Springer.

Verdière, N., Cantin, G., Provitolo, D., Lanza, V., Dubos-Paillard, E., Charrier, R., Aziz-Alaoui, M., & Bertelle, C. (2015). Understanding and simulation of human behaviors in areas affected by disasters: From the observation to the conception of a mathematical model. *Global Journal of Human-Social Science, 15*, 7–15.

Rogers, R. W. (1975). A protection motivation theory of fear appeals and attitude change1. *The Journal of Psychology, 91*(1).

Whytlaw, J. L., Nicole, H., Juita-Elena Y., Tihara R., Saige H., Taiwo O.-L., Pamela A.-N., Eduardo L., & Rafael D. (2021). Changing vulnerability for hurricane evacuation during a pandemic: Issues and anticipated responses in the early days of the COVID-19 pandemic. *International Journal of Disaster Risk Reduction, 61.*

Wim, K., Ruud, Z., & Philippe, D. M. (2012). The informed society: An analysis of the public's information-seeking behavior regarding coastal flood risks. *Risk analysis: An official publication of the Society for Risk Analysis, 32*(8).

Xie, W., Lee, E. W. M., Cheng, Y., Shi, M., Cao, R., & Zhang, Y. (2020). Evacuation performance of individuals and social groups under different visibility conditions: Experiments and surveys. *International Journal of Disaster Risk Reduction, 47*(prepublish).

Xuan, T. B., Thi, N. H., Thi, L. H., A., L. C., Quang, P. H., Gia, V. L., Thi T. L. X., Thanh N. T., Thi P. Q., Thi K. T. N., Thi N. Q., Ho, C. S. H., Ho, & R. C. M. (2020). Impact of COVID-19 on economic well-being and quality of life of the vietnamese during the national social distancing. *Frontiers in Psychology, 11.*

Yiannis, X., & Georgia, K. (2022). Prediction of humans' behaviors during a disaster: The behavioral pattern during disaster indicator (BPDI). *Safety Science, 152.*

Yongjun, H., Li, H., Tao, X., Aihua, Y., Yi, Y., Yanfei, F., Junshuai, C., Ling, D., Jiabao, Z., & Zengshu, Y. (2022). Smart community emergency evacuation management system and risk assessment based on mobile big data. *Wireless Communications and Mobile Computing, 2022.*

Yunong, H., Hung, W., & Tiong, T. N. (2015). Associations between economic loss, financial strain and the psychological status of Wenchuan earthquake survivors. *Disasters, 39*(4).

Chapter 9
Emergency Evacuation Choices and Reasons Under Pandemic Situation; Lessons from the Luding Earthquake

Abstract This chapter contributes to evacuation and emergency research by questioning the reasons behind evacuation and stay decisions of Sichuan residents during the Luding earthquake, 5th September 2022, whose lives have been affected by the Covid-19 epidemic and subsequent mobility limitations. To answer those questions, the study employed K-means clustering to recognize the patterns of evacuation reasons among the population and later attempted to identify how those reasons can vary upon the population by conducting parametric statistical tests. Our major results suggest that (1) people with similar evacuation decisions have shown similar evacuation reasons, (2) lack of disaster coping knowledge among the young population can result in cascading disasters in the case of emergencies, (3) the evacuation barriers were more influential on evacuation decision-making among the urban population than rural residents, (4) fear on infection of Covid-19 in the outdoors and mobility limitations due to pandemic control rules severely and negatively restricted the respondents' evacuation decisions, and (5) although the weak structures in rural areas resulted in the residents lower trust in structures' quake-resistance, lack of information of emergency facilities suffered the evacuation choices of the urban population. Insights of these aspects are expected to provide a deeper understanding of human perspective differences on evacuation motivations and barriers based on the individuals' characteristics, built environment, and pandemic situation by updating emergency plans and educating the public about evacuation and emergency management during double emergencies.

Keywords Travel-related CO_2 emission · Rural built environment · Travel attitudes · Structure equation modeling · Exploratory factor analysis · Rural China

9.1 Introduction

In the event of a disaster or emergency, the safety of individuals is of utmost importance. As evacuation is one of the most primary ways to save lives, understanding how people make decisions regarding evacuation can help emergency responders

Y. Ao and H. Bahmani, *Navigating Complexity: Understanding Human Responses to Multifaceted Disasters*, https://doi.org/10.1007/978-981-99-8207-3_9

and planners develop more effective strategies to ensure the safety of those affected by developing preparedness plans and allocating sufficient resources. According to Leonard, mass evacuation can be divided into "immediate evacuation" and "potential evacuation" (Pfefferbaum et al., 2016). While the first one occurs shortly after an emergency giving the population almost no time for preparation, potential evacuation provides more room for informing the people about the future hazard, making them more well-prepared. Notably, depending on the nature of the hazard, studies in the field of evacuation decision-making may choose the type of evacuation. In this study, to avoid the complexity, unless mentioned directly, the word evacuation refers to immediate evacuation. Human evacuation behavior and the factors affecting their risk perception, subject to several internal and external factors, have attracted considerable attention in the literature over the last decades (Bahmani et al., 2023a, 2023b). Studies in the field of social science attempted to find how individuals' evacuation choice varies per sociodemographic factors. Recent studies illustrated that age, income, housing type, drill experience, and knowledge of the shelter location affect people's evacuation behavior (Lee et al., 2018; Rodrigueza & Estuar, 2021; Wu et al., 2022). Moreover, an enormous amount of work has been carried out on the influence of the built environment on human evacuation (Ao et al., 2020a, 2020b, 2020c; Ionescu et al., 2021). Disaster type is another factor that influences evacuation choice (Alam et al., 2023). Although there have been several attempts in the literature to capture different aspects affecting human evacuation decision-making, the matter of evacuation decision-making has been chiefly explored by delving into evacuation decisions and relationships with individuals' social or demographic information. At the same time, the underlying reasons for human evacuation behavior and the impacts of principal factors on them were often neglected.

Sichuan province, permanent home to 83.675 million of China's population, ranked as the fifth largest population in China till May 2021 (Construction, 2013), located in the west of China, which was previously less economically developed than the eastern section. Moreover, ethnic and impoverished areas make this part of the country more vulnerable to disasters. This region is highly industrial, with the prevalence of machinery manufacture, metallurgy, chemical industries, and pharmacies (Jiang et al., 2008). The geological feature of this region has caused several natural disasters in history, among which the Wenchuan earthquake, May 12, 2008, with a magnitude of 8, was a great catastrophe that caused a death toll of more than 680,000 people and 130 billion US$ economic loss (Construction, 2013). However, the high casualties during the Wenchuan earthquake were not only caused by the massive destruction. Reports evidencing the fear of earthquake anxiety and lack of risk awareness caused some victims to jump out through buildings falling dead (You et al., 2009). On 2022 September 5, Sichuan province experienced the Luding earthquake, 6.8 magnitudes, whilst the Covid-19 epidemic crisis was at its peak, exacerbating the stress and panic among people whose evacuation choices have been seriously affected by the pandemic control regulations. Therefore considering the combined risks associated with disasters and global pandemics with ongoing economic and physical toll, people's evacuation decision-making may differ from what we see during a single risk scenario. Despite the need to study such a situation, studies linking disaster risk reduction, especially public evacuation, and biological hazards are missing in the literature.

Some recent studies attempted to address the relationship between the Covid-19 pandemic and evacuation during emergencies. However, the literature focused on whether massive population mobility and surge in Covid-19 infected cases (Page-Tan & Fraser, 2022; Sakamoto et al., 2020) or individuals' concerns about Covid-19 infection prohibit them from evacuation in case of emergencies (Alam et al., 2023; Santos-Reyes, 2020). Therefore, to our knowledge, there is no research on evacuation or stay decision reasons considering restrictions of pandemic time policies. Furthermore, this research aims to understand the evacuation decision-making of individuals with similar or dissimilar reasons motivating or placing an obstacle in the way of evacuation decision-making within the content of double whammy risks; the Luding earthquake that happened on September 5, 2022, when the increasing rate of Covid-19 infection in Sichuan province imposed a wide range of epidemic control policies on the residents to curb the pandemic. We have also delved into understanding how the residents' sociodemographic information, the built environment, and the epidemic may cause differences in (non) evacuation reasons. The findings of this study would be helpful for emergency managers, policy-makers, and researchers attempting to understand human evacuation behavior facing double emergencies. Researchers and emergency planners can gain valuable insights into the complex interplay of factors influencing individuals' decision-making and behaviors during evacuation scenarios by comprehensively exploring evacuation motivations and barriers. This understanding can inform the development of targeted interventions, communication strategies, and policies to improve evacuation preparedness and response to minimize risks and enhance the safety of communities facing hazardous situations. The remainder of the paper is structured as follows. The relevant studies on evacuation behavior and its influential factors are discussed in Sect. 9.2. An overview of research methods is presented in Sect. 9.3 by explaining sample and data collection methods, hypothesis construction, and analysis methods. The study's detailed results are given in Sect. 9.4, and Sect. 9.5 is allocated to the discussion, conclusions, policy implementations, and future research directions in Sect. 9.5.

In natural capital, compared with using past savings, the exchange of property for cash ($p = 0.000$, $B = -2.199$) is significantly negatively correlated with the distance between the house and the river and lake, which is consistent with the research conclusion of Baker (Bao et al., 2021). The closer the distance to the river, the greater the loss and damage caused by the flood, and selling the property can quickly solve the economic needs. In addition, farmers with drainage ditches are more likely to choose to rely on government or social organizations for relief funds ($p = 0.001$, $B = 0.817$), borrowing from banks or loan companies ($p = 0.03$, $B = 0.405$), selling property and livestock in exchange for cash ($p = 0.005$, $B = 0.103$). Compared with the adaptive livelihood strategy of improving drainage facilities, the choice of increasing house height ($p = 0.001$, $B = -0.454$) and moving ($p = 0.004$, $B = -0.634$) are significantly negatively correlated with the distance from the house to the river and lake, which is consistent with the research conclusions of Farman et al. (Xu et al., 2018) and Gioli et al. (King et al., 2014). Farmers with drainage ditches are more likely to choose to increase agricultural irrigation measures (p = 0.00, B = 0.087), and farmers with drainage ditches will choose a variety of flood disaster response livelihood strategies, but the choice of flood disaster adaptation livelihood strategies is limited.

9.2 Related Works

Multiple previous attempts have been made to understand the factors contributing to disaster evacuation decision-making, among which surveys and statistical tests such as regression methods have been widely applied to answer the research's questions. A study investigating factors contributing to evacuation behavior during a volcanic eruption, ethnicity, evacuation destination, transportation mode, residency period, risk awareness, income, and presence of children have shown a significant impact on households' evacuation decisions (Thakur et al., 2022). Using the "Protective Action Decision Model", alongside machine learning methods, research on household evacuation in the Attica wildfires, 2018, Greece revealed that age, risk perception, residency length, elderly number in family, information seeking before evacuation, gender, education, and income played critical roles in the residents' evacuation choice. Although the study has illustrated the importance of vehicle accessibility to evacuation behavior, since such disasters as earthquakes happen immediately, giving people almost no time for deliberate transportation mode, the type of disaster can significantly influence evacuation decision-making (Alam et al., 2023).

On the other hand, a study by Verma et al. figured out that the more people spent time looking at the hurricane information, the better their understanding of the hurricane statutes, and consequently, the more confident they became to evacuate (Verma et al., 2022). Another recent study has successfully proved that high evacuation intention would not necessarily lead to evacuation (Kakimoto & Yoshida, 2022). In fact, situation awareness errors affecting people's risk identification and perception are other reasons that should be considered in evacuation behavior modeling. Scholars have also found herding behavior (blind follow behavior), which indicates the evacuees' intention to follow the crowd's direction during evacuation (Shi et al., 2022), influences people's risk perception (Zhou & Wei, 2010) and subsequently affects their evacuation decision-making (Liu & Mao, 2022). The empirical studies observed the herding and route choice behavior of humans by changing the number of participants, urgency level, and imposing physical barriers, and identified the strong influence of high population density on herding behavior (Kakimoto & Yoshida, 2022). However, controversial findings did not find a relationship between herding behavior and students' evacuation decision-making, while the participant's tendency to seek additional emergency information affected their evacuation choice (Zheng et al., 2020).

Although disaster training for individuals who lacked real evacuation experience did not help them control their panic during emergencies (Chang et al., 2020), another study confirmed that those with disaster experiments performed better in virtual shooting incident experiments (Zhu et al., 2022). Recent studies have distinguished an ironic relationship between the built environment and evacuation behavior due to the influence of the former on people's risk perception and evacuation efficiency. The literature found a strong association between students backtracking behavior after their evacuation and their seat location in the classroom (Delcea & Cotfas, 2019). On the other hand, several researchers conducted studies on building designs and put forward design suggestions for safer evacuation (Alam et al., 2023; Delcea et al., 2020; Ionescu et al., 2021; Rostami & Alaghmandan, 2021). In the last decade, various studies have determined evacuation behavior and its influential factors, some of which are presented in Table 9.1; however, the growing body of literature absences research on the impact of underlying factors on the reasons behind human evacuation choice, whether evacuation or stay decision which is of great importance to perceiving the complex environment of human evacuation decision-making.

The complexity of understanding the mechanism of evacuation decision-making would increase in the face of double disasters, such as a combination of geological and biological hazards. With the surge of Covid-19 infection, which was first reported on December 31, 2019, in Wuhan, China, the World Health Organization (WHO) declared a Global Health Emergency on January 9, 2020 (Organization, 2020). Worldwide outbreaks and epidemic control measures have affected billions of people's daily life. The impact of Covid-19 went further by severe economic decline in 90 countries (Bank, 2022; Tran et al., 2020), increased family violence (Borah Hazarika & Das, 2021; Shah et al., 2021), food insecurity (Kumar et al., 2022), and increased stress and anxiety (Raza et al., 2020; Shah et al., 2021). On the other hand, the study of the Covid-19 pandemic and evacuation behavior is rarely reported. In general, prior work is limited to a subset of exploring the relationships between the risk of spreading Covid-19 and evacuation. A study found that the number of Covid-19 infected cases in evacuation destinations was more influential on the spread of the virus than population movement during an evacuation. However, the role of limiting population mobility in controlling the spread of Covid-16 is undeniable (Page-Tan & Fraser, 2022). Another study investigated the effectiveness of evacuation measures during the Covid-19 pandemic (Sakamoto et al., 2020) and suggested minimizing human contact during the evacuation and sheltering the population in destinations with lower infection rates (Pei et al., 2020). To wrap it all up, the existing literature suggests properly linking evacuation plans and real-time epidemic control information.

On the other hand, the researchers have examined the impact of the Covid-19 epidemic on public evacuation decision-making. Participants' likelihood to participate in evacuation drills during the Covid-19 epidemic in Mexico City was studied (Santos-Reyes, 2020). On the other hand, another research was conducted on Cyclone Amphan in 2020 when the region was experiencing the Covid-19 outbreak (Alam et al., 2023). This study found that in those areas, the residents' evacuation decision was negatively affected by fear of Covid-19 infection caused mainly by rummers. Moreover, the non-evacuation decision among the study's participants was triggered

Table 9.1 Influential factors on human evacuation behavior and the resources

Influential factors on evacuation behavior	Sources	Influential factors on evacuation behavior	Sources
Age	Bandecchi et al. (2019; Lee et al. (2018), Najmanová and Ronchi (2017)	Personal emotions	Mao et al. (2020)
Gender	Koshiba and Suzuki (2018), Lee et al. (2018), Najmanová and Ronchi (2017)	Disaster experience	Hasan et al. (2013), Lazo et al. (2015), Lee et al. (2018), Wu et al. (2022)
Job	Lee et al. (2018)	Family income	Lee et al. (2018)
Race (ethnicity)	Thakur et al. (2022; Whitehead et al. (2000)	Housing type	Lee et al. (2018)
Education	Katzilieris et al. (2022), Lee et al. (2018)	Housing structure	Ao et al. (2020a), (2020b), (2020c)
Marital statues	Lee et al. (2018)	House floor	Ao et al. (2020a), (2020b), (2020c)
Homeownership	Lee et al. (2018)	Presence of physical barriers or smoke	Gao et al. (2022), Wang et al. (2020), Xie et al. (2020)
Household size	Lee et al. (2018)	Choice of destination	Thakur et al. (2022)
Period of residency	Thakur et al. (2022)	Available transportation	Thakur et al. (2022)
Pet ownership	Lee et al. (2018), Whitehead et al. (2000)	Herding behavior	Ding and Sun (2020)
Cattle ownership	Paul (2012)	Situation awareness	Ao et al. (2020a), (2020b), (2020c), Verma et al. (2022)
Distance to shelter	Paul (2012)	Access to earthquake information sources	Kellens et al. (2012, Whitehead et al. (2000)
Presence of vulnerable groups in family	Lee et al. (2018), Whytlaw et al. (2021)	Trust in information sources	Paul (2012), Rainear and Lin (2021), Zheng et al. (2020)
Social ties between family members	Alawadi et al. (2020), Ding and Sun (2020), Xie et al. (2020)	Destination place attachement	Ariccio et al. (2020)

by overcrowding, insufficient public shelters, disbelief in emergency warnings, difficulty in transportation, intention for property protection, and sociocultural barriers. Finally, one of the most recent studies has focused on the Evacuation Decion-Making Mechanism of individuals during double-whammy situations using hierarchical logistic regressions (Bahmani et al., 2023a, 2023b). As the literature lacks considering the severe pandemic control legislation and condition on the evacuation behavior of the affected residents, this research has chosen to study the evacuation decision-making of the residents in Sichuan province during the Luding earthquake, September 5, 2022, who were experiencing mobility limitations imposed by the epidemic control policies to curb the growing rate of the Covid-19 infections.

9.3 3Materials and Methods

9.3.1 Study's Methodology and Hypothesis

Previous scholars applied several theoretical frameworks, among which Belief-Desire- Intention Model, Protection Motivation Theory, and Social Forces Model highly contributed to depicting a clear perspective of human evacuation behavior (Bratman et al., 1988; Helbing & Molnar, 1995; Rogers, 1975). In this study, evacuation motivations encompass the reasons and incentives that prompt individuals to consider evacuating from an area of potential danger. Evacuation barriers refer to the obstacles or challenges that hinder or impede individuals' ability or willingness to evacuate when faced with a hazardous event. These barriers can manifest at different levels, including personal, social, and logistical factors. This study, motivated by the research by Paul (Lazo et al., 2015), attempts to identify clusters of evacuation motivations and barriers among the respondents and understand how distinct background features across the population, as listed in Table 9.2, can cause differences in evacuation motivation and barriers. To answer the research questions, we designed and distributed an online questionnaire within eight days after the Luding earthquake through "Wen Juan Wang" platform. The survey was divided into four parts, covering different topics. The first part requested sociodemographic information, including age, gender, location during the earthquake, education, marital status, household size, presence of vulnerable groups, income, homeownership, community type, and earthquake experience. The second part focused on housing and built environment information, such as the type of housing, structure, and floor. The third part of the survey aimed to gather information about the respondents' experiences during the Covid-19 pandemic and lockdowns, including personal health code statutes, vaccination statuses, and lockdown policies. Lastly, the participants' evacuation choices and reasons were asked five scale Likert questions, strongly agree, agree, neutral, disagree, and strongly disagree. The validity detection test was performed by asking the respondents to choose "agree" if the respondents read the questionnaire carefully.

To ensure the consistency of the construct measurements, Cronbach's alpha was employed for "evacuation motivations" and "evacuation barriers". The overall Cronbach's alpha for evacuation motivations and barriers was 0.794 and 0.781, respectively, higher than a minimum value of 0.70, indicating an acceptable consistency among items in each scale (Taber, 2018). This study employed K-means clustering

Table 9.2 Study's hypothesizes

		Risk mitigation/reduction motivations (Cronbach's alpha = 0.794)				
Group	Factors	Prevention of loss of life or injuries (M1)	Prevention of financial loss (M2)	Prevention of further disasters (M3)	Intention to follow others (M4)	Received evacuation notice (M5)
Sociodemographic factors	Age		□	□	□	□
	Gender	□	□	□	□	□
	Community type	□	□	□	□	□
	Education	□	□	□	□	□
	Marriage	□	□	□	□	□
	Job	□	□	□	□	□
	Income		□			
	Household size	□	□	□	□	
	Vulnerable groups	□	□	□	□	□
	Earthquake experience	□	□	□	□	□
Dwelling features	Homeownership	□	□	□		□
	Housing type	□	□	□	□	□
	Housing structure	□	□	□		
	Housing floor	□	□	□	□	□
Pandemic features	Personal health code statues					□
	Vaccination statues					

(continued)

Table 9.2 (continued)

		Risk mitigation/reduction motivations (Cronbach's alpha = 0.794)				
Group	Factors	Prevention of loss of life or injuries (M1)	Prevention of financial loss (M2)	Prevention of further disasters (M3)	Intention to follow others (M4)	Received evacuation notice (M5)
	Epidemic control policies					□

		Risk mitigation/reduction barriers (Cronbach's alpha = 0.781)							
Group	Factors	Personal safety attitude, Covid-19 infection's concern (B1)	Lack of evacuation knowledge (B2)	Restrictions imposed by pandemic control rules (B3)	Lack of safe public shelter (B4)	Protection of property (B5)	Believe in earthquake-resistant house (B6)	Anxiety (B7)	Lack of situation awareness (B8)
Sociodemographic factors	Age	□	□	□		□	□	□	□
	Gender	□	□	□		□	□	□	□
	Community type	□	□	□	□	□	□	□	□
	Education	□				□	□	□	□
	Marriage	□		□		□		□	□
	Job	□	□	□		□	□	□	□

(continued)

Table 9.2 (continued)

Group	Factors	Risk mitigation/reduction barriers (Cronbach's alpha = 0.781)							
		Personal safety attitude, Covid-19 infection's concern (B1)	Lack of evacuation knowledge (B2)	Restrictions imposed by pandemic control rules (B3)	Lack of safe public shelter (B4)	Protection of property (B5)	Believe in earthquake-resistant house (B6)	Anxiety (B7)	Lack of situation awareness (B8)
	Income					☐		☐	☐
	Household size	☐		☐		☐		☐	☐
	Vulnerable groups	☐	☐	☐		☐		☐	☐
	Earthquake experience		☐			☐	☐	☐	☐
Dwelling features	Homeownership					☐	☐	☐	☐
	Housing type			☐	☐	☐	☐	☐	☐
	Housing structure						☐	☐	
	Housing floor				☐	☐	☐	☐	☐
Pandemic features	Personal health code statues	☐		☐				☐	☐
	Vaccination statues	☐						☐	
	Epidemic control policies	☐		☐		☐		☐	☐

to identify patterns in the reasons behind participants' evacuation decisions. K-means clustering is a widely used unsupervised machine learning algorithm that was employed in this study to identify inherent patterns or groupings within a dataset. It offers a systematic approach to partitioning data points into distinct clusters based on their similarity, aiming to discover underlying structures or relationships within the data (Ikotun et al., 2022). In addition, to test the hypotheses listed in Table 9.2, the chi-square test of independence and Welch's ANOVA were used. While the former has been employed to identify the association between evacuation reasons and population demographic features with two groups, the latter method has been applied to categorical independent variables with more than two groups to reveal the relationship with the dependent variable: evacuation motivations and barriers. The assumptions for running this statistical test were met as the observations were independent, values in the cells were mutually exclusive, and expected values were five or greater in at least 80% of cells. Because data violates the assumption of homogeneity of variance checked by Levene test, the study has benefited from using Welch's ANOVA to compare the effect of independent variables with more than two groups on dependent variables. At the same time, Shapiro–Wilk test proved that all data were normally distributed (Celik, 2022). In addition, we performed pairwise comparisons post hoc test, Games-Howell multiple comparisons method to identify significant differences between specific groups where differences existed. Overall, the use of k-means clustering, chi-square test of independence, and Welch's ANOVA in this study provided a robust scientific basis for the conclusions drawn and contributed to the rigor of the research. Note that our study primarily employed parametric tests and clustering methods to examine the relationships between sociodemographic factors, built environment conditions, and the pandemic's influence on evacuation response and reasons. While we recognize the valuable insights that spatial analysis could offer, conducting a comprehensive spatial analysis would have required additional data collection and extensive spatial modeling techniques, exceeding the scope of our study. The overall methodology applied in this research is presented in Fig. 9.1.

9.3.2 Case Study

Due to its history of deadly earthquakes, including the 2008 Wenchuan earthquake, Sichuan province is considered one of China's most earthquake-prone provinces (Huang et al., 2011). Our study is focused on the Luding earthquake, which occurred on September 5, 2022, in Sichuan province. The earthquake's epicenter was located in Luding county, Ganzi prefecture, with coordinates of 29.59 degrees north latitude and 102.08 degrees east longitude and a focal depth of 16 km. This earthquake had a magnitude of 6.8 Richter and affected a wide area, causing 93 deaths and 25 loss of contact (Administration, 2022; Los, 2022). The Luding earthquake occurred during a challenging time when most parts of Sichuan province faced lockdown due to the Covid-19 pandemic. Chengdu, the capital of Sichuan, reported 1332 local Covid-19

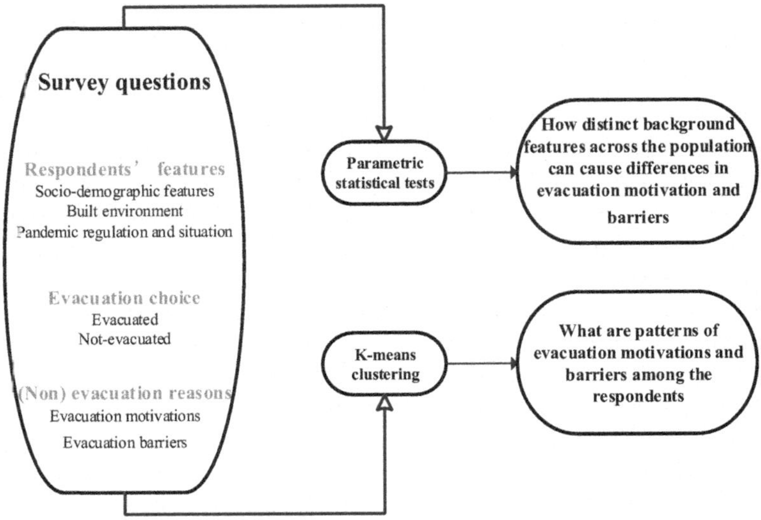

Fig. 9.1 Research framework

cases between August 12 and September 4 (Government, 2022). The main restrictions during the lockdown time imposed by the provincial and municipal epidemic prevention and control bureau were as follows:

- Restrictions were imposed on shops and public areas, and they stopped working unless they provided essential services.
- Residents were required to undergo daily nucleic acid tests and were allocated health codes based on their Covid-19 status, test results, and travel history. People with a red health code, indicating a confirmed or suspected Covid-19 case, close contact, or coming from a high-risk area, were not allowed to enter public places. Travel restrictions were imposed for those with a yellow health code due to respiratory symptoms or contact with Covid-19 patients. Residents not belonging to the last two categories were assigned green health codes and were required to provide nucleic acid tests to enter public areas.
- Non-emergency vehicles were not allowed to travel on the roads during this period.
- Strict lockdowns were imposed on high-risk areas, preventing residents from leaving their apartments. The "people are not allowed to leave the communities" rule was enforced in medium-risk areas.
- Individuals outside the restricted areas were advised to remain at home and were strictly prohibited from traveling, except for one household member per day who could go to nearby markets to purchase essentials within a limited time.

To that end, the cities/counties affected by the Luding earthquake, as indicated in Fig. 9.2, have been selected for the survey. To ensure data integrity, the study selected one representative per household and monitored respondents' geographical

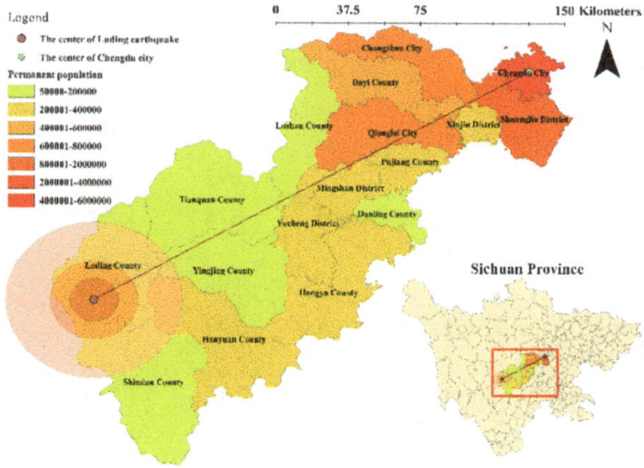

City/ district/ county	Num. of participants	City/ district/ county	Num. of participants
Chengdu city	679	Tianquan county	72
Shuangliu district	93	Yingjing county	32
Chongzhou city	103	Hanyuan county	120
Dayi county	54	Shimian county	54
Qionglai county	66	Luding county	64
Pujiang county	53	Yingjing County	32
Mingshan county	7	Leshan city	14
Danling county	51	Dongpo District	13
Lushan county	36	Renshou County	16
Yucheng district	58	Other areas	52
Hongya county	57	**Total number**	**1257**

Fig. 9.2 Location map of the sample villages, counties, and cities

locations during the earthquake. Through a public survey, 1512 questionnaires were collected, and after validity checks, 1257 valid questionnaires remained for analysis.

9.4 Results

9.4.1 Descriptive Analysis

A public survey was conducted in the affected areas from September 13 to September 21, resulting in a database of 1512 questionnaires. A validity detection test was conducted, asking respondents to confirm that they had carefully read the questionnaire by choosing the "agree" option, excluding 254 unacceptable questionnaires. Additionally, one filled questionnaire was excluded from the sample during the first round of data screening as it was located outside our study area. Therefore, the final valid sample size consisted of 1257 questionnaires. While most respondents were female, the population age was almost equally distributed among the age groups. The proportion of married respondents was higher than single individuals, and 59.3% of the study's participants resided in urban areas. 75.3% of the study's population owned a house, while nearly 90% of the respondents lived in a household whose size was smaller than five people. The presence of vulnerable groups, i.e., children below 12, elderly over 65 years old, and disabled people, have not been reported frequently, as only one-third of our respondents lived with at least one of those groups. Moreover, 95.6% of the study's population had experience dealing with earthquakes. In terms of the built environment, 81.5% of the participants live in non-single buildings (which means residing in other buildings rather than a residential complex), and more than 60% of the respondents' houses had less than seven stories. Steel and concrete was the most widely observed structure. At the time of the earthquake, most people held

Table 9.3 Study's population ranking for evacuation motivations and barriers

Evacuation motivations and barriers	Mean	Std. Deviation
I wanted to save my family's and myself's lives (M1)	4.44	1.141
I did not want to suffer from financial loss (M2)	3.92	1.488
I did not want to be stuck in the area after the event or experience aftershocks (M3)	4.28	1.236
I saw others evacuating/frightening (M4)	3.48	1.536
I received an evacuation notification from the government's official sources/warning system (M5)	3.49	1.622
I felt unsafe outdoors (the risk of covid-19 transmission is high) (B1)	2.14	1.375
I do not know how to evacuate (B2)	1.86	1.274
The lockdown policies limited my choices (B3)	2.15	1.461
There is no safe place for sheltering in my residential area (B4)	2.39	1.510
I wanted to protect my home and belongings (B5)	1.70	1.145
I believe my house is strong and earthquake-resistant (B6)	3.26	1.481
I was afraid to evacuate because of the earthquake panic (B7)	1.75	1.236
I didn't realize there was an earthquake (B8)	1.75	1.288

Note M1-M5 are evacuation motivations that have been asked from 813 people who evacuated, while B1-B8 are evacuation barriers answered ranked by 444 respondents who did not choose to evacuate

green health codes, although 41.4% of the participants had experienced moderate epidemic control rules; they were allowed to leave their houses but could not freely go out of their communities. Eight percent of the study's population were under strict lockdown in their houses, hotels, or the government's central lockdown facilities and were not allowed to leave their rooms, and fifty percent were not restricted by lockdowns as home isolation was not strictly implemented, and they could move freely in their district. The detailed findings of the descriptive analysis of the respondents are given in the appendix, and Table 9.3 summarizes evacuation motivation and barriers' mean score and standard deviation.

9.4.2 K-means Clustering

An exploratory k-means cluster analysis was conducted to investigate the grouping of evacuation decision-making considering evacuation motivations and barriers. The sample consisted of 1257 participants, and five evacuation motivations and eight evacuation barriers were included in the analysis. Prior to clustering, the variables were standardized to ensure comparability. In the K-means clustering analysis, the initial cluster center is randomly chosen. The Euclidean distance between each data point and the centroids was Calculated using the Euclidean distance by the formula below:

$$\text{Euclidean distance} = \sqrt{(x_2 - x_1)^2 + (y_2 - y_1)^2}$$

In this formula, (x2 – x1) represents the difference in x-coordinates between the two points, and (y2 – y1) represents the difference in y-coordinates.

We Updated the centroid positions by calculating the mean of all the data points assigned to each cluster by running the algorithm 10 times from random initial positions. The maximum number of iterations within each algorithm was 300.

The silhouette method combining separation and cohesion was used to determine the number of clusters (Saputra et al., 2020) where Higher values indicate better-defined clusters by use of the formula below:

$$\text{Silhouette Coefficient} = \frac{(b - \mathrm{a})}{\text{Max }(a, b)}$$

where

"a" is the average distance from the data point to other points within the same cluster.

"b" is the average distance from the data point to the points in the nearest neighboring cluster.

The quality of the clusterings was assessed using the silhouette coefficient, which indicated a strong separation between the clusters. The average silhouette coefficient for non-evacuees and evacuees groups was 0.743 and 0.555, respectively, suggesting well-defined clusters with minimal overlap. As the p-value from the ANOVA test is below chosen significance level (e.g., $\alpha = 0.05$), we can reject the null hypothesis and conclude that there are significant differences in the silhouette values among the clusters of both groups indicating that the clustering algorithm successfully differentiated the data points, into distinct clusters (for non-evacuees ANOVA = 10,400.634, *p*-value = 0.000, N = 1257 and for evacuees ANOVA = 5353.225 p-value = 0.000, N = 1257). The Chi-square independence test has also revealed a significant relationship between the clusters and the independent variables, as shown in Table 9.4.

- **Clusters of evacuees**:

In cluster 1, owning 310 members, the primary evacuation reason was preventing injuries/death and secondary disasters, as over 60 and 40% of the population chose those evacuation reasons, respectively. In this cluster, preventing financial loss and following others during the quake and evacuation due to formal notice have gained 27% agreement. However, in cluster 2, representing 503 members, a high proportion of people stated that the prevention of human and financial loss as well as the adverse effects of aftershocks, were their evacuation reasons. In contrast, this cluster owned a lower ranking for evacuation due to herding behavior and official evacuation notice. A closer look at the characteristics of the individuals in these two clusters revealed that the second cluster has a higher proportion of people with earthquake experience

Table 9.4 Statistical values of K-means clustering

	Chi-square value	df		Chi-square value	df
B1	1593.13*	10	M1	1829.9*	10
B2	1460.45*	10	M2	2138.38*	10
B3	1629.04*	10	M3	2116.13*	10
B4	1981.4*	10	M4	1643.74*	10
B5	1407.37*	10	M5	1597.74*	10
B6	1790.86*	10			
B7	2058.1*	10			
B8	2058.1*	10			

* Indicates the significant value < 0.05

and also a higher number of its members had a vulnerable group at the time of the quake. Additionally, the majority of the population in cluster 2 is aged between 26 and 33, whereas cluster 1 has owned younger population aged between 11 and 25 years old. (see Tables 9.5 and 9.6; Fig. 9.3).

- **Clusters of non-evacuees**:

Cluster 1 has 326 people, among which three fourth of the population strongly disagree that five of the evacuation barriers impede their decision-making. Those evacuation barriers can be grouped to built environment impact on people's risk perception and the individuals' risk management and perception abilities. In this cluster, the rankings for one barrier significantly differ from the others as most respondents believed their intention to save assets was the primary reason for their non-evacuation decision. On the contrary, in cluster 2, holding 118 people, the distribution of the opinions regarding non-evacuation reasons was mainly neutral and showed a comparatively more moderate perspective towards the evacuation barriers. In this cluster, lack f evacuation knowledge and property protection intention have gained the highest votes as evacuation barriers, while inadequate public shelter obtained the lowest ranking. The comparative feature statistics of the populations in the non-evacuees clusters illustrated that people in the second cluster were mostly married, and the average monthly salary of cluster one was higher than the second cluster. Additionally, most respondents assigned to cluster 1 resided in rural areas and lived in high-raised buildings, and their experience dealing with earthquakes showed a higher proportion than the first cluster. (see Fig. 9.4).

9.4.3 Evacuation Motivations

As the results in Table 9.7 show, age has a significant and moderate effect on M1-M3. The Game Howell multiple comparisons illustrate that the mean value for people

Table 9.5 Feature statistics of evacuees clusters

Cluster 1	Proportion of major group (%)	Cluster 2	Proportion of major group (%)
Jender (female)	63.87	Jender (female)	62.03
Marital status (married)	61.29	Marital status (married)	64.21
Homeownership (house-owner)	75.48	Homeownership (house-owner)	74.35
Building type (complex residential buildings)	81.29	Building type (complex residential buildings)	80.52
Earthquake experience (yes)	94.84	Earthquake experience (yes)	97.42
Vulnerable groups (yes)	30.32	Vulnerable groups (yes)	35.39
Rural resident	57.42	Rural resident	51.89
Health code (green)	97.42	Health code (green)	98.41
Vaccinated (yes)	97.74	Vaccinated (yes)	99.60
People around at the time of the earthquake (3–4 people)	23.23	People around at the time of the earthquake (3–4 people)	27.44
Income (3000–5000 RMB)	27.74	Income (3000–5000 RMB)	28.63
Structure of house (steel and concrete)	55.48	Structure of house (steel and concrete)	51.69
Housing floor (1-3th floor)	41.29	Housing floor (1-3th floor)	39.96
Risk area (low risk)	85.81	Risk area (low risk)	85.69
Age (11–25 y.o)	30.65	Age (26–33 y.o)	28.63
Education (bachelor)	65.48	Education (bachelor)	67.00
Pandemic isolation (home isolation)	51.29	Pandemic isolation (home isolation)	57.46
Job (full-time)	30.32	Job (full-time)	35.79

aged more than 41 years old is lower than the rest of the three other groups, indicating that compared with the younger people, participants more than 41 years old were less motivated to evacuate to save their lives, which is in line with the findings previous studies (Lazo et al., 2015; Stein et al., 2010) (mean difference between people aged more than 41 years old and 11–25, 26–33, and 34–41 years old is 0.410, 0.563, 0.453, respectively). Similarly, people aged between 26–33 years old were more willing to prevent financial loss and further disasters than those over 41 (mean difference of M2 = 0.37, mean difference of M3 = 0.451). Marital status has a small but significant impact on M1-M3. We found that relatively more unmarried participants chose prevention of financial loss as their primary evacuation motivation compared to 65.76% of married people (mean value unmarried individuals = 3.97 vs. mean value married individuals = 3.88). Among evacuees motivated to save lives, the mean

Table 9.6 Feature statistics of non-evacuees clusters

Cluster 1	Proportion of major group (%)	Cluster 2	Proportion of major group (%)
Jender (female)	55.52	Jender (female)	63.56
Marital status (married)	59.20	Marital status (married)	71.19
Homeownership (house-owner)	72.09	homeownership (house-owner)	87.29
Building type (complex residential buildings)	84.66	Building type (complex residential buildings)	74.58
Earthquake experience (yes)	95.09	Earthquake experience (yes)	92.37
Vulnerable groups (no)	67.79	vulnerable groups (no)	63.56
Rural resident	74.83	Rural resident	54.24
Health code (green)	98.16	Health code (green)	94.92
Vaccinated (yes)	97.55	Vaccinated (yes)	98.31
People around at the time of the earthquake (3–4 people)	25.46	People around at the time of the earthquake (2 people)	28.81
Income (> 9000 RMB)	34.97	Income (3000–5000 RMB)	27.97
Structure of house (steel and concrete)	61.04	Structure of house (steel and concrete)	59.32
Housing floor (11-15th floor)	31.90	Housing floor (1-3th floor)	39.83
Risk area (low risk)	80.37	Risk area (low risk)	83.90
Age (11–25 y.o)	30.67	Age (11–25 y.o)	32.20
Education (bachelor)	70.25	Education (bachelor)	59.32
Pandemic isolation (home isolation)	47.85	Pandemic isolation (home isolation)	48.31
Job (full-time)	34.66	Job (others)	30.51

value for unmarried respondents is 4.52, while the married population owns a mean value of 4.39. Moreover, unmarried individuals had a higher tendency to evacuate to avoid the risk of upcoming disasters (mean value unmarried individuals = 4.37 vs. mean value married individuals = 4.23). The job category has shown a significant moderate impact on herding behavior, and a significant difference exists between evacuation decisions of self-employed people and those working for governmental organizations (mean difference = 0.733). Additionally, the results indicated that the individuals without earthquake experience were more likely to follow the evacuating crowd than those with experience (mean value with earthquake experience = 3.49 vs. mean value without earthquake experience = 3.3). The statistical tests showed that community type significantly impacts M1 and M5, with the mean value of 4.4 and 3.63 for rural communities vs. 4.47 and 3.36 for urban communities, respectively.

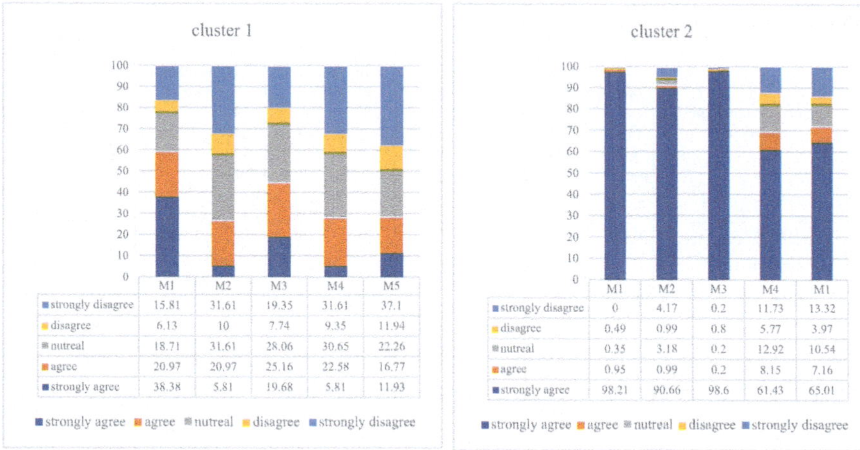

Cluster 1:

	M1	M2	M3	M4	M5
strongly disagree	15.81	31.61	19.35	31.61	37.1
disagree	6.13	10	7.74	9.35	11.94
nutreal	18.71	31.61	28.06	30.65	22.26
agree	20.97	20.97	25.16	22.58	16.77
strongly agree	38.38	5.81	19.68	5.81	11.93

Cluster 2:

	M1	M2	M3	M4	M1
strongly disagree	0	4.17	0.2	11.73	13.32
disagree	0.49	0.99	0.8	5.77	3.97
nutreal	0.35	3.18	0.2	12.92	10.54
agree	0.95	0.99	0.2	8.15	7.16
strongly agree	98.21	90.66	98.6	61.43	65.01

Fig. 9.3 Evacuees clustering and their rankings for evacuation barriers. Lef) cluster 1; right) cluster 2

Cluster 1:

	B1	B2	B3	B4	B5	B6	B7	B8
strongly disagree	77.3	65.34	60.12	86.2	28.53	74.66	90.8	90.8
disagree	5.83	8.28	7.06	7.36	4.6	7.91	1.53	1.53
nutreal	9.2	12.58	9.51	1.84	17.48	5.99	1.23	1.23
agree	4.29	6.13	9.51	0.92	18.4	1.53	1.84	1.84
strongly agree	3.37	7.67	13.8	3.68	30.98	9.91	4.6	4.6

Cluster 2:

	B1	B2	B3	B4	B5	B6	B7	B8
strongly disagree	21.19	22.03	6.78	9.32	5.08	17.8	9.32	9.32
disagree	19.49	15.25	14.41	22.03	6.78	20.34	22.88	22.88
nutreal	32.2	25.42	50	51.69	43.22	34.74	38.98	38.98
agree	10.17	9.32	8.47	6.78	26.27	16.95	10.17	10.17
strongly agree	16.95	27.97	20.34	10.17	18.64	10.17	18.64	18.64

Fig. 9.4 Non-evacuees clustering and their rankings for evacuation barriers. (Left) cluster 1; (right) cluster 2

Results also indicate that the housing floor significantly affects M1. Although the mean score of M1 increased with each additional residential floor up to the 10th floor, suggesting that individuals living at higher elevations were more motivated to save their lives, a significant difference in attitude toward life protection was observed only between residents living on the 1st-3rd floors and those on the 7th-10th floors (mean difference = 0.308). On the other hand, community type shows a small and moderate effect on M1 and M5. 86.07% of the urban residents agreed that their evacuation decision was motivated by their intention to save lives, whereas 81.33% of the rural residents held the same perspective (mean value rural communities = 4.4

Table 9.7 Statistical test results for M1–M5

Factors/ evacuatuion motivations	Prevention of loss of life or injuries (M1)		Prevention of financial loss (M2)		Prevention of further disasters (M3)		Intention to follow others (M4)		Received evacuation notice (M5)	
	F value/ Chi-square	Eta-squared	F value/ Chi-Square	Eta-squared	F value/ Chi-Square	Eta-squared	F value/ Chi-Square	Eta-squared	F value/ Chi-Square	Eta-squared
Age a	7.996	0.036*	3.004	0.01*	4.714	0.02*	2.49	0.01	1.346	0
Gender b	0.648	0	3.426	0	2.861	0	7.877	0	5.172	0
Community type b	13.185	0.001*	5.352	0	4.69	0	7.349	0.01	11.602	0.01*
Education a	1.451	0.006	0.692	0	0.945	0	1.017	0	0.97	0
Marriage b	22.077	0.003*	19.69	0.00*	17.737	0.00*	7.451	0	7.858	0
Job a	1.638	0.022	1.158	0.01	1.645	0.01	2.544	0.02*	0.466	0
Income a	-	-	0.602	0	-	-	-	-	-	-
Household size a	0.74	0.005	0.91	0.01	1.409	0.01	1.77	0.01	-	-
Presence of vulnerable groups b	5.011	0	3.708	0	8.87	0	5.312	0	5.584	0
Earthquake experience b	5.843	0.001	4.11	0	4.512	0	16.395	0.00*	3.492	0
Homeownership b	2.291	0	0.802	0	3.169	0	-	-	6.2	0
Housing type b	6.53	0.002	0.73	0	2.919	0	2.234	0	3.257	0
Housing structure a	1.68	0.005	0.107	0	1.605	0.01	1.263	0.01	1.453	0.01
Housing floor a	2.702	0.011*	0.204	0	1.386	0.01	-	-	0.436	0.001

*Eta-squared values marked by (*) are significant at the level of 0.05.

[a] and [b] indicate the application of Welch's ANOVA and Chi-square test of independence, respectively.

Small size effect
Medium size effect
Large size effect

vs. mean value urban communities = 4.47). Evacuation notice in rural communities was the primary reason for 59.2% of the population to evacuate, whereas 52.51% of the residents in urban areas evacuated because of the evacuation notice (mean value rural communities = 3.63 vs. mean value urban communities = 3.36). This research found no statistical difference in the motivation for preventing financial loss among individuals with different income levels, which aligns with the findings of some previous studies (Lazo et al., 2015; Smith & McCarty, 2009). However, contradictory findings exist regarding the association between income level and intention to prevent financial loss, as some studies illustrated the less significant impact of disaster information on individuals with high income (Verma et al., 2022). Additionally, no significant variation in peoples' tendency to prevent human loss is found based on the presence or absence of vulnerable groups in their families, consistent with the previous literature (Stein et al., 2010).

9.4.4 Evacuation Barriers

The results signify that age can moderately influence the individuals' non-evacuation decisions affected by their insufficient evacuation knowledge and high trust in the quake-resistant house. The post hoc test further illustrated that among those who did not evacuate, people aged 11–26 strongly believed in their houses' quake resistance, contrary to those over 41 who held a less optimistic perspective toward their houses' quake resistance (mean difference = 0.622). On the other hand, regarding the trust in the safety of the structures, the youngest participants, 11–25 years old, have significantly lower mean scores than those aged 26–33 (the mean difference = -0.522), which might imply a lower level of evacuation preparedness and knowledge among this critical age group. Results also show that gender significantly impacts non-evacuation decisions due to fear of getting infected by Covid-19 (mean value men = 2.01 vs. mean value women = 2.23), the intention for property protection (mean score of 1.71 for women vs. mean score of 1.69 for men), and anxiety during the quake (mean score of 1.48 for men vs. 1.95 for women). Marital status significantly and negatively affected the quake anxiety preventing the individuals from evacuation (mean value unmarried individuals = 1.56 vs. mean value married individuals = 1.87). The property protection intention obstacle to the evacuation decision significantly differed across different household sizes as those with more than seven family members showed a significant difference in mean score compared with those with zero, two, and five to six family members (mean differences are − 0.568, − 0.510, and − 0.980, respectively). The study also found that anxiety prohibited evacuation of the participants with five or six members in the family than those with four or fewer family members. The study identified that participants with vulnerable members in their families were more likely to suffer from a lack of evacuation knowledge, with 16.22% of them indicating this as the reason for not evacuating, compared to 11.15% of those without vulnerable family members (mean value absence of vulnerable people = 1.74 vs. mean value presence of vulnerable people = 2.1).

The respondents' non-evacuation decisions in urban and rural communities differed significantly per five of the evacuation barriers. Since almost one-fourth of the rural residents answered that epidemic control policies and quake anxiety restricted their evacuation, which is higher than the similar belief among the urban residents, it can be concluded that the evacuation decision was more negatively affected by the epidemic control policies and anxiety in the rural communities (mean value epidemic-rural = 2.25 vs. mean value epidemic-urban = 2.11 and, mean value anxiety-rural = 2.03 vs. mean value anxiety-urban = 1.63). 82.47% of urban residents refused to evacuate due to the intention to keep the property safe, which was higher than the recognized pattern in rural communities, with 64.7% of people with the same perspective. On the other hand, 28.57% of urban residents agreed that lack of public shelter affected their evacuation choice, whereas 16.8% of the participants in rural areas reported the same answer (mean value rural areas = 2.29 vs. mean value urban areas = 2.44). Moreover, Covid-19 infection concern is the primary evacuation obstacle selected by 19.48% of urban residents, while only 11.03% of rural residents' concern about getting infected by the Covid-19 virus prohibited them from evacuation (mean value rural areas = 2.07 vs. mean value urban areas = 2.17). We have also identified that 67.64% of the rural residents did not evacuate because they did not recognize the earthquake. In comparison, more than 80% of the residents in urban areas reported the same (mean value rural areas = 2.07 vs. mean value urban areas = 1.61).

With the significant moderate effect of education on B1, the mean score difference between postgraduates and undergraduates is significant (mean score difference = 0.487. On the other hand, among those who refused to evacuate to protect their properties, individuals with high school degrees were more than undergraduates and postgraduates (mean difference of 0.601 and 0.655 for those group comparisons, respectively). The moderate impact of education on non-evacuation decisions due to quake anxiety is significant, and the groups' difference is significant between those with high school education and undergraduates and postgraduates (mean differences are 0.626, 0.70, respectively). Having an earthquake experience impacts the respondents' evacuation knowledge and intention to save properties; 50% of those without earthquake experience stated that they did not evacuate because they did not know how to evacuate. In contrast, only 10.71% of respondents with earthquake experience expressed the same concern (mean value without earthquake experience = 2.87 vs. mean value with earthquake experience = 1.8). 25% of those without experience stated they did not evacuate to save their properties. However, the intention to protect properties decreased to 6.3% among those with earthquake experience (mean value without earthquake experience = 2.42 vs. mean value with earthquake experience = 1.68). Moreover, those with earthquake experience reported a lower level of trust in their house's quake resistance than those without earthquake experience (mean value without earthquake experience = 3 vs. mean value with earthquake experience = 3.28). However, earthquake experience significantly assisted earthquake recognition by the participants. 11.33% of those with previous experience dealing with an earthquake reported situation unawareness. In contrast, this proportion is triple for those without earthquake experience (mean value without earthquake experience = 2.37 vs. mean value with earthquake experience = 1.72).

The statistical tests proved job categories' significant effect on non-evacuation decisions caused by the four evacuation obstacles. Farmers' concern for their health safety in outdoor areas has greatly prohibited them from evacuation since there is a significant mean difference between farmers and professionals (mean difference = 2.585). Lack of evacuation knowledge prohibiting evacuation significantly differed among job categories, and the difference existed between unemployed individuals and the government and full-time employees (the mean difference is − 1.07 and − 1.96, respectively). Additionally, pandemic regulations significantly affected non-evacuation decisions among job categories; and significant differences between mean scores of farmers and unemployed or retired people and most of the other jobs exist, among which the highest mean differences belong to farmers and part-time employees (2.766), farmers and professionals (2.626), farmers and government employees (2.625), and unemployed or retired and government employees (1.198). The earthquake anxiety among the unemployed or retired individuals affected their evacuation choice more than professionals and those with part-time jobs (mean differences are 1.126 and 1.278, respectively).

A significant effect of housing type on the non-evacuation decision due to restrictions imposed by pandemic control rules, anxiety, and failure in earthquake recognition was identified. The respondents living in single buildings were more affected by pandemic regulation than those in complex residential buildings (mean value complex residential buildings = 2.05 vs. mean value single residential buildings = 2.64). On the other hand, the non-evacuation decision due to situation unawareness and quake anxiety is significantly higher for the residents living in single buildings than those residing in complex residences (mean value complex residential buildings-situation unawareness = 1.71 vs. mean value single residential buildings-situation unawareness = 1.94 and mean value complex residential buildings-anxiety = 1.66 vs. mean value single residential buildings-anxiety = 2.64). Housing structure also has a medium effect on peoples' quake-resistance understanding of their houses, as residents of steel and concrete buildings own higher mean scores than those living in brick and concrete structures (mean difference = 0.429. The quake anxiety has shown a negative and medium impact on evacuation decision-making of house owners (mean value house owner = 1.83 vs. mean value non-house owner = 1.5 for non-owners). Moreover, although the housing floor impacts the property protection attitudes of the residents, the significant difference is only between those who lived on the 1st–3rd and 4th–6th floors, as the former group showed a higher mean score (mean difference = 0.470), indicating among those who did not evacuate, those in lower floors were more willing to save their assets. The housing floor also moderately affected the residents' anxiety preventing their evacuation; People living on the 4th–6th floors reported that their non-evacuation decision was less affected by quake anxiety than those living on the 1st–3rd and 11th–15th floors.

The health code status had a significant impact on the concern of Covid-19 infection as the mean value for those without green health code was significantly different for those holding green-health code at the time of the earthquake (mean value non-green health code holders = 3.45 vs. mean value green health code holders = 2.1). In addition, respondents with yellow or red health codes were less likely to evacuate due to pandemic regulation, with a mean value of 3.18 and 2.13 for non-green and green-health codes, respectively. Similarly, those holding yellow or red health codes during the earthquake were less aware of the situation (mean value non-green health code holders = 3.18 vs. mean value green health code holders = 1.71). The results also illustrate the significant and moderate effect of the pandemic regulations on people's health safety concerns; the stricter the pandemic control rules, the more concerned people were about the risk of Covid-19 infections, which negatively affected their evacuation decision (mean difference between strict and moderate pandemic control = 0.645, mean difference between strict and no pandemic control = 1.045). Areas with strict and moderate pandemic isolation had a higher impact on the non-evacuation decision than those without specific isolation, with mean differences of 1.292 and 1.0709, respectively. Finally, people strictly influenced by the epidemic control regulations showed significantly higher intention to save their properties than those not limited by the epidemic control regulations (mean difference = 0.551). The results of statistical tests for evacuation barriers are given in Table 9.8.

Table 9.8 Statistical test results for B1-B8

Factors/ evacuetuion motivations	Personal safety attitude, Covid-19 infection's concern (B1)		Lack of evacuation knowledge (B2)		Restrictions imposed by pandemic control rules (B3)		Lack of safe public shelter (B4)		Protection of property (B5)		Believe in the earthquake-resistant house (B6)		**Anxiety (B_7)**		**Lack of situation awareness (B_8)**	
	F value/ Chi-square	Eta-squared	F value/ Chi-Square	Eta-squared	F value/ Chi-square	Eta-squared	F value/ Chi-Square	Eta-squared	F value/ Chi-square	Eta-squared	F value/ Chi-Square	Eta-squared	F value/ Chi-square	Eta-squared	F value/ Chi-Square	Eta-squared
Age a	0.874	0.01	5.274	0.03*	1.271	0.01	-	-	0.298	0	3.465	0.02*	0.921	0.01	0.501	0
Gender b	9.877	0.01*	5.726	0	1.414	0	-	-	9.818	0.00*	3.697	0	18.903	0.03*	7.366	0
Community type	17.968	0.00*	3.775	0	15.96	0.00*	21.323	0.00*	19.292	0.02*	5.995	0	15.938	0.02*	29.234	0.03*
Education a	2.773	0.02*	0.344	0	-	-	1.496	0.01	3.714	0.03*	1.055	0.01	3.36	0.03*	0.653	0
Marriage b	2.563	0	-	-	2.465	0	-	-	8.256	0.01	1.305	0.02	9.543	0.01*	6.886	0
Job a	4.963	0.06*	7.669	0.06*	6.85	0.07*	-	-	1.984	0.04	3.465	0.02	2.526	0.07*	1.639	0.02
Income a	-	-	-	-	-	-	-	-	0.526	0	-	-	1.354	0.01	0.307	0
Household size a	2.455	0.03*	-	-	2.182	0.04	-	-	4.485	0.02*	-	-	4.2	0.07*	1.104	0.01
Presence of vulnerable groups b	2.431	0	11.17	0.02*	5.084	0	-	-	1.651	0	-	-	8.462	0.01	7.443	0
Earthquake experience b	-	-	40.198	0.04*	-	-	-	-	35.751	0.00*	10.166	0.00*	7.212	0.01	19.731	0.01*
Homeownership b	3.561	0	-	-	-	-	-	-	7.184	0.01	1.869	0	12.396	0.01*	7.926	0.01
Housing type b	-	-	-	-	22.35	0.02*	3.209	0	5.946	0.01	4.92	0	34.875	0.02*	28.151	0.00*
Housing floor a	-	-	-	-	-	-	1.076	0.01	3.987	0.03*	0.607	0.01	4.276	0.04*	1.095	0.01
Housing structure a	-	-	-	-	-	-	-	-	-	-	4.513	0.02*	0.813	0.01	-	-
Personal health code statues b	19.4	0.03*	-	-	17.892	0.01*	-	-	-	-	-	-	7.586	0.02	34.51	0.03*
Epidemic control policies a	10.719	0.05*	-	-	30.055	0.11*	-	-	4.327	0.02*	-	-	0.202	0	1.708	0.01
Vaccination statues b	5.121	0.015	-	-	-	-	-	-	-	-	-	-	2.885	0.047	-	-

*Eta-squared values marked by (*) are significant at the level of 0.05.

[a] and [b] indicate the application of Welch's ANOVA and Chi-square test of independence, respectively.

Small size effect
Medium size effect
Large size effect

9.5 Discussion and Concluding Remarks

Apart from the growing literature to understand who evacuates and who stays, this study attempted to provide a deeper understanding of the reasons for the evacuation decision of the residents in Sichuan province who experienced the Luding earthquake on September 5, 2022, while the region was facing a severe covid-19 pandemic. This research's findings are expected to assist governments and emergency managers optimize emergency management plans to be more effective in addressing double emergencies when one limits human mobility. Considering the rise of epidemics worldwide, modification of emergency plans and improvement of evacuation facilities and public emergency awareness are of great importance.

The K-means clustering results revealed that the participants could be grouped into four clusters according to their similarities in holding reasons for (non) evacuation decisions, among which some clusters represent strongly opposing or supporting perspectives toward evacuation decisions, while others showed moderately distributed voting across evacuation barriers and motivations. Property protection and prevention of human loss were the primary reasons for the stay and evacuation decisions; however, the proportion of people voting for them differed among clusters. In the second stage, we attempted to see if the evacuation barriers and motivations differ across the population by running parametric statistical tests. We observed that as age increased, people's intention to save lives and prevent injuries, avoid financial loss and stay safe from the danger of further disaster decreased. However, aging has shown a less substantial impact on the two latter evacuation motivations. On the other hand, statistics have also shown that some evacuation barriers were significantly different per age group. First, as the age increased, fewer participants reported non-evacuation decisions due to a lack of evacuation information. The knowledge gap shows a significant difference between those 11–25 years old and 26–33 years old, which is a critical finding illustrating the vulnerability of the former group, primarily students. Second, the youngest population's trust in quake-resistance buildings is significantly higher than those aged over 40. This difference may be derived from more earthquake experience among elders, which led to a more pessimistic perspective toward the quake resistance of the structures. Considering that 16.1% of Sichuan's province population is below 14 years old (Ning, 2021), inappropriate evacuation choices made by this group during a severe emergency would result in a great catastrophe. Therefore, more attention should be given to the disaster preparedness of the young generation.

Anxiety caused by the earthquake was a vital evacuation barrier whose impact on women was sufficiently higher than on men, corresponding with similar findings (Norris et al., 2002). Data analysis also proved that women were more concerned about Covid-19 infection outdoors and tended more to protect their assets leading to non-evacuation choices, which is different from the findings of research conducted in Bangladesh, possibly due to cultural differences (Mallick et al., 2009). However, gender did not cause a significant difference in evacuation motivation, as in our study, men and women were equally motivated to protect health and safety. The study also demonstrated that couples were also less motivated than single individuals to save lives and assets and stay safe from further potential disasters, contrary to the findings of a study on the motivational role of solid social ties and emotional support among couples in evacuation decision-making, (Lee et al., 2018; Li et al., 2020). One explanation for our finding would be lower responsibilities among single persons (Alam et al., 2023) and the sense of place attachment among couples (Qing et al., 2022), which can negatively affect their evacuation choice. Additionally, earthquake anxiety influencing the respondents' evacuation decisions played a different role among married and unmarried individuals. Contrary to many studies on the positive impact of marriage or couplehood on evacuation decision-making (Alawadi et al., 2020), our results illustrated that anxiety caused by the quake would be more restrictive for couples than for single people on their evacuation. This research's findings also confirmed that the impact of job categories on evacuation barriers was tangible in four cases, which corresponds to the results by Lee et al. (2018). Farmers' answers demonstrated that their evacuation choice had been limited due to pandemic rules and reat concern for their safety in outdoor areas regarding Covid-19 infection which is considerably higher than the other job categories. On the other hand, earthquake anxiety has strongly influenced the retired and unemployed individuals' evacuation choice, and this negative impact is significantly higher than in most other job categories. In summary, it has been proved that the evacuation barriers would be more restrictive among farmers, retired, and unemployed populations. Additionally, while some studies did not reach a robust conclusion on education's impact on evacuation choice (Xu et al., 2017), some research still proves the relationship between education and evacuation behavior (Lee et al., 2018). This study's sample showed that higher education has a significant positive impact on reducing earthquake anxiety and the intention to protect properties. However, as different attitudes about Covid-19 infection concern in the outdoors was only recognized between respondents with a postgraduate degree and undergraduates, the influence of education on this matter remains questionable.

Previous research has found that disaster preparedness behavior in China can differ between rural and urban residents (Ao et al., 2020a, 2020b, 2020c; Chai et al., 2021). According to our results, rural residents' intention to save lives as a primary reason for evacuation was higher than urban residents', possibly related to social norms' differences in an urban and rural lifestyle. In line with the last finding, attempting to save assets was more observed among urban residents than rural ones, possibly due to the considerable difference between living expenses in urban and rural regions in China. On the other hand, although the rural residents have better perceived the occurrence of the earthquake and better responded to the official evacuation notice (possibly because of which could be pertinent to stronger quakes in those regions (Administration, 2022) or poorer quality buildings in rural areas (Ao et al., 2020a, 2020b, 2020c) and difference in trust in the government and perceived potential risks (Han et al., 2017; Niu & Zhao, 2018), non-evacuation decisions because of insufficient evacuation information and quake anxiety were higher in rural zones, indicating the need to improve their disaster preparedness knowledge. Moreover, as has been studied by the last researchers, lower construction quality in Sichuan's rural areas (Ao et al., 2020a, 2020b, 2020c) and insufficient disaster preparedness (Hu et al., 2018) played a critical role in the evacuation decision-making of rural communities. Lower trust in the quake-resistant house in rural neighborhoods was highly reported, which stands higher than the reported number in the urban communities. However, the lack of access to public shelters hindered more people in urban communities than rural areas, which contradicts the existing literature. It is therefore suggested, to tackle this issue, the government should provide a safeguard for urban residents to guarantee lower losses in case of a destructive disaster. Emergency lost insurance would be a practical solution in this matter. Moreover, evacuation notice systems and facilities in urban communities need to be improved. Although Sichuan province, after the Wenchuan earthquake, has seen a significant improvement in evacuation facilities such as shelters, there is a need to educate the public in urban districts on how and when to use those facilities.

The participants' evacuation choices with prior experience in dealing with an earthquake have been less affected by the herding behavior than those without earthquake experience, which confirms the findings by Ge et al. (2011). However, there was no significant difference between those with or without earthquake experience and their motivation to prevent financial and health damages, which is a notable finding. Non-evacuation decisions among the individuals without earthquake experience were made because of lower evacuation knowledge, higher intention to save properties, and higher possibility for situation unawareness. The results also confirmed that stay choice due to trust in quake-resistant buildings among those who did not experience an earthquake owned a higher proportion than those with earthquake experience, partially confirming the existing studies' results (Shoji et al., 2020; Zhu et al., 2022). Moreover, this study has demonstrated that compared to smaller households, families with 5–6 members might experience more anxiety and be more willing to save their belongings, which can cause a refusal to evacuate. On the other hand, households with vulnerable groups would be more likely to fail the evacuation because of insufficient evacuation knowledge. The last findings could partially explain the conclusions drawn by previous scholars (Tanim et al., 2022). Therefore, care should be paid to the household features to improve the coping capabilities of families to be prepared for disasters. It is crucial to provide disaster and evacuation training for those without prior experience dealing with emergencies, rural residents, students, females gender, people with lower education, and families with children under 12 and elders above 65 years old.

Based on the study's results, we can substantially argue that living on the higher floors can both negatively and positively affect the evacuation choice; as the buildings' elevation increases, people's motivation to save lives plays a more critical role in motivating them to evacuate, while anxiety and intention for property protection would be unavoidable for evacuation. Our findings have also illustrated that for those residents living on the upper floors, the intention to save their lives is higher. This argument is consistent with the results of (Ao et al., 2020a, 2020b, 2020c), who believed that the higher risk perception of the higher floors residents would motivate them to evacuate while contradicting the conclusions regarding the evacuation difficulty of the residents on the upper floor (Putri et al., 2022). In addition, house owners' evacuation decision has been negatively affected by the earthquake anxiety as compared to those who did not own the property living on. Other built environment factors, housing type and structure, have significantly impacted the respondents' non-evacuation reasons. First, the residents in complex residential buildings who did not evacuate experienced lower anxiety and less difficulty recognizing the quake, which could be due to more communication space in complex residentials resulting in better informal information sharing (Asiamah et al., 2020). The type of structure has also affected people's evacuation refusal due to their belief in quake-resistant houses. Steel and concrete structures may positively influence the residents' risk perception by causing more trust in their safety in case of earthquakes. In contrast, the participants living on brick and concrete structures' trust in their structure's safety was lower, which was previously observed after the Bam earthquake in Iran in 2008, when the construction industry and building code significantly upgraded, resulting in enhanced people's trust in the safety of their homes (Arefian, 2018). The built environment can motivate evacuation decisions or obstacles by affecting people's risk perception. Hence, it is necessary to pay enough care to design safe houses that follow the construction safety standards and correspond with the people's sense of safety.

Lastly, as the study hypothesized, the pandemic control situation caused a significant negative impact on the evacuation decisions of non-evacuees. Among those who did not evacuate, individuals assigned by non-green health codes and those living in communities with stricter epidemic control rules were potentially more concerned about Covid-19 infection risk and felt more considerable limitations by the epidemic control situation. In addition, the study has found that an extremely more proportion of those with non-green codes failed to recognize the occurrence of the earthquake recognition, which may indicate their limitation in accessing earthquake information which could be possibly limited by being in isolation. We have also recognized that people's tendency to save their belongings was more remarkable in areas with strict pandemic control policies. The non-evacuees in urban regions were more concerned about getting infected by Covid-19 outside their houses, although more respondents in rural areas reported that their non-evacuation decision was restricted due to epidemic control rules, which might imply harsher pandemic control legislation implemented in rural communities. All in all, not only did the infection concern prohibit the residents from evacuation, lockdowns and health codes have posed critical obstacles to public evacuation decision-making. Evacuation and emergency management protocols should be improved and updated with real-time pandemic information, avoiding severe mobility restrictions for those in danger of emergencies. The public should also be provided with relevant and accurate information about double disaster preparedness to control their anxiety when facing such a situation. Moreover, referring to the existing studies on the impact of evacuation and people's mobility on the infection rate, there is a need to bring more flexibility and insight into pandemic control rules since the occurrence of an emergency would cause a greater risk to public safety rather than a pandemic.

The study attempted to collect primary data immediately after the Luding earthquake. However, challenges were encountered while collecting the information during the lockdown, such as difficulty in reaching people in different counties due to mobility limitations and the concern of losing the participants' memory, which urged us to organize the research basis within a short time. Considering the specific features of the case study, the methodology applied in this study can be replicated in other cases by careful adjustment. In addition, future studies can focus on evacuation behavior during different emergencies, such as fire accidents and study vulnerable groups' behavior during epidemics to recognize their needs and limitations on the way to a safe emergency evacuation.

Appendix 1

	Options	Frequency	Percentage
Age	11–25 years old = 1	347	27.6
	26–33 years old = 2	328	26.1
	34–41 years old = 3	272	21.6
	41–73 years old = 4	310	24.7
Gender	Female = 1	768	61.1
Community type	Rural areas = 1	746	59.3
Job	Unemployed and retired = 1	84	6.7
	Farmer = 2	25	2.0

(continued)

(continued)

	Options	Frequency	Percentage
	Professional (such as teacher/doctor/lawyer, and others) = 3	104	8.3
	Civil servant/Public institution worker/Government worker = 4	267	21.2
	Full-time employee of state-owned enterprises = 5	422	33.6
	Part-time worker = 6	46	3.7
	Self-employed individual = 7	49	3.9
	Others = 8	260	20.7
Education level	Below high school/Technical secondary school/Higher vocational school = 1	130	10.3
	High school = 2	147	11.7
	Undergraduate/Junior college = 3	842	67.0
	Postgraduate = 4	138	11.0
Marriage statues	Married = 1	792	63.0
Homeownership	Owner = 1	947	75.3
Number of people at home at the time of the earthquake	0 person = 1	285	22.7
	1 people = 2	263	20.9
	2 people = 3	264	21.0
	3–4 people = 4	315	25.1
	5–6 people = 5	88	7.0
	More than 7 people = 6	42	3.3
Presence of vulnerable groups	Yes = 1	421	33.5
Income	Less than 3000 RMB = 1	164	13.0
	3000–5000 RMB = 2	337	26.8
	5000–7000 RMB = 3	255	20.3
	7000–9000 RMB = 4	142	11.3
	More than 9000 RMB = 5	359	28.6
Earthquake experience	Yes = 1	1206	95.9
Housing type	Non-single building = 0	1024	81.5
	Single building = 1	233	18.5
Housing structure	Brick and concrete = 1	492	39.1
	Steel and concrete = 2	702	55.8
	Others = 3	63	5.0
Housing floor	1–3 floors = 1	481	38.3
	4–6 floors = 2	347	27.6
	7–10 floors = 3	175	13.9

(continued)

(continued)

	Options	Frequency	Percentage
	11–15 floors = 4	130	10.3
	Above 15th floor = 5	124	9.9
Personal health code statues	Green = 1	1232	98.0
Vaccination statues	Vaccination = 1	1241	98.7
Epidemic control policies	Strict isolation ((home or government places, cannot leave at all) = 1	101	8.0
	Moderate isolation (home) = 2	520	41.4
	No isolation = 3	636	50.6

References

Administration, S. E. (2022). *The emergency management department issued the intensity map of sichuan luding M6.8 earthquake*. Retrieved August 16 from http://www.china.com.cn/txt/2022-09/14/content_78419390.htm

Alam, M. S., Chakraborty, T., Hossain, M. Z., & Rahaman, K. R. (2023). Evacuation dilemmas of coastal households during cyclone Amphan and amidst the COVID-19 pandemic: A study of the Southwestern region of Bangladesh. *Natural Hazards, 115*(1), 507–537.

Alawadi, R., Murray-Tuite, P., Marasco, D., Ukkusuri, S., & Ge, Y. (2020). Determinants of full and partial household evacuation decision making in hurricane Matthew. *Transportation Research Part D: Transport and Environment, 83*, 102313.

Ao, Y., Martek, I., Ao, Y., & Martek, I. (2020). Seismic evacuation preparedness behavior of rural residents. *Rural Built Environment of Sichuan Province, China*, 317–340.

Ao, Y., Huang, K., Wang, Y., Wang, Q., & Martek, I. (2020a). Influence of built environment and risk perception on seismic evacuation behavior: Evidence from rural areas affected by Wenchuan earthquake. *International Journal of Disaster Risk Reduction, 46*, 101504.

Ao, Y., Zhou, X., Ji, F., Wang, Y., Yang, L., Wang, Q., & Martek, I. (2020c). Flood disaster preparedness: Experience and attitude of rural residents in Sichuan, China. *Natural Hazards, 104*, 2591–2618.

Arefian, F. F. (2018). *Organising post-disaster reconstruction processes*. Springer. https://doi.org/10.1007/978-3-319-70911-6

Ariccio, S., Petruccelli, I., Cancellieri, U. G., Quintana, C., Villagra, P., & Bonaiuto, M. (2020). Loving, leaving, living: Evacuation site place attachment predicts natural hazard coping behavior. *Journal of Environmental Psychology, 70*, 101431.

Asiamah, N., Kouveliotis, K., Eduafo, R., & Borkey, R. (2020). The influence of community-level built environment factors on active social network size in older adults: Social activity as a moderator. *International Quarterly of Community Health Education, 41*(1), 77–87.

Bahmani, H., Ao, Y., Li, M., Yang, D., & Wang, D. (2023). Dual disasters: Seismic evacuation decision-making during COVID-19 lockdown: A case study of Luding earthquake, Sichuan Province. *Journal of Transport Geography*, 103622.

Bahmani, H., Ao, Y., Yang, D., & Wang, D. (2023). Students' evacuation behavior during an emergency at schools: A systematic literature review. *International Journal of Disaster Risk Reduction*, 103584.

Bandecchi, A. E., Pazzi, V., Morelli, S., Valori, L., & Casagli, N. (2019). Geo-hydrological and seismic risk awareness at school: Emergency preparedness and risk perception evaluation. *International Journal of Disaster Risk Reduction, 40*, 101280.

Bank, W. (2022). *Emerging risks to the recovery*. In.

Bao, X., Zhang, F., Deng, X., & Xu, D. (2021). Can trust motivate farmers to purchase natural disaster insurance? Evidence from earthquake-stricken areas of Sichuan, China. *Agriculture, 11*(8), 783.

Borah Hazarika, O., & Das, S. (2021). Paid and unpaid work during the Covid-19 pandemic: A study of the gendered division of domestic responsibilities during lockdown. *Journal of Gender Studies, 30*(4), 429–439.

Bratman, M. E., Israel, D. J., & Pollack, M. E. (1988). Plans and resource-bounded practical reasoning. *Computational Intelligence, 4*(3), 349–355.

Celik, N. (2022). Welch's ANOVA: Heteroskedastic skew-t error terms. *Communications in Statistics-Theory and Methods, 51*(9), 3065–3076.

Chai, L., Han, Y., Han, Z., Wei, J., & Zhao, Y. (2021). Differences in disaster preparedness between urban and rural communities in China. *International Journal of Disaster Risk Reduction, 53*, 102020.

Chang, D., Cui, L., & Huang, Z. (2020). A cellular-automaton agent-hybrid model for emergency evacuation of people in public places. *IEEE Access, 8*, 79541–79551.

Construction, S. P. D. o. H. a. (2013). *5.12 Urban and rural planning practice of post-disaster reconstructio.*

Delcea, C., & Cotfas, L.-A. (2019). Increasing awareness in classroom evacuation situations using agent-based modeling. *Physica A: Statistical Mechanics and Its Applications, 523*, 1400–1418.

Delcea, C., Cotfas, L.-A., Craciun, L., & Molanescu, A. G. (2020). An agent-based modeling approach to collaborative classrooms evacuation process. *Safety Science, 121*, 414–429.

Ding, N., & Sun, C. (2020). Experimental study of leader-and-follower behaviours during emergency evacuation. *Fire Safety Journal, 117*, 103189.

Gao, D. L., Xie, W., & Lee, E. W. M. (2022). Individual-level exit choice behaviour under uncertain risk. *Physica A: Statistical Mechanics and Its Applications, 604*, 127873.

Ge, X.-X., Dong, W., & Jin, H.-Y. (2011). Study on the social psychology and behaviors in a subway evacuation drill in China. *Procedia Engineering, 11*, 112–119.

Government, C. M. P. s. (2022). *Latest ! Notice of novel coronavirus Pneumonia Prevention and Control Headquarters in Chengdu* Retrieved August 16 from https://difang.gmw.cn/sc/2022-08/30/content_35986742.htm

Han, Z., Lu, X., Hörhager, E. I., & Yan, J. (2017). The effects of trust in government on earthquake survivors' risk perception and preparedness in China. *Natural Hazards, 86*, 437–452.

Hasan, S., Mesa-Arango, R., & Ukkusuri, S. (2013). A random-parameter hazard-based model to understand household evacuation timing behavior. *Transportation Research Part c: Emerging Technologies, 27*, 108–116.

Helbing, D., & Molnar, P. (1995). Social force model for pedestrian dynamics. *Physical Review E, 51*(5), 4282.

Hu, J., Chen, C., & Kuai, T. (2018). Improvement of emergency management mechanism of public health crisis in rural China: A review article. *Iranian Journal of Public Health, 47*(2), 156.

Huang, Y., Zhou, L., & Wei, K. (2011). 5.12 Wenchuan earthquake recovery government policies and non-governmental organizations' participation. *Asia Pacific Journal of Social Work and Development, 21*(2), 77–91.

Ikotun, A. M., Ezugwu, A. E., Abualigah, L., Abuhaija, B., & Heming, J. (2022). K-means clustering algorithms: A comprehensive review, variants analysis, and advances in the era of big data. *Information Sciences.*

Ionescu, Ş, Nica, I., & Chiriţă, N. (2021). Cybernetics approach using agent-based modeling in the process of evacuating educational institutions in case of disasters. *Sustainability, 13*(18), 10277.

Jiang, L., Wang, J., & Liu, L. (2008). Providing emergency response to Wenchuan earthquake. *Technical Assistance Consultant's Report*

Kakimoto, R., & Yoshida, M. (2022). Evacuation action during torrential rain considering situation awareness error using protection motivation theory. *International Journal of Disaster Risk Reduction, 82*, 103343.

Katzilieris, K., Vlahogianni, E. I., & Wang, H. (2022). Evacuation behavior of affected individuals and households in response to the 2018 Attica wildfires: From empirical data to models. *Safety Science, 153*, 105799.

Kellens, W., Zaalberg, R., & De Maeyer, P. (2012). The informed society: An analysis of the public's information-seeking behavior regarding coastal flood risks. *Risk Analysis: An International Journal, 32*(8), 1369–1381.

King, D., Bird, D., Haynes, K., Boon, H., Cottrell, A., Millar, J., Okada, T., Box, P., Keogh, D., & Thomas, M. (2014). Voluntary relocation as an adaptation strategy to extreme weather events. *International Journal of Disaster Risk Reduction, 8*, 83–90.

Koshiba, Y., & Suzuki, Y. (2018). Factors affecting post-evacuation behaviors following an earthquake: A questionnaire-based survey. *International Journal of Disaster Risk Reduction, 31*, 548–554.

Kumar, A., Mishra, A. K., Saroj, S., & Rashid, S. (2022). Government transfers, COVID-19 shock, and food insecurity: Evidence from rural households in India. *Agribusiness, 38*(3), 636–659.

Lazo, J. K., Bostrom, A., Morss, R. E., Demuth, J. L., & Lazrus, H. (2015). Factors affecting hurricane evacuation intentions. *Risk Analysis, 35*(10), 1837–1857.

Lee, D., Yoon, S., Park, E.-S., Kim, Y., & Yoon, D. (2018). Factors contributing to disaster evacuation: The case of South Korea. *Sustainability, 10*(10), 3818.

Li, L., Ding, N., Ma, Y., Zhang, H., & Jin, H. (2020). Social Relation Network and group behavior based on evacuation experiments. *Physica A: Statistical Mechanics and Its Applications, 545*, 123518.

Liu, Y., & Mao, Z. (2022). An experimental study on the critical state of herd behavior in decision-making of the crowd evacuation. *Physica A: Statistical Mechanics and Its Applications, 595*, 127087.

Los, U. M. D. E. C. D. E. (2022). *The Luding earthquake in Sichuan has resulted in 93 deaths and 25 loss of contact*. Retrieved August 16 from https://m.guancha.cn/politics/2022_09_12_657547.shtml

Mallick, B. J., Witte, S. M., Sarkar, R., Mahboob, A. S., & Vogt, J. (2009). Local adaptation strategies of a coastal community during cyclone Sidr and their vulnerability analysis for sustainable disaster mitigation planning in Bangladesh. *Journal of Bangladesh Institute of Planners, 2*, 158–168.

Mao, Y., Yang, S., Li, Z., & Li, Y. (2020). Personality trait and group emotion contagion based crowd simulation for emergency evacuation. *Multimedia Tools and Applications, 79*, 3077–3104.

Najmanová, H., & Ronchi, E. (2017). An experimental data-set on pre-school children evacuation. *Fire Technology, 53*(4), 1509–1533.

Ning, J. (2021). *The main data of the seventh national population census of Sichuan Province*.

Niu, G., & Zhao, G. (2018). Identity and trust in government: A comparison of locals and migrants in urban China. *Cities, 83*, 54–60.

Norris, F. H., Friedman, M. J., Watson, P. J., Byrne, C. M., Diaz, E., & Kaniasty, K. (2002). 60,000 disaster victims speak: Part I. An empirical review of the empirical literature, 1981–2001. *Psychiatry, 65*(3), 207–239.

Organization, W. H. (2020). *COVID-19 strategy update—14 April 2020*. https://fctc.who.int/publications/i/item/covid-19-strategy-update---14-april-2020

Page-Tan, C., & Fraser, T. (2022). COVID-19 to go? The role of disasters and evacuation in the COVID-19 pandemic. *Global Environmental Change, 73*, 102471.

Paul, B. K. (2012). Factors affecting evacuation behavior: The case of 2007 Cyclone Sidr, Bangladesh. *The Professional Geographer, 64*(3), 401–414.

Pei, S., Dahl, K. A., Yamana, T. K., Licker, R., & Shaman, J. (2020). Compound risks of hurricane evacuation amid the COVID-19 pandemic in the United States. *GeoHealth, 4*(12), e2020GH000319.

Pfefferbaum, B., Jacobs, A. K., Van Horn, R. L., & Houston, J. B. (2016). Effects of displacement in children exposed to disasters. *Current Psychiatry Reports, 18*, 1–5.

Putri, S., Haerdy, R., & Nurdini, A. (2022). Evacuation access simulation during earthquake disaster in vertical public housing, Rancacili bandung as a case study. In *IOP conference series: Earth and environmental science.*

Qing, C., Guo, S., Deng, X., Wang, W., Song, J., & Xu, D. (2022). Stay in risk area: Place attachment, efficacy beliefs and risk coping. *International Journal of Environmental Research and Public Health, 19*(4), 2375.

Rainear, A. M., & Lin, C. A. (2021). Communication factors influencing flood-risk-mitigation motivation and intention among college students. *Weather, Climate, and Society, 13*(1), 125–135.

Raza, S. H., Haq, W., & Sajjad, M. (2020). COVID-19: A psychosocial perspective. *Frontiers in Psychology, 11*, 554624.

Rodrigueza, R. C., & Estuar, M. R. J. E. (2021). A spatial agent-based model for preemptive evacuation decisions during typhoon. In *Proceedings of the 2021 IEEE/ACM international conference on advances in social networks analysis and mining.*

Rogers, R. W. (1975). A protection motivation theory of fear appeals and attitude change1. *The Journal of Psychology, 91*(1), 93–114.

Rostami, R., & Alaghmandan, M. (2021). Performance-based design in emergency evacuation: From maneuver to simulation in school design. *Journal of Building Engineering, 33*, 101598.

Sakamoto, M., Sasaki, D., Ono, Y., Makino, Y., & Kodama, E. N. (2020). Implementation of evacuation measures during natural disasters under conditions of the novel coronavirus (COVID-19) pandemic based on a review of previous responses to complex disasters in Japan. *Progress in Disaster Science, 8*, 100127.

Santos-Reyes, J. (2020). Factors motivating Mexico City residents to earthquake mass evacuation drills. *International Journal of Disaster Risk Reduction, 49*, 101661.

Saputra, D. M., Saputra, D., & Oswari, L. D. (2020). Effect of distance metrics in determining k-value in k-means clustering using elbow and silhouette method. In *Sriwijaya international conference on information technology and its applications (SICONIAN 2019).*

Shah, M., Rizzo, S., Percy-Smith, B., Monchuk, L., Lorusso, E., Tay, C., & Day, L. (2021). Growing up under COVID-19: Young People's agency in family dynamics. *Frontiers in Sociology, 6*, 722380.

Shi, X., Xue, S., Shiwakoti, N., Li, D., & Ye, Z. (2022). Examining the effects of exit layout designs on children pedestrians' exit choice. *Physica A: Statistical Mechanics and Its Applications, 602*, 127654.

Shoji, M., Takafuji, Y., & Harada, T. (2020). Behavioral impact of disaster education: Evidence from a dance-based program in Indonesia. *International Journal of Disaster Risk Reduction, 45*, 101489.

Smith, S. K., & McCarty, C. (2009). Fleeing the storm (s): An examination of evacuation behavior during Florida's 2004 hurricane season. *Demography, 46*, 127–145.

Stein, R. M., Dueñas-Osorio, L., & Subramanian, D. (2010). Who evacuates when hurricanes approach? The role of risk, information, and location. *Social Science Quarterly, 91*(3), 816–834.

Taber, K. S. (2018). The use of Cronbach's alpha when developing and reporting research instruments in science education. *Research in Science Education, 48*, 1273–1296.

Tanim, S. H., Wiernik, B. M., Reader, S., & Hu, Y. (2022). Predictors of hurricane evacuation decisions: A meta-analysis. *Journal of Environmental Psychology, 79*, 101742.

Thakur, S., Ranjitkar, P., & Rashidi, S. (2022). Investigating evacuation behaviour under an imminent threat of volcanic eruption using a logistic regression-based approach. *Safety Science, 149*, 105688.

Tran, B. X., Nguyen, H. T., Le, H. T., Latkin, C. A., Pham, H. Q., Vu, L. G., Le, X. T. T., Nguyen, T. T., Pham, Q. T., Ta, N. T. K., & Nguyen, Q. T. (2020). Impact of COVID-19 on economic well-being and quality of life of the Vietnamese during the national social distancing. *Frontiers in Psychology, 11*, 565153.

Verma, R., Shen, J., Benedict, B. C., Murray-Tuite, P., Lee, S., & Ukkusuri, S. V. (2022). Progression of hurricane evacuation-related dynamic decision-making with information processing. *Transportation Research Part D: Transport and Environment, 108*, 103323.

Wang, Q., Song, W., Zhang, J., Ye, R., & Ma, J. (2020). Experimental study on knee and hand crawling evacuation for different age group students. *International Journal of Disaster Risk Reduction, 48*, 101613.

Whitehead, J. C., Edwards, B., Van Willigen, M., Maiolo, J. R., Wilson, K., & Smith, K. T. (2000). Heading for higher ground: Factors affecting real and hypothetical hurricane evacuation behavior. *Global Environmental Change Part b: Environmental Hazards, 2*(4), 133–142.

Whytlaw, J. L., Hutton, N., Yusuf, J. E. W., Richardson, T., Hill, S., Olanrewaju-Lasisi, T., Antwi-Nimarko, P., Landaeta, E., & Diaz, R. (2021). Changing vulnerability for hurricane evacuation during a pandemic: Issues and anticipated responses in the early days of the COVID-19 pandemic. *International Journal of Disaster Risk Reduction, 61*, 102386.

Wu, J., Yang, X., Deng, X., & Xu, D. (2022). Does disaster knowledge affect residents' choice of disaster avoidance behavior in different time periods? Evidence from China's earthquake-hit areas. *International Journal of Disaster Risk Reduction, 67*, 102690.

Xie, W., Lee, E. W. M., Cheng, Y., Shi, M., Cao, R., & Zhang, Y. (2020). Evacuation performance of individuals and social groups under different visibility conditions: Experiments and surveys. *International Journal of Disaster Risk Reduction, 47*, 101527.

Xu, D., Peng, L., Liu, S., Su, C., Wang, X., & Chen, T. (2017). Influences of sense of place on farming households' relocation willingness in areas threatened by geological disasters: Evidence from China. *International Journal of Disaster Risk Science, 8*, 16–32.

Xu, D., Peng, L., Liu, S., & Wang, X. (2018). Influences of risk perception and sense of place on landslide disaster preparedness in southwestern China. *International Journal of Disaster Risk Science, 9*, 167–180.

You, C., Chen, X., & Yao, L. (2009). How China responded to the May 2008 earthquake during the emergency and rescue period. *Journal of Public Health Policy, 30*, 379–394.

Zheng, L., Guo, Y., Peeta, S., & Wu, B. (2020). Impacts of information from various sources on the evacuation decision-making process during no-notice evacuations in campus environment. *Journal of Transportation Safety & Security, 12*(7), 892–923.

Zhou, Y., & Wei, J. Y. (2010). Risk Analysis and Evacuation Research on Crowded Students for Colleges and Universities' Emergencies. *Key Engineering Materials, 439*, 766–771.

Zhu, R., Lucas, G. M., Becerik-Gerber, B., Southers, E. G., & Landicho, E. (2022). The impact of security countermeasures on human behavior during active shooter incidents. *Scientific Reports, 12*(1), 929.

Chapter 10
Post-earthquake Resettlement Choices in Rural Sichuan, China

Abstract In this chapter this reason. Random sampling and face-to-face questionnaire surveys were combined with factor analysis and binary logistic regression to understand the resettlement modes desired by the residents and the influencing factors. The results show that residents who have lived in their current places long and whose houses were not built recently are more likely to choose the in-situ resettlement. Accessibility to employment and public services has a significant impact on residents' choice of in-situ resettlement or reallocated resettlement, so does the previous resettlement experience. The research results can provide useful suggestions for Chinese rural area post-earthquake resettlement planning following a human-centric approach with empirical data.

Keywords Post-earthquake resettlement mode · Influencing factor · Factor analysis · Binary logistic regression · Wenchuan earthquake · Lushan earthquake · Changning earthquake

10.1 Introduction

China is a country that suffers from severe earthquakes (Guo et al., 2014; YX, 2008). Sichuan is one of the most known earthquake regions in the world (Ajzen, 1988; Xu et al., 2019). The earthquake distribution map of Sichuan region is shown in Fig. 10.1, from which it can be seen that earthquakes have occurred frequently in Yibin City, Ya'an City, and Aba Tibetan and Qiang Autonomous Prefecture for the past years. This map is generated by ArcMap using Chinese seismological network data for earthquakes of magnitude 4 or above, from 2012 to 2020. The Wenchuan earthquake (in Aba Tibetan and Qiang Autonomous Prefecture) in 2008 caused 69,227 people dead, 374,644 injured and 17,923 missing; the Lushan earthquake (in Ya'an city) on April 20, 2013, killed 196 people, left 21 missing and 11,470 injured; the Changning (in Yibin city) Earthquake on June 17, 2019, resulted in 11 deaths and 122 injuries (CNSB, 2020). These earthquakes changed the natural environment of the disaster areas and left the environment there more disaster-prone and fragile.

Y. Ao and H. Bahmani, *Navigating Complexity: Understanding Human Responses to Multifaceted Disasters*, https://doi.org/10.1007/978-981-99-8207-3_10

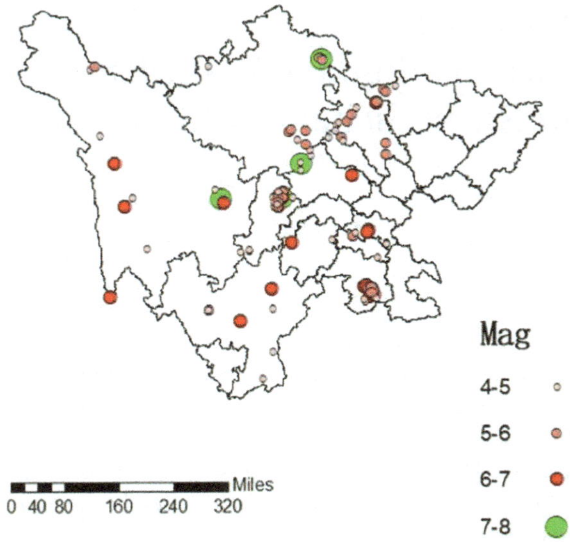

Fig. 10.1 Earthquake distribution map of Sichuan region

The residents are severely impacted by such disasters: their homes and farmland are buried and abandoned and their industries are relocated. They face the choices of resettlement, which directly affects their quality of life (MY, 2009). Therefore, post-disaster resettlement planning for resilient rural areas for reducing disasters' impacts is particularly important for local residents (Davis & Alexander, 2015). Post-earthquake resettlement planning shall follow the principles of putting people first, with scientific overall planning guidance and step-by-step implementations which combines self-reliance, state support and social assistance (Regulations on post Wenchuan earthquake recovery & reconstruction, 2008).

The current research on resettlement choices after disasters mainly focuses on the relationship between residents' willingness to relocate and its influencing factors but lacks systematic understanding on the choice made for specific post-disaster resettlement modes.

There are normally two modes for resettlement after a strong earthquake: in-situ resettlement and reallocated resettlement. In-situ resettlement refers to the resettlement of residents in their original place of residence (Jha et al., 2010). Reallocated resettlement refers to the relocation of residents to areas that are less prone to earthquakes. These two resettlement modes both have their unique advantages. In-situ resettlement does not require mobilization and land acquisition and therefore does not cause much social tension (Jha et al., 2010). Reallocated resettlement offers better employment, environmental and public services, and increases the chances of livelihoods (Iuchi, 2010).

In addition, according to the existing research, post-disaster resettlement modes include centralized resettlement and scattered resettlement. Developing concentrated rural settlements in a village is a feasible way to achieve sustainable development and improve the resilience of villages after disasters. Shortcomings in scattered resettlement include wasting land resources, poor living conditions and environmental degradation. Centralized resettlement is considered an effective means of utilizing rural land, improving infrastructure, public services, and rural living conditions (Alaci, 2010; Peng et al., 2012). However, in rural China, the scattered villages are the basic forms of villages (Yang et al., 2017). Therefore, it is still desired by some residents after earthquakes.

Furthermore, previous studies on post-disaster resettlement behavior mostly focused on developed countries (Durage et al., 2013; Lazo et al., 2015; Tobin et al., 2011), on how urban families cope with hurricanes, floods and other natural disasters (Huang et al., 2012; Lim et al., 2015; Mortreux & Barnett, 2009; Do, 2019). There is little research on how rural families in developing countries and poor areas cope with geological disasters like earthquakes. Knowing who is at higher risk due to their resettlement mode choices and the reasons behind their choices could help policymakers design targeted policies to reduce loss of life and damage for such understudied areas (Xu et al, 2017).

Recently, identifying impacting factors of farmers' resettlement choices has become a hot topic in academic research. Numerous studies show that the main factors affecting rural residents' choice of resettlement mode are on two levels: natural environment, resettlement planning, transportation conditions, government policies and other macro factors; family structure, financial conditions, education level, disaster risk perception and other personal and family characteristics on the micro level (Yang, 2009). Studies have been conducted on the willingness of residents to move after disasters in the United States. The results show that the main migration groups caused by natural disasters are the elderly who have lived in a place for a short period of time and families with low education levels (Morrow-Jones & Morrow-Jones, 1991). Lu's research also finds that people who have lived in a place for years may not choose to move their jobs or homes elsewhere due to strong local dependency (Lu et al., 2018). An investigation into the migration of different populations after the 1970 Earthquake in Peru shows that many of the victims fell into poverty after the earthquake and many young people prefer to seek employment opportunities in big cities (Osterling, 1979). The housing conditions also impact the willingness to move. Studies find that when the existing housing cannot meet the needs of residents, the willingness to move will increase (Li J, 2008). However, the longer the residents have lived in the local area, they become more dependent on the local area, leading to their reluctance to move. The appeal of a location increases with the accessibility to workplaces and everyday destinations (Iuchi, 2010). Therefore, convenient and diverse means of transportation will significantly affect residents' willingness to relocate. In the Chinese countryside, motorbikes are a good substitute for cars as farmers with motorbikes can easily reach the surrounding areas (Ao et al., 2019). Rural residents' satisfaction with the living environment also has an impact on rural residents' final choices. Studies show that people prefer to live in places that are comfortable, accessible and free from disasters (Bhat & Guo, 2004; Liu et al., 2017; Peng et al., 2013; Wijegunarathna et al., 2018). Similarly, whether people will resettle is largely related to their risk perception of disasters (Lindell and

Perrry, 2000; Wachinger et al., 2013). Facing potential disaster risks, some people living in risky areas will decide to resettle, while others are reluctant to move. For rural residents who have experienced post-disaster resettlement, the experience of previous post-disaster resettlement is also an important factor affecting later choices (Lu et al., 2018). As illustrated, much research has been performed to identify and analyze impacting factors but these factors are not grouped under themes to be used for systematic analysis for a region in China, which makes it difficult for the local government to design specific resettlement options based on the preferences of the local residents.

To summarize, previous studies do not specify different modes of resettlement for rural residents after earthquakes and are mainly executed in developed countries for various impacting factors. It is needed to study rural residents' post-earthquake resettlement mode choices in developing countries like China as the rural residents suffer a lot from severe earthquakes and their quality of life should be improved with empirical studies for better planning. And it is also important to group influencing factors under themes to study each theme separately to holistically understand the impacts. Therefore, the purpose of this study is to explore the impact of factors under different themes on residents' choice of resettlement modes in Chinese rural earthquake-stricken areas. The rural residents affected by the three major earthquakes in Sichuan are contacted to study their post-disaster resettlement mode choices and influencing factors. Two groups of four post-disaster resettlement modes ([in-situ resettlement, reallocated resettlement] and [centralized resettlement, scattered resettlement]) are set up. The differences of residents' willingness to choose resettlement mode between the two groups after the earthquake have been compared. The results will shed light upon further resettlement planning practice after earthquakes for the Chinese rural areas.

The following chapters are organized as follows: Chapter 2 explains the case areas and the identified impacting factor themes accordingly; Chap. 3 illustrates the data collection and data analysis methods step by step. Chapter 4 discusses the analysis results and Chap. 5 summarizes the whole research with future research directions.

10.2 Case Areas and Impact Factor Themes

10.2.1 Sample Village Selection

More than 80 percent of China's earthquakes of magnitude 5 or greater occur in the countryside (Ha et al., 2015). Therefore, this study takes the rural residents who suffered the most from the three major earthquakes in Sichuan (Wenchuan 5.12 earthquake, Lushan 4.29 earthquake and Changning 6.17 earthquake) as the research objects. Random sampling method was adopted to randomly select three villages in each of the rural areas severely affected by the three earthquakes. A total of nine villages have been selected as sample villages, namely Xing Wenping village, Oil

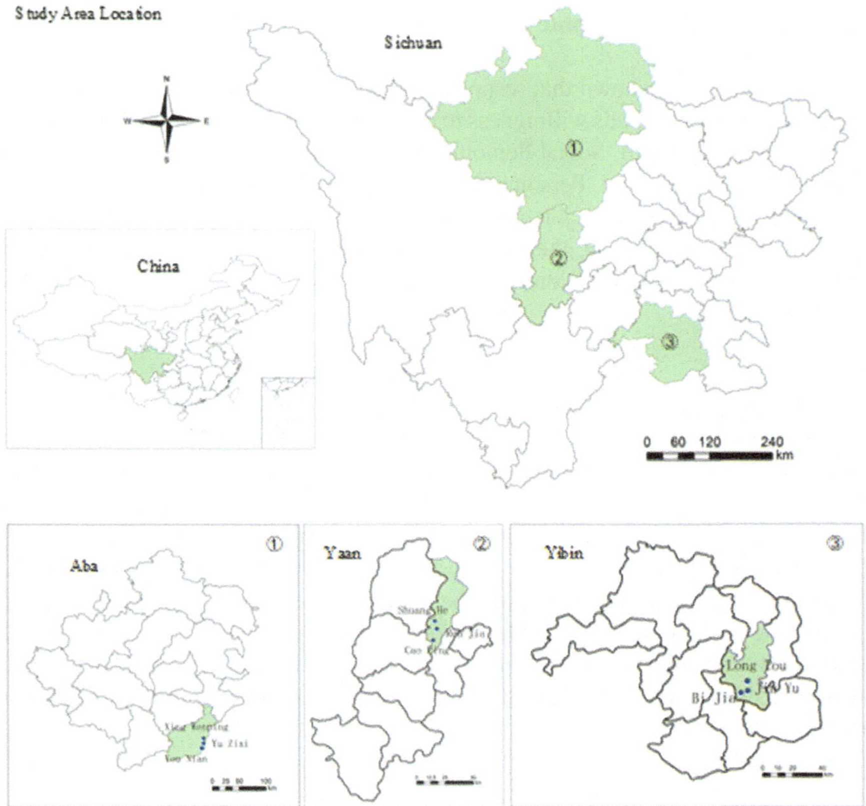

Fig. 10.2 Sample villages distribution

Mill village, Yuzixi village in Wenchuan; Caocao village, Renjia village, Shuanghe village in Lushan; Bijia village, Goldfish village, Dragon village in Changning. The location distribution of sample villages is shown in Fig. 10.2. These villages' contexts are used to combine with literature to identify four themes of impacting factors in 2.2.

10.2.2 Influencing Factor Themes

Based on the literature review, we have identified four themes for grouping influencing factors on local residents post-disaster resettlement choices. The data for all the factors under each theme is gathered for further analysis.

10.2.2.1 Social Demographic Factors

Existing studies have shown that respondents' social demographic background can significantly affect people's willingness to relocate after an earthquake (Bukvic et al., 2015; Shah et al., 2020). Social demographic factors can be divided into personal factors and family factors. Personal factors include gender, age, household registration type, education level, job status, and current residence duration. And family factors include family type, the number of migrant workers, the main source of income, whether there is a motorcycle (scooter), and the number of motorcycles (scooters).

10.2.2.2 Residential Factors

Residential factors are mainly considered from the perspective of residents' living conditions. Existing literature shows that the condition of houses, such as the year of completion, has an impact on the selection of post-disaster resettlement mode (Li J, 2008). Therefore, this paper also considers the factors like the year of the completion of the current house, whether the quality inspection has been carried out after the building is finished, and whether there is a shelter (open space) in the location. At the same time, the daily destination accessibility of local residents is also added as it significantly affects residents' willingness to choose among post-disaster resettlement modes (Iuchi, 2010).

10.2.2.3 Previous Recovery Experience Factor

Existing studies have shown that post-earthquake resettlement experience can affect their resettlement choice behavior again (Lu et al., 2018). Liu et al. (2017) have studied the satisfaction of residents under the two post disaster resettlement modes, namely the in-situ resettlement and the reallocated resettlement. The research shows that the overall satisfaction of reallocated resettlement is higher than that of in-situ resettlement. The same study group has also evaluated the satisfaction level of the previous recovery experience. In addition, in terms of neighborhood relations, family relations and other home related factors, the effect of in-situ resettlement is much better than that of reallocated resettlement (Liu et al., 2017). This paper considers the impact of resettlement mode after the last earthquake experience as another factor theme.

10.2.2.4 Built Environment Perception Factors

Residents' perception of local risks may have an impact on their choices of resettlement modes. In the face of potential disasters, some people will make migration decisions, while others are unwilling to migrate for various reasons (Wachinger et al.,

2013). Based on this, this study includes the local residents' perception of the built environment and explores their impacts on the choice made. There are many possible factors in this theme and therefore it is needed to find the most relevant ones for further analysis.

10.3 Data Collection and Analysis Methods

According to the identified themes of impacting factors, questionnaires have been developed and validated by experts. Trained surveyors have been recruited and fieldwork has been performed to gather data from local residents using the questionnaires (3.1). The descriptive statistics is illustrated in 3.2. With the collected data, we firstly have explored the most important factors for the built environment factor theme, using exploratory factor analysis. Based on the exploratory analysis, two binary logistic regression models have been constructed to identify the important factors under all themes that influence local residents' resettlement choices (3.3).

10.3.1 Data Collection

To collect valid data, the research team has recruited 18 experienced researchers, including two faculty members, 14 graduate students, and two undergraduate students. The researchers are all from rural areas, including 16 from rural Sichuan. The surveyors all have the experience of a face-to-face questionnaire survey in villages and households. They are familiar with the rural environment and can effectively communicate with rural residents. Before the study, all 18 researchers received professional training in etiquette, questioning techniques and the logical consistency of questions. In addition, the questionnaire was explained and discussed. If there are questions raised by the researchers on the questionnaire, the questionnaire is modified to address the questions. Investigators carried 10–20 questionnaires to one sample village for a preliminary survey. After the preliminary survey, researchers have adjusted the survey plan accordingly. By ensuring all issues resolved, a formal survey is finally conducted. The survey was conducted between January 5th, 2020, and January 10th, 2020. The households have been selected by random sampling. If the households are not willing to participate in the questionnaire survey, the next household will be randomly selected. All the personal sensitive data is anonymized and all the respondents have agreed upon participation.

10.3.2 Descriptive Statistics

The research group has distributed and assisted in filling out 900 questionnaires face-to-face, and finally completed 688 questionnaires effectively. After removing the questionnaires with too much missing content or lack of untruthful answers, 663 valid questionnaires have been obtained, with an effective rate of 96.37%. The distance from the epicenter of various villages and the number of valid questionnaires are shown in Table 10.1.

According to the statistics of 663 valid questionnaires, 404 people (60.94%) chose to settle in the original place, and the rest 259 people (39.06%) chose to settle in other places; 464 people (69.98%) chose centralized resettlement, while the rest 199 people (30.02%) chose scattered resettlement.

10.3.2.1 Social Demographic Theme

The basic information of interviewees is shown in Table 10.2. According to the survey data, more women have been surveyed than men, which can be explained by the fact that male laborers go out to work, while females stay at home to deal with housework. Respondents over 40 years old accounted for 72.70% of the total. The employment situation in and around the surveyed villages cannot meet the needs of young people, so more of them choose to work or study elsewhere. The proportion of rural registered permanent residence is 85.37%, and some residents in the rural–urban fringe have changed to urban registered permanent residence, which indicates that the nature of Rural China is gradually changing. The majority of people have had primary or junior high school education. The main types of personal income are

Table 10.1 Distance to epicenter and number of valid questionnaires

Sample villages		Distance from village center to epicenter (km)	Number of valid questionnaires	Number of households in the village	Total number of villages
Wen Chuan	Xing Wenping	11	74	200	800
	Yu Zixi	1.8	68	400	1200
	You Nian	8.2	62	128	307
Lu Shan	Shuang He	16	97	424	1500
	Ren Jia	9.4	85	1090	3346
	Cao Ping	4.3	68	854	3704
Changning	Long Tou	6.1	86	340	1000
	Jin Yu	0.1	71	390	1200
	Bi Jia	1.1	52	210	600

[a] This distance is the driving distance measured by Baidu navigation application

farming and working outside. The number of years of living in the current place is generally long. The main types of families are parents and children living together and three generations living under the same roof. 58.82% of families have means of transportation for short distance.

10.3.2.2 Residential Factor Theme

The variables of the current residential building year are classified as follows: "before 2000 (1)", "2000–2004 (2)", "2005–2009 (3)", "2010–2014 (4)" and "2015–2020 (5)";Whether the quality complies with the requirements variables are classified as "none (0)" and "yes (1)" after the building is built; Whether there is shelter (open space) on the location is classified as "none (0)" and "yes (1)"; Various distance variables are classified as: "[0, 1) KM (1)", "[1, 2)KM (2)", "[2, 3) KM (3)", "[3, 4) KM (4)", "[4, up) KM (5)". The researchers have used Ove software to locate points on sites and measure distances from homes to the nearest school, health center (hospital), county center and bus stops. After preliminary statistical analysis, the average value of these variables and the data sources are shown in Table 10.3.

10.3.2.3 Previous Recovery Experience Theme

The two resettlement modes are classified as "in-situ resettlement (0)" and "reallocated resettlement (1)" or "centralized resettlement (0)" and "scattered resettlement (1)". The quantified mean of statistical data and their sources are shown in Table 10.4.

10.3.2.4 Built Environment Perception Theme

Combined with the existing research, in the design of the built environment perception questionnaire, the residents' satisfaction with their own environment and their risk perception in the future will have an impact on the expected results. The questionnaire asked 27 questions about the perception of the built environment. Using Likert Scale 5, 1 means "totally disagree" and 5 means "totally agree". Respondents need to evaluate these 27 statements according to their own judgment. In order to determine important factors, factor analysis has been used to reduce the dimension of built environment perception factors for later analysis.

Factor analysis was first proposed by Chales Spearman in 1904 (XW, 2008). Factor analysis is a statistical method to simplify and analyze high-dimensional data. Its principle is dimensionality reduction, which condenses complex variables into a few factors (Zeng et al., 2019). This method causes the least information loss and can ensure the integrity of original information to the maximum extent (Zhang YW, 2019). Since under built environment theme, there are many factors involved, we have applied factor analysis to explore the most significant factors for further logistic regression analysis. The general form of factor analysis is:

Table 10.2 Social demographic theme

Personal factor		Number	Proportion (%)	Family property		Number	Proportion (%)
Gender	Male	303	45.70	Homestyle	Live alone	31	4.68
	Female	360	54.30		Two people live together	103	15.54
Age	[0,20) year	28	4.22		Parents live with their children	230	34.69
	[20–30) year	75	11.31		Of three generations under one roof	261	39.37
	[30,40) year	78	11.76		Of four generations under one roof	27	4.07
	[40,50) year	146	22.02		Other	11	1.66
	[50,60) year	175	26.40	Migrant workers	0 people	264	39.82
	[60,∞) year	161	24.28		1 people	164	24.74
Type of registered permanent residence	Rural household registration	566	85.37		2 people	179	27.00
	Urban registration	97	14.63		3 people	30	4.52
Schooling	Nothing	78	11.76		4 or more people	26	3.92
	Primary school	255	38.46	The main source of income	Agricultural industry	61	9.20
	Junior high school	220	33.18		Out-migrantion for work	398	60.03
	Senior high school	71	10.71		Shop management	98	14.78
	Bachelor degree or above	39	5.88		Domestic workshop	6	0.90
Working condition	No job	248	37.41		House for rent	3	0.45

(continued)

Table 10.2 (continued)

Personal factor		Number	Proportion (%)	Family property		Number	Proportion (%)
	Farming	103	15.54		Agritainment	4	0.60
	Out-migrantion for work	137	20.66		Other	93	14.03
	In business	83	12.52	Motorcycle or not	No	273	41.18
	Governmental service	17	2.56		Yes	390	58.82
	Other	75	11.31	Number of Motorcycles	0	273	41.18
Length of residence in current place	[0,10) year	165	24.89		1	346	52.19
	[10,20) year	109	16.44		2	40	6.03
	[20,30) year	105	15.84		3	3	0.45
	[30,40) year	70	10.56		4 or more	1	0.15
	[40,∞) year	214	32.28				

Table 10.3 Residential condition items and sources

Items	Average (KM)	Source
The year of construction of the present residence	3.62	Li J, (2008)
Is there any quality inspection after the building is built?	0.44	Committee (2008)
Is there a shelter (open space)?	0.89	Shiwakoti (2016)
The distance from the present residence to the nearest school	1.70	Eliasson (2010a, 2010b)
The distance from the present residence to the nearest health centre (hospital)	1.71	Eliasson (2010a, 2010b)
The distance from the present residence to the center of the county	3.63	Eliasson (2010a, 2010b)
The distance from the present residence to the nearest bus station	1.52	Eliasson (2010a, 2010b)

Table 10.4 Previous recovery experiences items and sources

Items	AVG	Source
The last time after the earthquake to choose the In-situ resettlement or reallocated resettlement?/The last time after the earthquake to choose centralized resettlement or scattered resettlement?	0.34/ 0.33	Authors

$$x_i = a_{i1}F_1 + a_{i2}F_2 + \cdots + a_{in}F_n + \varepsilon_i (i = 1, 2, \ldots, p) \quad (10.1)$$

where $\boldsymbol{a}_{ij}$ *is the correlation coefficient between the common factor* $\boldsymbol{F}_i$ *and the variable* $\boldsymbol{x}_i$ *and* $\boldsymbol{\varepsilon}_i$ *is the special factor, representing the influencing factors other than the common factor* (XW, 2008)

Before the factor analysis, Kjeldahl Meyer Olgin (KMO) and Bartlett test are carried out. In general, if the KMO value is greater than 0.70 and the p value of Bartlett variance homogeneity test is less than 0.05, the data is considered suitable for factor analysis. The test results are shown in Table 10.5. The KMO value is 0.879 and the p value is 0.000. The test effect is significant, indicating that there is a certain correlation, which is suitable for factor analysis.

The results of factor analysis are shown in Table 10.6. In general, if the contribution rate of accumulation variance of each factor reaches more than 60%, the scale is

Table 10.5 KMO and Bartlett's test results

KMO and Bartlett test		Built environment perception
KMO Sampling suitability quantity		0.879
Bartlett test of sphericity	Approximate chi square	9632.906
	Freedom	351
	Significance	0.000

considered to have good validity. The total variance interpretation reaches 69.933%, that is, the factor analysis can explain 69.933% of the variables, and the effect of factor analysis is good. The closer the load coefficient is to 1, the more relevant the index is to the common factor. Generally, we can ignore load coefficients less than 0.4. After removing the factors whose factor load is less than 0.4, seven main factors are obtained.

10.3.3 Binary Logistic Regression Analysis

To understand all the factors identified that impact resettlement mode choices, binary logistic regression analysis has been performed. Binary Logistic regression model does not require the distribution of explanatory variables, nor does it require a linear relationship between explanatory variables and the explained variables. This makes it suitable for analyzing problems with classified variables. In the general regression model, the dependent variable of the function is usually interval variable, and the dependent variable should meet the assumption conditions of normal distribution. Logistic regression is different from general regression models. The purpose of logistic regression is to predict the probability of each classification of a certain classification variable, so the dependent variable must be a classification variable, and logistic regression has no special requirements for independent variables, which can be interval variables, categorical variables or a mixture of the two variables. Meanwhile, according to the number of dependent variable values, logistic regression model is further divided into binary logistic regression model and multivariate logistic regression model. In the multivariate logistic regression model, the dependent variable can take multiple values. In the binary logistic regression model, the dependent variable is a binary value, set as $\boldsymbol{Y}$ and follows a binomial distribution with values of 0 and 1, and the independent variables are $\boldsymbol{X}_1, \boldsymbol{X}_2, \ldots, \boldsymbol{X}_n$. The binary logistic regression model corresponding to independent variables is:

$$P(Y=1) = \frac{EXP(\beta_0 + \beta_1 X_1 + \beta_2 X_2 + \cdots + \beta_n X_{\boldsymbol{m}})}{1 + EXP(\beta_0 + \beta_1 X_1 + \beta_2 X_2 + \cdots + \beta_n X_n)} \tag{10.2}$$

$$P(Y=0) = \frac{1}{1 + EXP(\beta_0 + \beta_1 X_1 + \beta_2 X_2 + \cdots + \beta_n X_n)} \tag{10.3}$$

$$\boldsymbol{logit\, P}(\boldsymbol{Y}=1) = \boldsymbol{\beta}_0 + \boldsymbol{\beta}_1 \boldsymbol{X}_1 + \boldsymbol{\beta}_2 \boldsymbol{X}_2 + \cdots + \boldsymbol{\beta}_n \boldsymbol{X}_n \tag{10.4}$$

Similar to the linear regression model, $\boldsymbol{\beta}_0$ is the constant term (or intercept), and $\boldsymbol{\beta}_i$ is the partial regression coefficient corresponding to $\boldsymbol{X}_i (\boldsymbol{i} = 1, 2, \ldots, \boldsymbol{m})$ (Wang & Guo, 2001). For our study, binary logistic regression analysis has been conducted as we have only two options for resettlement modes for each model.

Taking all the above factors as independent variables, selecting set [in-situ resettlement (0), reallocated resettlement (1)], [centralized resettlement (0), scattered

Table 10.6 EFA results of BE perception: factor component matrix

	Component							Source
	Satisfaction with current housing	Satisfaction with current residence	Sense of security and belonging to the current residence	Satisfaction with government subsidies	Impact of the last earthquake	Interpersonal relationship	Risk perception of earthquake recurrence	
	1	2	3	4	5	6	7	
Satisfaction with house power supply	0.730							Bhat and Guo (2004), Liu et al. (2017), Peng et al. (2013)
Satisfaction with house gas supply	0.733							Bhat and Guo (2004), Liu et al. (2017), Peng et al. (2013)
Satisfaction with house drainage	0.740							Bhat and Guo (2004), Liu et al. (2017), Peng et al. (2013)
Satisfaction with housing structure	0.812							Bhat and Guo (2004), Liu et al. (2017), Peng et al. (2013)

(continued)

Table 10.6 (continued)

	Component							Source
	Satisfaction with current housing	Satisfaction with current residence	Sense of security and belonging to the current residence	Satisfaction with government subsidies	Impact of the last earthquake	Interpersonal relationship	Risk perception of earthquake recurrence	
Satisfaction with housing area	0.778							Bhat and Guo (2004), Liu et al. (2017), Peng et al. (2013)
Satisfaction with house orientation and daylighting	0.788							Bhat and Guo (2004), Liu et al. (2017), Peng et al. (2013)
Satisfaction with the quality of life in the place of residence		0.589						Bhat and Guo (2004), Liu et al. (2017), Peng et al. (2013)
Satisfaction with residence economy		0.701						Bhat and Guo (2004), Liu et al. (2017), Peng et al. (2013)

(continued)

Table 10.6 (continued)

	Component							Source
	Satisfaction with current housing	Satisfaction with current residence	Sense of security and belonging to the current residence	Satisfaction with government subsidies	Impact of the last earthquake	Interpersonal relationship	Risk perception of earthquake recurrence	
Satisfaction with the ecological environment of the residence		0.425						Bhat and Guo (2004), Liu et al. (2017), Peng et al. (2013)
Satisfaction with the educational environment of the residence		0.737						Bhat and Guo (2004), Liu et al. (2017), Peng et al. (2013)
Satisfaction with medical conditions in residence		0.762						Bhat and Guo (2004), Liu et al. (2017), Peng et al. (2013)
Satisfaction with residential infrastructure		0.753						Bhat and Guo (2004), Liu et al. (2017), Peng et al. (2013)

(continued)

Table 10.6 (continued)

	Component							Source
	Satisfaction with current housing	Satisfaction with current residence	Sense of security and belonging to the current residence	Satisfaction with government subsidies	Impact of the last earthquake	Interpersonal relationship	Risk perception of earthquake recurrence	
Satisfaction with employment opportunities in residence		0.815						Bhat and Guo (2004), Liu et al. (2017), Peng et al. (2013)
Sense of belonging to the place of residence			0.803					Liu et al. (2017)
Security of place of residence			0.788					Liu et al. (2017)
Road connections to safe locations			0.693					Bhat and Guo (2004)
Adaptability to local customs			0.556					Li and Guo (2016), Peng et al. (2013)
Satisfaction with the amount of government subsidies				0.860				Liu et al. (2017), Lu et al. (2018)

(continued)

Table 10.6 (continued)

	Component							Source
	Satisfaction with current housing	Satisfaction with current residence	Sense of security and belonging to the current residence	Satisfaction with government subsidies	Impact of the last earthquake	Interpersonal relationship	Risk perception of earthquake recurrence	
Satisfaction with government subsidies				0.895				Liu et al. (2017), Lu et al. (2018)
Satisfaction with the transparency of government subsidy information				0.848				Liu et al. (2017), Lu et al. (2018)
The degree of damage to the building structure caused by the last earthquake					0.893			Li and Guo (2016), Peng et al. (2013)
The damage degree of the last earthquake to the road					0.821			Li and Guo (2016), Peng et al. (2013)
The impact of the last earthquake on you and your family					0.850			Jianguang (1994)

(continued)

Table 10.6 (continued)

	Component							Source
	Satisfaction with current housing	Satisfaction with current residence	Sense of security and belonging to the current residence	Satisfaction with government subsidies	Impact of the last earthquake	Interpersonal relationship	Risk perception of earthquake recurrence	
Neighborhood relations						0.858		Liu et al. (2017)
Family relations						0.853		Liu et al. (2017)
There is a great possibility of another destructive earthquake							0.829	Lindell and Perry (2000), Wachinger et al. (2013)
When another destructive earthquake occurred, the family was greatly affected							0.737	Lindell and Perry (2000), Wachinger et al. (2013)
Inclusion statistics								
characteristic value	4.024	3.789	2.846	2.545	2.318	1.991	1.369	
Percentage variance	14.904	14.032	10.542	9.426	8.586	7.374	5.069	

(continued)

Table 10.6 (continued)

	Component							Source
	Satisfaction with current housing	Satisfaction with current residence	Sense of security and belonging to the current residence	Satisfaction with government subsidies	Impact of the last earthquake	Interpersonal relationship	Risk perception of earthquake recurrence	
Cumulative variance percentage	14.904	28.936	39.478	48.904	57.490	64.864	69.933	
Extraction method: principal component analysis								
Rotation method: Kaiser standardized orthogonal rotation method								
The rotation has converged after 6 iterations								

resettlement (1)] as dependent variables, binary logistic regression analysis has been performed using SPSS 20.0 software. The results of binary logistic regression are shown in Table 10.7. And the model fitting results show that the chi square scores are 9.188 and 8.574 with the degrees of freedom of 8, and the Cox-Snell R2 values are 0.317 and 0.330 respectively. The fitting effect of the regression model is good, and the data can be further correlated. When there are many independent factors in the actual problem, there may be a certain correlation between two or more independent variables, which is called multicollinearity. When the collinearity trend of independent variables is very obvious, it will seriously affect the fitting of the model (Zhang, 1994). In this study, variance expansion factor (VIF) has been used to test multicollinearity. The greater the variance expansion factor, the stronger the multicollinearity. When the VIF value is greater than 10, it is considered that there is strong multicollinearity between variables and it is unacceptable (S, 2014). The VIF values of the independent variables in this research are less than 10 (see Table 10.7), indicating that there is no multicollinearity between the independent variables.

10.4 Results

10.4.1 Social Demographic Factor Theme Results

For personal factors, gender has no obvious impact on the resettlement modes choice. The type of household registration has no obvious impact on whether you want to settle in the original place or in another place; Education level and working conditions have no obvious impact on whether they want centralized resettlement or decentralized resettlement. The older the residents are and the longer they live in their current residence, the more inclined they are to choose the centralized in-situ resettlement. The reason may be the strong habit and dependence on local facilities. The more educated (B = 0.372, P = 0.007) the residents are, the more they want to resettle in other places. In the study of Zorrilla and Sandberg, it is also found that residents with higher education are also more likely to choose to migrate outward under natural disasters (Saldaña-Zorrilla & Sandberg, 2009), which is consistent with the results of this study. Residents with rural household registration account type (B = – 0.554, P = 0.084) show that residents with rural household registration prefer scattered resettlement. The data results of work (farming) ($B_1 = -0.651$, $P_1 = 0.076$) show that villagers prefer to choose the original site for resettlement.

In the family attribute variables, the main source of income, family type and the number of migrant workers have no obvious impact on whether they want centralized resettlement or decentralized resettlement. Compared with families whose main source of income is migrant workers, people who earn money by running shops and renting houses are more inclined to choose reallocated resettlement. Families with a large number of migrant workers (B = 0.203, P = 0.038) prefer to relocate to other places, where they can find better employment opportunities. Families living

Table 10.7 Results of binary logistic regression analysis

Variable	In-situ resettlement (0) reallocated resettlement (1)			Centralized resettlement (0) scattered resettlement (1)		
	B	*P*-value	VIF	B	*P*-value	VIF
Personal attributes						
Gender	–0.262	0.229	1.117	0.242	0.306	1.106
Age	–0.360	0.000	2.031	–0.220	0.043	2.019
Length of current residence	–0.214	0.005	1.652	–0.170	0.043	1.519
Account type	–0.203	0.480	1.101	–0.554	0.084	1.101
Education level	0.372	0.007	1.881	0.209	0.164	1.897
Working condition (no working)		0.320	1.202		0.369	1.209
Working conditions (farming)	–0.651	0.076		–0.187	0.649	
Working conditions (working outside)	–0.221	0.621		–0.147	0.763	
Working conditions (business)	–0.160	0.684		–0.106	0.817	
Working conditions (government services)	–0.629	0.204		0.712	0.173	
Working conditions (others)	–0.010	0.988		–0.956	0.258	
Family attributes						
Main source of income (migrant workers)		0.149	1.123		0.986	1.123
Main source of income (agricultural production)	0.577	0.237		0.049	0.917	
Main source of income (shop operation)	0.841	0.018		–0.074	0.838	
Main source of income (family workshop)	0.745	0.114		–0.389	0.419	
Main source of income (house rental)	3.044	0.019		–0.595	0.620	
Main source of income (farmhouse)	–19.191	0.999		–21.267	0.999	
Main source of income (other)	0.898	0.508		–18.324	0.999	
Family type	–0.347	0.002	1.082	0.044	0.704	1.089

(continued)

Table 10.7 (continued)

Variable	In-situ resettlement (0) reallocated resettlement (1)			Centralized resettlement (0) scattered resettlement (1)		
	B	*P*-value	VIF	B	*P*-value	VIF
Number of migrant workers	0.203	0.038	1.116	–0.119	0.285	1.117
Is there a motorcycle (battery car)	0.368	0.389	4.151	1.332	0.003	4.146
Number of motorcycles (battery cars)	0.324	0.340	4.083	1.161	0.001	4.083
Residential properties						
Year of completion of current residence	0.294	0.006	1.166	–0.065	0.531	1.163
Is there any quality acceptance after the house is built	–0.159	0.494	1.308	–0.640	0.013	1.335
Is there a shelter (open space) in the location	0.228	0.520	1.200	–0.689	0.058	1.176
Distance to nearest school	0.596	0.032	8.323	0.362	0.225	8.313
Distance to nearest health center (hospital)	0.508	0.040	7.687	0.486	0.058	7.710
Distance to town center	0.031	0.695	1.473	0.112	0.181	1.516
Distance to nearest bus stop (bus stop)	0.268	0.054	1.777	0.051	0.747	1.728
Past recovery experience						
After the last earthquake, choose the In-situ resettlement or reallocated resettlement./centralized resettlement or scattered resettlement	–1.853	0.000	1.399	–2.358	0.000	1.361
Built environment perception						
Satisfaction with current housing	0.057	0.596	1.164	0.036	0.761	1.180
Satisfaction with current residence	–0.283	0.009	1.180	0.170	0.152	1.172
Sense of security and belonging to the place of residence	–0.020	0.841	1.088	–0.223	0.046	1.078
Satisfaction with government subsidies	0.002	0.985	1.055	–0.085	0.476	1.056
Damage of houses and roads and its impact on families	0.268	0.028	1.355	–0.408	0.001	1.315
Interpersonal relationship	0.113	0.268	1.077	0.071	0.539	1.069
Risk perception	0.205	0.066	1.158	–0.494	0.000	1.146
Constant	–1.575	0.192		0.083	0.949	

together for several generations ($B = -0.347$, $P = 0.002$)in the local area prefer to choose the in-situ resettlement, possibly because of their strong attachment to the local area. The existence and number of motorcycles significantly affect whether residents choose centralized resettlement or scattered resettlement. Families with more motorcycles ($B = 1.161$, $P = 0.001$)prefer scattered resettlement.

10.4.2 Residential Factor Theme Results

In the residential factor theme, the completion year and accessibility of the current housing significantly affect the residents' choice of in-situ resettlement or reallocated resettlement. To be more specific, the earlier the current residence ($B = 0.294$, $P = 0.006$) is built, the weaker their intention is to move, which is consistent with the analysis in the personal factor theme. That is: the longer the residence time, the more desire for residents to choose the in-situ resettlement. The closer the residence is to the nearest school ($B = 0.596$, $P = 0.032$), health center (hospital) ($B = 0.508$, $P = 0.040$) and bus stop ($B = 0.268$, $P = 0.054$), the more likely it is for them to choose the in-situ resettlement. Furthermore, whether the quality acceptance is carried out after the house is built ($B = -0.640$, $P = 0.013$), whether there is a shelter (open land)) ($B = -0.689$, $P = 0.058$) on the location and the distance from the hospital ($B = 0.486$, $P = 0.058$) near the residence significantly affect the residents' choice between centralized resettlement and scattered resettlement.

10.4.3 Previous Recovery Experience Theme Results

Based on the valid 464 questionnaire results answering questions in this theme, the resettlement experience after the last earthquake significantly affects the resettlement mode choice made in the future. It can be seen from Table 10.7 that when the resettlement mode selected after the last earthquake is in-situ, residents may prefer to choose non-local resettlement in the future resettlement ($B = -1.853$, $P = 0.000$). For people with the centralized resettlement ($B = -2.358$, $P = 0.000$) after the last earthquake, they may be more willing to choose scattered resettlement in the future. This shows that residents prefer to choose the opposite resettlement mode in the future. This finding indicates that in the rural areas of Sichuan, there have been not effective and satisfactory resettlement planning which could result in consistent choice for desired resettlement from the local residents in the future.

10.4.4 Built Environment Perception Theme Results

Overall, the impact of built environment perception on the choice of villagers' post disaster resettlement mode is limited. Among the perceived variables of built environment, satisfaction with the current residence has a significant impact on residents' choice of in-situ resettlement or reallocated resettlement. Residents who are satisfied with their current residence (B = − 0.283, $P = 0.009$) are more likely to choose the in-situ resettlement. In fact, environmental factors have been proved to be an important consideration in site selection decision-making, and people prefer to live in a place with good environment (Klaiber, 2014; Tacoli, 2009). The more sense of security and belonging to the place of residence, the more residents are willing to choose centralized resettlement (B = − 0.223, $P = 0.046$). The more serious the damage of houses and roads and the impact on families (B = − 0.408, $P = 0.001$), the higher the risk perception (B = − 0.494, $P = 0.000$), the more residents prefer centralized resettlement. A study in Vietnam shows similar results (Koubi et al., 2016). We can also see that there are some factors identified in literature do not have impacts on residents post-disaster resettlement choices in rural Sichuan. Therefore, it indicates that it is needed to follow context-specific data analysis for better village design in the post-disaster era.

10.5 Discussion

In this study, the rural areas where three major earthquakes occurred in Sichuan Province in recent years were selected as the research object, and nine sample villages were selected in a reasonable way for investigation and relevant data were collected. By using binary logistics regression and factor analysis method, this paper makes an empirical analysis of the research objectives, and explores the influence of personal attributes, family attributes, residential attributes, the last post-disaster resettlement mode selection and the perception of living environment on residents' choice of resettlement mode in earthquake-stricken areas.

In personal attributes, the older the residents are, the longer they live in their current residence, and they are more inclined to choose the original site for centralized resettlement. Residents with higher education level are more likely to relocate. Lu et al. (2018) suggested in the research that people who have lived in the local area for many years may not choose to move their jobs or residences to other places because of their strong dependence on the local area. The villagers in farming are more willing to choose original resettlement, which is consistent with the research of Chen Yang (2009) that original resettlement is more conducive to villagers' farming. Rural residents with registered permanent residence are more inclined to choose decentralized resettlement. Combined with the actual situation of Chinese rural areas, the external form of decentralized traditional villages is the basic pattern of the external form of Traditional Villages in China (Yang et al., 2017), that is to say, rural residents with registered permanent residence are more willing to choose the traditional rural resettlement. This is consistent with the results of this study.

Among the family attributes, the families whose main source of income is shop operation and house rental are more inclined to choose resettlement in different places than those whose main source of income is going out to work. Families who have lived together for generations in the local area prefer to be resettled in the original place. Families with a large number of migrant workers are more willing to choose resettlement in different places. The object of this study is rural China and the income in rural areas is generally not high. Young villagers prefer to seek better employment opportunities in other places, because there are few employment opportunities in rural areas and the employment scope is narrow. If this trends continues, the rural area will be with more elderly and children which makes them more fragile in disasters and therefore, a better designed centralized settlement that can attract these residents to relocate might be preferred. On the other hand, in the future, the development of characteristic industries in rural areas and surrounding areas might enable young people to obtain a certain income without going out to work, which will affect the choice of post-earthquake resettlement models for rural residents. This is consistent with the proposal of Kniveton et al. that different employment opportunities will have an impact on residents' choice of relocation (Kniveton et al., 2011). Families with more motorcycles (battery cars) are more willing to choose scattered resettlement. It may be that families living in scattered places have a relatively long travel distance. In order to reduce the travel time, it is more likely to own motorcycles. Generally speaking, family travel tools are used to reduce travel time. According to the actual situation in rural China, families with motorcycles (battery cars) are relatively convenient to travel (Bhat & Guo, 2007), so they are more willing to choose decentralized resettlement, which is consistent with the results of this study.

Among the properties of housing, the earlier the year of completion of the existing housing, the weaker the intention of moving. Generally speaking, the longer a house exists, the worse its function will be. When the existing houses can't meet the needs of residents more and more, it promotes the residents' willingness to move to a certain extent (Iuchi, 2010). However, the research in this paper shows that the earlier the completion year of the current residence, the longer the residents live in the current residence, the stronger their dependence on the local area, and the weaker their intention to move. This is consistent with Lu et al. research. The closer you are to the nearest school, health center (hospital), and bus stop (bus stop), the more likely you are to choose the original site for resettlement. In transportation related research, the choice behavior of residence or work place is usually related to public transport accessibility, travel cost, travel mode, traffic congestion and departure time (Arentze & Timmermans, 2007; Nurlaela & Curtis, 2012). In general, the attractiveness of a place increases with the accessibility of the workplace and daily life. Therefore, convenient and diverse means of transportation will significantly affect residents' willingness to move (Iuchi, 2010). This supports the findings in this research as well. Whether the house has been inspected in terms of quality after being built, whether there is a shelter (open space) in the place where it is located, and the distance from the hospital near the place of residence significantly affect residents' choice of centralized or decentralized resettlement. This finding aligns with the conclusion of Lindell et al. that residents prefer to choose centralized resettlement when they think their residential conditions are relatively safe (Lindell and Perry, 2000).

In the last choice of post-disaster resettlement mode, residents would prefer to choose the opposite resettlement mode in the future. For rural residents who have experienced a post-disaster resettlement, the experience of the last post-disaster resettlement is also an important factor affecting the final mode choice (Lu et al., 2018). This finding also aligns with another survey made by the same group of researchers to evaluate the satisfaction levels of residents' previous resettlement experience and the results show that residents are not happy with their previous experience. Therefore, this study is in need for better design future resettlement based on residents' preferences, using empirical data.

In the perception of living environment, residents who are satisfied with their current residence are more inclined to choose the original site for resettlement. The better the sense of security and belonging to the place of residence, the more residents are willing to choose centralized resettlement. This is consistent with the research result of Bhat and Guo (2004) that is, people prefer to live in places with comfortable environment, harmonious neighborhood and far away from disasters. The more serious the road damage and its impact on families, the higher the risk perception, and the more residents tend to be resettled in different places. Lindell and Perry (2000) research shows that whether people will make the decision to move when a disaster occurs is largely related to the risk perception of the disaster. Some people have a strong perception of the disaster risk, some people have a weak perception of the disaster risk, and some people are even indifferent to the disaster risk. Therefore, in the face of potential disaster risks, some people with strong risk perception will make relocation decisions, which is consistent with the research results of this paper.

The main limitation of this study is that in terms of social demographic variables, this study does not consider wealth but those with more financial means may choose differently. However, we do have selected motorbike possession as one of the variables which can reflect to some extent the wealth of residents. In terms of space, this study does not analyze the specific relocation orientation, resulting in not in-depth analysis of the influencing factors, and it only preliminary discusses the relationship between rural residents' relocation intention and its influencing factors. In the future research, we shall consider the wealth factor and analyze the difference of residents' choice of post disaster resettlement modes. we could also analyze the specific relocation orientation of residents with different relocation intentions.

10.6 Conclusions and Limitations

Facing severe earthquake threats in Rural China, it is important for the government to design built environment for better resettlement in a human-centric and resilient way to improve quality of life and public health. The current studies focus mainly on other disasters than earthquakes in developed countries and analyze multiple factors in a non-systematic way. Therefore, the main purpose of this research is to explore the influencing factors under different themes on the resettlement mode choices made in rural China, using empirical data and systematic analysis. We have chosen case study areas first and then identified four important themes affecting post-earthquake resettlement mode choices based on literature and the context in rural China. Questionnaires are developed accordingly and used for gathering data for these four themes' factors. For reducing factor numbers in themes, factor analysis has been applied. For analyzing all the factors in different themes, two binary logistic regression models with identified factors are built.

Through investigation and research, it is found that the factors affecting the choice of residents' resettlement modes are context-specific. Not all the identified factors have significant impacts for each resettlement mode. The systematic analysis of the main influencing factors provides a better understanding of the motivation of residents' choice of resettlement modes in Rural China, which can not only guide rural residents to reasonably choose resettlement modes, but also place people in the center for post disaster resettlement design (Maly, 2018). This analysis also provides reference for the planning and construction of future resilient villages in disaster-prone areas. In addition, among the influencing factors of choosing in-situ resettlement or reallocated resettlement, the negative correlation of satisfaction with the current residence is obvious, indicating that satisfaction with the current residence is still the main factor causing residents' migration. Therefore, in the overall construction and planning of the village in the future, residents satisfaction should be put in center. Possible actions could be: the traffic roads in the village should be actively improved and the construction of infrastructure and other public service facilities in the village shall be carried out simultaneously with the residential construction. The local economy should be improved so that young people would stay in the village and take better care of the elderly and children, and making them less vulnerable facing disasters.

References

Ajzen, I. (1988). *Attitudes personality, and behavior*. Open University Press.

Alaci, D. S. A. (2010). Regulating urbanisation in Sub-Saharan Africa through cluster settlements: Lessons for urban managers in Ethiopia. *Theor Empir Res Urban Manag, 5*(14), 20–34.

Ao, Y., Yang, D., Chen, C., & Wang, Y. (2019). Exploring the effects of the rural built environment on household car ownership after controlling for preference and attitude: Evidence from Sichuan, China. *Journal of Transport Geography, 74*, 24–36. https://doi.org/10.1016/j.jtrangeo.2018.11.002

Arentze, T., & Timmermans, H. (2007). Congestion pricing scenarios and change of job or residential location: Results of a stated adaptation experiment. *Journal of Transport Geography, 15*(1), 56–61. https://doi.org/10.1016/j.jtrangeo.2006.02.013

Bhat, C. R., & Guo, J. (2004a). A mixed spatially correlated logit model: Formulation and application to residential choice modeling. *Transportation Research Part b: Methodological, 38*(2), 147–168. https://doi.org/10.1016/s0191-2615(03)00005-5

Bhat, C. R., & Guo, J. (2007). A comprehensive analysis of built environment characteristics on household residential choice and auto ownership levels. *Transportation Research Part b: Methodological, 41*(5), 506–526. https://doi.org/10.1016/j.trb.2005.12.005

Bukvic, A., Smith, A., & Zhang, A. (2015). Evaluating drivers of coastal relocation in Hurricane Sandy affected communities. *International Journal of Disaster Risk Reduction.* https://doi.org/10.1016/j.ijdrr.2015.06.008

Chen, Y. (2009). Preliminary discussion on disaster and migration. *Journal of Catastrophology, 24*(2), 138–144.

CNSB. (2020). *China yearbook of household survey in 2019*. China Statistical Press.

Committee, C. C. (2008). *Unified design description of farmers' self built houses after earthquake.*

Davis, I., & Alexander, D. (2015). *Recovery from disaster*. Routledge.

Do, X. B. (2019). Fukushima Nuclear Disaster displacement: How far people moved and determinants of evacuation destinations. *International Journal of Disaster Risk Reduction, 33*, 235–252. https://doi.org/10.1016/j.ijdrr.2018.10.009

Durage, S. W., Kattan, L., Wirasinghe, S. C., & Ruwanpura, J. Y. (2013). Evacuation behaviour of households and drivers during a tornado. *Natural Hazards, 71*(3), 1495–1517. https://doi.org/10.1007/s11069-013-0958-6

Eliasson, J. (2010). The influence of accessibility on residential location. In F. Pagliara, J. Preston, & D. Simmonds (Eds.), *Residential location choice: Models and applications* (pp. 137–164). Springer Berlin Heidelberg. https://doi.org/10.1007/978-3-642-12788-5_7

Eliasson, J. (2010). The influence of accessibility on residential location. *Advances in Spatial Science.*

Guo, S., Liu, S., Peng, L., & Wang, H. (2014). The impact of severe natural disasters on the livelihoods of farmers in mountainous areas: A case study of Qingping Township, Mianzhu City. *Natural Hazards, 73*(3), 1679–1696. https://doi.org/10.1007/s11069-014-1165-9

Ha, H., Zhu, Z., Hu A. Z., Xu, C. J., Ji, S. G., Chen, Y. S., & Zhao, G. W. 2015. Summary of China's rural earthquake safety policies and regulations. *Recent Developments in World Seismology*, 03, 3–19. https ://doi.org/10 .3969 / j.issn.0235-4975.2015.03.003

Huang, S. K., Lindell, M. K., Prater, C. S., Wu, H. C., & Siebeneck, L. K. (2012). Household evacuation decision making in response to Hurricane ike. *Natural Hazards Review, 13*(4), 283–296. https://doi.org/10.1061/(asce)nh.1527-6996.0000074

Iuchi, K. (2010). Redefining a place to live: Decisions, planning processes, and outcomes of resettlement after disasters. *Dissertations & Theses Gradworks.*

Jha, A., Duyne Barenstein, J., Phelps, P., Pittet, D., & Sena, S. (2010). Safer homes, stronger communities: a handbook for reconstructing after natural disasters. *World Bank Publications*, 1–370 (370). 18361

Jianguang, Z. (1994). Environmental Hazards in the Chinese Public's Eyes. *Risk Analysis an Official Publication of the Society for Risk Analysis, 14*(2), 163–167. https://doi.org/10.1111/j.1539-6924.1994.tb00041.x

Klaiber, H. A. (2014). Migration and household adaptation to climate: A review of empirical research. *Energy Economics, 46*, 539–547. https://doi.org/10.1016/j.eneco.2014.04.001

Kniveton, D., Smith, C., & Wood, S. (2011). Agent-based model simulations of future changes in migration flows for Burkina Faso. *Global Environmental Change, 21*, S34–S40. https://doi.org/10.1016/j.gloenvcha.2011.09.006

Koubi, V., Spilker, G., Schaffer, L., & Bernauer, T. (2016). Environmental stressors and migration: Evidence from Vietnam. *World Development, 79*, 197–210. https://doi.org/10.1016/j.worlddev.2015.11.016

Lazo, J. K., Bostrom, A., Morss, R. E., Demuth, J. L., & Lazrus, H. (2015). Factors affecting hurricane evacuation intentions. *Risk Analysis, 35*(10), 1837–1857. https://doi.org/10.1111/risa.12407

Li J, L. X. 2008. Analysis on Influencing Factors of rural residents' willingness to move. *Economic Geography, 28*(3), 454–459. https://doi.org/10.15957/j.cnki

Li, Y., & Guo, Y. (2016). Be proactive for better decisions: Predicting information seeking in the context of earthquake risk. *International Journal of Disaster Risk Reduction, 19*, 75–83. https://doi.org/10.1016/j.ijdrr.2016.08.008

Lim, M. B. B., Lim, H. R., Piantanakulchai, M., & Uy, F. A. (2015). A household-level flood evacuation decision model in Quezon City, Philippines. *Natural Hazards, 80*(3), 1539–1561. https://doi.org/10.1007/s11069-015-2038-6

Lindell, M. K., & Perry, R. W. (2000a). Household adjustment to earthquake hazard: A review of research. *Environment and Behavior, 32*(4), 461–501.

Liu, H., Zhang, D., Wei, Q., & Guo, Z. (2017). Comparison study on two post-earthquake rehabilitation and reconstruction modes in China. *International Journal of Disaster Risk Reduction, 23*, 109–118. https://doi.org/10.1016/j.ijdrr.2017.04.016

Lu, Q. C., Zhang, J., & Rahman, A. S. (2018). Location choice behavior adapting to flood and cyclone hazards. *International Journal of Disaster Risk Reduction, 27*, 189–198. https://doi.org/10.1016/j.ijdrr.2017.10.006

Maly, E. (2018). Building back better with people centered housing recovery. *International Journal of Disaster Risk Reduction, 29*, 84–93. https://doi.org/10.1016/j.ijdrr.2017.09.005

Morrow-Jones, H. A., & Morrow-Jones, C. R. 1991. Mobility due to natural disaster: theoretical considerations and preliminary analyses. *Disasters*, 15 (2).

Mortreux, C., & Barnett, J. (2009). Climate change, migration and adaptation in Funafuti, Tuvalu. *Global Environmental Change, 19*(1), 105–112. https://doi.org/10.1016/j.gloenvcha.2008.09.006

MY, S. (2009). Study on migration of affected population in wenchuan earthquake area. *Social Science Research* (04), 1–7.

Nurlaela, S., & Curtis, C. (2012). Modeling household residential location choice and travel behavior and its relationship with public transport accessibility. *Procedia—Social and Behavioral Sciences, 54*, 56–64. https://doi.org/10.1016/j.sbspro.2012.09.725

Osterling, J. P. (1979). Peruvian disaster and the spontaneous relocation of some of its victims: Ancashino peasant migrants in Huayopampa. *Mass Emergencies, 4*, 117–120.

Peng, Y., Shen, L., Tan, C., Tan, D., & Wang, H. (2012). Critical determinant factors (CDFs) for developing concentrated rural settlement in post-disaster reconstruction: A China study. *Natural Hazards, 66*(2), 355–373. https://doi.org/10.1007/s11069-012-0488-7

Peng, Y., Shen, Q., Shen, L., Lu, C., & Yuan, Z. (2013). A generic decision model for developing concentrated rural settlement in post-disaster reconstruction: A China study. *Natural Hazards, 71*(1), 611–637. https://doi.org/10.1007/s11069-013-0924-3

Regulations on post Wenchuan earthquake recovery and reconstruction. (2008). *China Disaster Reduction* (06), 60–64.

S, W. (2014). SPSS statistical analysis.

Saldaña-Zorrilla, S. O., & Sandberg, K. (2009). Impact of climate-related disasters on human migration in Mexico: A spatial model. *Climatic Change, 96*(1–2), 97–118. https://doi.org/10.1007/s10584-009-9577-3

Shah, A. A., Ye, J., Shaw, R., Ullah, R., & Ali, M. (2020). Factors affecting flood-induced household vulnerability and health risks in Pakistan: The case of Khyber Pakhtunkhwa (KP) Province. *International Journal of Disaster Risk Reduction*, 42. https://doi.org/10.1016/j.ijdrr.2019.101341

Shiwakoti, N. (2016). Understanding differences in emergency escape and experimental pedestrian crowd egress through quantitative comparison. *International Journal of Disaster Risk Reduction, 20*, 129–137. https://doi.org/10.1016/j.ijdrr.2016.11.002

Tacoli, C. (2009). Crisis or adaptation? Migration and climate change in a context of high mobility. *Environment and Urbanization, 21*(2), 513–525. https://doi.org/10.1177/0956247809342182

Tobin, G. A., Whiteford, L. M., Jones, E. C., Murphy, A. D., Garren, S. J., & Padros, C. V. (2011). The role of individual well-being in risk perception and evacuation for chronic vs. acute natural hazards in Mexico. *Applied Geography, 31*(2), 700–711. https://doi.org/10.1016/j.apgeog.2010.12.008

Wachinger, G., Renn, O., Begg, C., & Kuhlicke, C. (2013). The risk perception paradox–implications for governance and communication of natural hazards. *Risk Analysis, 33*(6), 1049–1065. https://doi.org/10.1111/j.1539-6924.2012.01942.x

Wang, J. C., & Guo, Z. G. (2001). Logistic regression model: method and application.

Wijegunarathna, E. E., Wedawatta, G., Prasanna, L. J., & Ingirige, B. (2018). Long-term satisfaction of resettled communities: An assessment of physical performance of post-disaster housing. *Procedia Engineering, 212*, 1147–1154. https://doi.org/10.1016/j.proeng.2018.01.148

Xu, D., Yong, Z., Deng, X., Liu, Y., Huang, K., Zhou, W., & Ma, Z. (2019). Financial preparation, disaster experience, and disaster risk perception of rural households in earthquake-stricken areas: Evidence from the Wenchuan and Lushan earthquakes in China's Sichuan Province. *International Journal of Environmental Research and Public Health, 16*(18). https://doi.org/10.3390/ijerph16183345

Xu, D., Peng, L., Liu, S., Su, C., Wang, X., & Chen, T. (2017). Influences of sense of place on farming households' relocation willingness in areas threatened by geological disasters: Evidence from China. *International Journal of Disaster Risk Science, 8*(1), 16–32. https://doi.org/10.1007/s13753-017-0112-2

XW, M. (2008). Multicollinearity diagnosis method in linear regression equation and its empirical analysis. *Journal of Huazhong Agricultural University (Social Sciences Edition)* (02), 78–81+85. https://doi.org/10.13300/j.cnki.hnwkxb.2008.02.005

Yang, L., Sheng, F., Chen, W., & Ma, S. (2017). Types and transect differentiation of external forms of Chinese traditional villages. *Journal of Hengyang Normal University, 38*(03), 8–12. https://doi.org/10.13914/j.cnki.cn43-1453/z.2017.03.002

Yang, C. (2009). Preliminary discussion on disaster and migration. *Journal of Catastrophology, 24*, 138–144.

YX, L. (2008). Wenchuan earthquake in Sichuan and national earthquake relief. *Chinese Geological Education* (02), 16–20. https://doi.org/10.16244/j.cnki.1006-9372.2008.02.006

Zeng, W., Yan, S., & Tian, J. (2019). Quality evaluation of new urbanization in Wuhan based on factor analysis. *Statistics & Decision, 35*(02), 114–117. https://doi.org/10.13546/j.cnki.tjyjc.2019.02.026

Zhang YW, Y. H., Ma K. (2019). Research on residents' happiness of Tonghua City based on factor analysis. *Technology Wind* (03), 232. https://doi.org/10.19392/j.enki.1671-7341.201903206

Zhang, J. (1994). Environmental hazards in the Chinese public's eyes. *Risk Analysis, 14*(2), 163–167.

Chapter 11
Factors Affecting Villagers' Post-disaster Reconstruction Satisfaction: A Decade After the Wenchuan Earthquake

Abstract This chapter takes ten randomly selected villages in the five hardest-hit areas of Wenchuan earthquake as research objects, and obtained 483 valid sample data through the field questionnaire survey. Villagers are randomly sampled and descriptive statistical analysis, factor analysis and ordered logistic regression are used to explore the influencing factors and relationships of villagers' satisfaction with post-disaster reconstruction in Wenchuan earthquake-stricken areas. The results show that: (1) The more rural residents know about post-disaster reconstruction, the higher their satisfaction; (2) The more the annual income increase of families after resettlement, the higher the satisfaction of rural residents with the post-disaster reconstruction; (3) Six public factors, namely the village committee acts as, housing construction quality, public service, policy of benefiting farmers, cultural environment and hardware environment, all significantly affect the overall satisfaction of residents in post-earthquake reconstruction positively. This study thus enriches the theory of residents' satisfaction studies and the practice of post-earthquake reconstruction.

Keywords Earthquake-stricken area · Post-disaster reconstruction · Satisfaction analysis · Factorial analysis · Ordered Logistic regression

11.1 Introduction

Since the twentieth century, there have been nearly a thousand earthquakes of magnitude 6 or above in China, and the seismic activities are characterized by high frequency, high intensity and wide distribution (Ao et al., 2021; Li & Tian, 2005). Among them, the Wenchuan earthquake occurred in the Longmenshan seismic belt, which was in the northeast-southwest direction, and there were six earthquakes of magnitude six or above, and the largest earthquake was the Wenchuan earthquake of magnitude 8.0 in 2008 (Xue, 2009). According to statistics, the Wenchuan earthquake was an unprecedented disaster, with the hardest hit area exceeding 100,000 square kilometers, involving six cities and counties, 88 counties and cities, 1204 towns and villages, and 27.92 million people. In Sichuan Province alone, more than four million

Y. Ao and H. Bahmani, *Navigating Complexity: Understanding Human Responses to Multifaceted Disasters*, https://doi.org/10.1007/978-981-99-8207-3_11

houses collapsed and were damaged, with percentages of houses in Wenchuan area collapsing as high as 90%, and infrastructure for water, electricity and transportation suffered serious damage (Center, 2008).

The research on the damage caused by earthquake to human production and life and the coping strategies has always been the focus of scholars all over the world (Bryant, 1991). High-intensity earthquakes can destroy all kinds of urban and rural construction in disaster areas to varying degrees. In order to restore life order in disaster areas as soon as possible and explore scientific and efficient reconstruction mode, post-disaster reconstruction has become an important concern of experts and scholars from all over the world (Shi et al., 2021). At the same time, it is also an issue of people's livelihood that needs special attention to clarify the status quo of post-disaster reconstruction of residential areas and strive to improve the production and life satisfaction of rural residents. At present, the research on post-disaster reconstruction settlements mostly focuses on the research of big cities, but less attention is paid to the satisfaction of rural post-disaster reconstruction settlements (Yang & Jie, 2017).

Satisfaction is an individual's subjective experience of his or her own quality of life, an individual's comprehensive cognitive judgment of life, which reflects an individual's general evaluation of overall life and is influenced by both his own factors and environmental factors (Song et al., 2019). Individual differences of rural residents will lead to differences in perception of centralized living style of post-disaster reconstruction settlements, which will lead to different degrees of acceptance, thus showing differences in satisfaction with post-disaster reconstruction settlements (Peng et al., 2018). Professor Edward Blakely pointed out that the post-disaster reconstruction of residential areas is not only disaster prevention and emergency rescue, but also the re-planning and reconstruction of a region. The first thing is definitely the emergency planning when a disaster occurs, but more importantly, the functional allocation of the region, such as infrastructure, environmental planning, economic development planning, new residence resettlement and so on (Z, 2008). These living environment factors will affect rural residents' satisfaction with the post-disaster reconstruction of residential areas to a certain extent.

Most of the previous studies were analyzed from the perspective of the government. This study analyzes the construction of post-disaster reconstruction settlements from the perspective of individual villagers, understands the status quo of rural residents' satisfaction in post-disaster reconstruction settlements in Sichuan Province, and explores the impact of rural residents on the overall satisfaction of post-disaster reconstruction settlements. Influence factors and their influence relationships, and put forward variable factors on the basis of existing scholars' research, establish a system of factors influencing rural residents' satisfaction with post-earthquake reconstruction of residential settlements, find out significant influencing factors, and provide theoretical support for post-disaster reconstruction practice.

The structure of this paper is shown in Fig. 11.1: Sect. 11.2 reviews the literature, Sect. 11.3 describes the research methods and data sources, Sect. 11.4 presents the empirical results and discusses the results, Sect. 11.5 summarizes the full text and proposes research prospects, and the last section presents the limitations of this study.

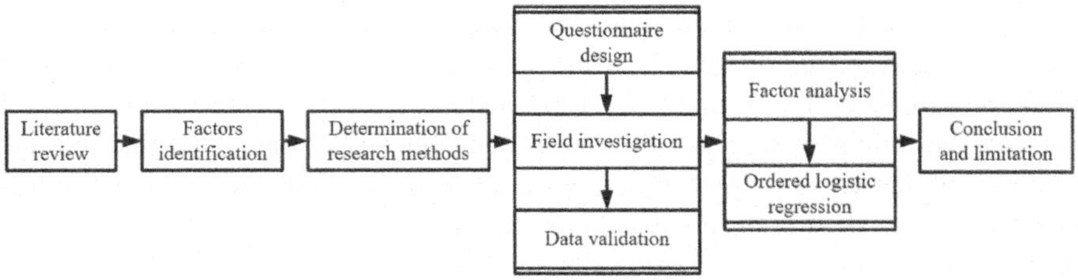

Fig. 11.1 Technology roadmap

11.2 Literature Review

China has adopted different reconstruction policies for rural post-earthquake reconstruction, including three main ways of overall construction, overall self-construction and self-construction and maintenance (Peng et al., 2013). Under the joint action of the family characteristics and production and management goals of rural residents, different types of farmers will be formed, and different types of farmers will have different positioning of maximizing the comprehensive benefits of the family, which will directly affect the decision-making behavior and mode selection of rural settlements post-disaster reconstruction (Steinberg, 2007).

The theoretical research on post-earthquake reconstruction emphasizes keeping pace with the times and combining corresponding development policies for reconstruction areas of different regional levels. For urban post-disaster reconstruction, Zhou Jian and Xia Nankai believed that the post-earthquake reconstruction planning should follow the development strategy of "shaping an international tourist city", relocating urban functional departments and residents from severely damaged old cities to new areas, and relocating old urban areas to new areas. It will be transformed into an economic development functional area with cross services of multiple industries. Yu Miao et al. proposed that rural social relations are complex, and it is necessary to pay more attention to social connection and regional capacity building when planning post-earthquake reconstruction in rural areas, and to maximize the flexibility of reconstruction strategies to form an adaptive mechanism for post-disaster recovery. When verifying the role and impact of participation in post-earthquake reconstruction, Wang Zhuo believed that participatory post-disaster rural reconstruction could provide a basis for targeted poverty alleviation, rural revitalization and urban and rural development.

Post-disaster reconstruction is a common behavior of social norms and government intervention. Guo et al. Y and J (2019) study the satisfaction of rural residents with basic public services in rural areas. The results show that the satisfaction of rural residents is influenced by gender, age, culture, income level, farmers' participation and understanding of basic public services. Different individual and family characteristics of rural residents for post-disaster reconstruction residential satisfaction exist. The satisfaction of post disaster reconstruction and resettlement shows a downward trend with the increase of age. The higher the education level of rural residents, the higher the satisfaction of post disaster reconstruction settlements (F, 2014). In addition, the difference of income sources will also have an impact on villagers' satisfaction. Sun and Chen (2016) study the influencing factors of rural residents' resettlement satisfaction in M Town, Jiangsu Province, and find that the farmers with larger proportion of non-agricultural income after resettlement in this area are less satisfied with this resettlement. The reason may be that for this kind of farmers, the compensation cost of this resettlement is less attractive to them and their overall satisfaction is not high. Peng et al. (2018) point out that the earthquake caused serious damage to cultivated land, land consolidation and reclamation directly affected farmers' economic income, and the process of rebuilding residential areas was an important opportunity to improve economic development, and the diversification of rural residents' income ways should be increased to make farmers' income no longer unitary.

In addition to the above-mentioned self-factors, the satisfaction of post-disaster reconstruction of residential areas is also affected by living environment factors (Song et al., 2019; Yang et al., 2021). Jansen (2014) believes that living satisfaction depends on personal expectations, and when housing does not meet the needs of residents, it will directly reduce their living satisfaction. Cao (2016) analyses the living satisfaction of the living environment, including housing orientation, supporting facilities, housing quality, and neighborhood relationship. On this basis, Aulia and Ismail (2013) further refines the external influencing factors into natural environment, equipment conditions, property services and traffic conditions to analyze the influencing factors of living satisfaction.

The actions of village committees or communities will have an impact on villagers' satisfaction with reconstruction. In the study of new rural construction, Hu puts forward that the working ability of village grassroots village Committee cadres and rural residents' awareness of rural construction policies will affect rural residents' satisfaction with village-level democratic system construction and cultural construction. In the process of resettlement housing allocation, Xiao et al. (2014) evaluated the satisfaction of the rebuilt residential areas after the earthquake from the perspective of the affected residents, and found that the disclosure of the information channel of the housing redistribution system in the rebuilt residential areas had a great influence on the satisfaction of the resettled residents. Hu Yizhi once made a comparative study on the reconstruction of New Orleans in the United States and the reconstruction after Wenchuan earthquake in China. He found that the communication with indigenous residents should be emphasized at the beginning of the reconstruction policy formulation process, and they should be invited to participate in the reconstruction process and listen to their opinions, which is not only conducive to the application of the reconstruction policy, but also to rebuild the confidence of indigenous people (Z, 2008). Paying attention to the rural housing reconstruction of farmers is the key to realize sustainable recovery. The reconstruction policy has different decisions on the reconstruction of village houses damaged to varying degrees, and the publicity of the reconstruction policy can clarify the affected people's perception of post-disaster reconstruction (Peng et al., 2018). Wang pointed out that earthquake disaster prevention knowledge dissemination is an important factor affecting residents' life satisfaction.

Housing construction is the top priority of post-earthquake reconstruction. According to the disaster reduction plan and regulations of China Earthquake Administration (CEA) from 2007 to 2019, local rural housing construction planning should be supervised in terms of site selection, avoidance and construction quality to ensure housing safety, and skilled technicians should be trained to master earthquake knowledge (Wu & Wu, 2020). Cassidy (2007) proposed that when a disaster occurs, the maximum duration of temporary shelter provided by the government is 5 years, so the construction period of post-earthquake reconstruction settlements should not exceed the maximum duration that victims can bear. At the same time, post-disaster reconstruction projects should have clear start and completion dates (Davidson et al., 2007).

Steinberg (2007) summarized the experience of post-disaster reconstruction in Aceh and Nias in Indonesia and pointed out that the construction of residential buildings in post-disaster reconstruction was only the first step of reconstruction, and the construction of surrounding environment and public facilities was also the focus of post-disaster reconstruction. Transportation is very important in post disaster reconstruction. If roads are blocked, much of the transport of reconstruction materials will be affected. Therefore, post-disaster reconstruction of traffic is a basic and pilot project (Kun, 2013). At the same time, public facilities such as water, electricity and communication are also the focus of the reconstruction of post-disaster settlements, and whether the quality of drinking water, the convenience of water use, power supply and communication meet the needs of residents also affect the satisfaction of post-disaster reconstruction settlements (Curti et al., 2008).

The existing research shows that the relevant policies of benefiting farmers will also have an impact on villagers' satisfaction with reconstruction. Developing characteristic agriculture to diversify agriculture, developing agricultural training and introducing non-agricultural industries will enable young laborers to engage in non-agricultural work, which will increase farmers' income, thus making up for the increased living cost after the disaster, which will improve rural residents' satisfaction with the reconstruction of settlements after the disaster. In addition, as the ownership of land belongs to rural collectives, after many rounds of discussion by the village committee, even the land adjustment will not encounter any difficulties, so it makes the residents believe that the problems of land reclamation or cultivated land demand can be well solved after the disaster (Peng et al., 2013). At the same time, the human environment is also the key point that cannot be ignored. Earthquake disasters in Sichuan Province mostly occur in areas with superior natural conditions and profound cultural heritage. Post-earthquake reconstruction should pay attention not only to the protection of natural environment but also to the reconstruction of national cultural traditions (Li X J, 2008).

Therefore, it is of great theoretical and practical significance to systematically explore residents' satisfaction degree in post-disaster reconstruction to further improve residents' quality of life, enhance public participation in post-disaster reconstruction and post-disaster reconstruction effect.

11.3 Methodology

11.3.1 Questionnaire Design

This study designs a questionnaire according to the existing related research and the current situation of post-earthquake reconstruction of residential areas. The questionnaire includes two parts: social demographic information and post-disaster reconstruction satisfaction survey scale.

Personal basic information includes personal basic information of respondents (shown in Table 11.1) and family information (shown in Table 11.2).

The satisfaction scale measures respondents' satisfaction with 28 elements of post-disaster reconstruction, which is coded by Likert's five-point scale, with the lowest satisfaction set at 1 and the highest satisfaction set at 5 (Yang et al., 2020c). Details of post-disaster reconstruction satisfaction scale and its measurement instructions are shown in Table 11.3:

11.3.2 Model Specification

Academics who study satisfaction have used a variety of methods and models, including multiple regression model (Yang et al., 2022), structural equation model, CCSI model, ordered Logistic regression analysis, Bayesian multilevel ordinal response model. However, there are limited researches on the influencing factors of residents' satisfaction and their relationship in post-disaster reconstruction, which restricts the rationality of the formulation of post-disaster reconstruction policies. Therefore, factor analysis is mainly used to reduce the dimension of influencing factors (Ao et al., 2020), and ordered Logistic regression model is used to analyze the relationship between the influencing factors and the satisfaction of post-earthquake reconstruction settlements.

In Logistic regression analysis, when the variable level is greater than 2 and it is an ordered variable, the ordered Logistic regression analysis can be used. The dependent variable of this study is satisfaction, and the options are completely dissatisfied, not very satisfied, generally satisfied, comparatively satisfied and very satisfied, which is suitable for ordered Logistic regression analysis. The Logistic regression model used in this study is expressed as follows:

$$Ln[\frac{p(y \leq j)}{1 - p(y \leq \text{j})}] = \alpha_j + \sum_{i=1}^{n} \beta_i x_i \tag{11.1}$$

Table 11.1 Description of basic personal information of respondents

Variable	Variable declaration	Variable type
Gender	1 = Man; 2 = Woman	Categorical variable
Age	The corresponding numerical value is the corresponding age For example: 25 = 25 years old	Continuous variable
Education level	1 = Uneducated; 2 = Primary School; 3 = Junior High School; 4 = Senior High School; 5 = University and above	Categorical variable
Is the current place of residence the birthplace?	Yes = 1, No = 0	Binary variable
Participate in the reconstruction decision-making process?	Yes = 1, No = 0	Binary variable
Have you received education on disaster prevention and mitigation?	Yes = 1, No = 0	Binary variable
Have you experienced secondary disasters after the earthquake?	Yes = 1, No = 0	Binary variable
Understanding of post-disaster reconstruction management regulations	Very little understanding = 1, A little understanding = 2, General understanding = 3, Better understanding = 4, Very understanding = 5	Sequence variable
Understanding of seismic fortification level of buildings	Very little understanding = 1, A little understanding = 2, General understanding = 3, Better understanding = 4, Very understanding = 5	Sequence variable
Understanding of post-disaster reconstruction methods	Very little understanding = 1, A little understanding = 2, General understanding = 3, Better understanding = 4, Very understanding = 5	Sequence variable

where, $j = 1, 2, 3, 4, 5$, which represents five levels of satisfaction; y is residents' satisfaction with reconstruction; x_i is the explanatory variable and control variable that affects farmers' life satisfaction; α_j is the intercept parameter; β_j is the regression coefficient, which indicates the direction and degree of influence of explanatory variables on the explained variables.

Table 11.2 Description of respondents' family information

Variable	Variable declaration	Variable type
Number of residential floors	1 = 1, 2 = 2, 3–6 = 3, Layer 7 and above = 4	Sequence variable
Is the rebuilt house completed on time?	Yes = 1, No = 0	Binary variable
Post-earthquake reconstruction	Overall construction = 1, overall self-construction = 0	Binary variable
Annual household income after earthquake	() Ten thousand yuan	Continuous variable
Stability of main household income after earthquake resettlement	Unstable = 1, stable = 0	Binary variable
Changes of annual household income after resettlement	Significant increase = 1, some increase = 2, no change = 3, some decrease = 4, significant decrease = 5	Sequence variable
Main income sources of families before the earthquake	Farming/fruit and vegetable planting, poultry/aquaculture, farmhouse tourism, land circulation, working outside, and others	Categorical variable
Main source of family income after earthquake resettlement	Farming/fruit and vegetable planting, poultry/aquaculture, farmhouse tourism, land circulation, working outside, and others	Categorical variable

11.3.3 Sample Selection and Data Collection

According to the degree of the Wenchuan earthquake disaster, the population, economy, industry and employment status, combined with the vigilance of rural residents and the degree of cooperation reflected, this study randomly selected 10 sample villages in five hardest hit areas in Sichuan Province, and the geographical location of each village is shown in Fig. 11.2.

The field investigation of this study was conducted from January 1st to January 5th, 2019. The research team is divided into five groups, each of which is responsible for data collection of two sample villages. The research team entered the village and randomly selected the residents in the village to carry out the questionnaire survey. If the residents did not accept the survey invitation, the researchers randomly selected the next household. In this study, 516 face-to-face questionnaires were conducted, and 33 questionnaires with missing information and internal inconsistency were excluded. 483 valid questionnaires were finally recovered. Table 11.4 shows the sample villages and the questionnaires collected information, and Table 11.5 shows the Statistical table of basic information of respondents. The change trend of the

Table 11.3 Post-disaster reconstruction satisfaction survey scale variable

Variable		References	Satisfaction		
			Minimum	Minimum	Variable type
F1	Subsidy guarantee	Alparslan et al. (2008)[4] Morimoto (2012)	1	5	Sequence variable
F2	Information channel	Xiao et al. (2014)	1	5	Sequence variable
F3	Education and publicity of disaster prevention and mitigation	Wang Z (2012)	1	5	Sequence variable
F4	Reconstruction policy propaganda	Peng et al. (2018)	1	5	Sequence variable
F5	Reconstruction decision-making participation	Li X J (2008)	1	5	Sequence variable
F6	Type of layout of an apartment	Cinicioglu et al. (2007)	1	5	Sequence variable
F7	House safety	Ergonul (2005), Wu and Wu (2020)	1	5	Sequence variable
F8	Quality of building materials	Ergonul (2005), Wu and Wu (2020) Cinicioglu et al. (2007)	1	5	Sequence variable
F9	Technology of constructors	Cinicioglu et al. (2007), Ergonul (2005), Wu and Wu (2020)	1	5	Sequence variable
F10	Reconstruction duration	Davidson et al. (2007), Ergonul (2005)	1	5	Sequence variable
F11	Drinking water quality	Curti et al. (2008)	1	5	Sequence variable
F12	Water convenience	Curti et al. (2008)	1	5	Sequence variable
F13	Power supply demand	Curti et al. (2008)	1	5	Sequence variable
F14	Communication requirements	Curti et al. (2008)	1	5	Sequence variable

(continued)

Table 11.3 (continued)

Variable		References	Satisfaction		
			Minimum	Minimum	Variable type
F15	Planting space around housing	Li X J (2008)	1	5	Sequence variable
F16	Cultivated land distance	Li X J (2008)	1	5	Sequence variable
F17	Land reclamation	Ansal et al. (2009), Peng et al. (2013)	1	5	Sequence variable
F18	Agricultural training	Ansal et al. (2009), Peng et al. (2013)	1	5	Sequence variable
F19	Agricultural diversification	Ansal et al. (2009), Peng et al. (2013)	1	5	Sequence variable
F20	Non-agricultural industry introduction	Ansal et al. (2009), Peng et al. (2013)	1	5	Sequence variable
F21	Talent education	Speare (1974)	1	5	Sequence variable
F22	Policies and systems	Li X J (2008)	1	5	Sequence variable
F23	Cultural tradition	Li X J, (2008), Speare (1974)	1	5	Sequence variable
F24	Earthquake shelter	Wen (2001)	1	5	Sequence variable
F25	Road planning	Kun (2013)	1	5	Sequence variable
F26	Road quality	Kun (2013), Steinberg (2007)	1	5	Sequence variable

(continued)

Table 11.3 (continued)

Variable		References	Satisfaction		
			Minimum	Minimum	Variable type
F27	Sanitary environment/ Village appearance	Inneke et al. (2013) MacAskill and Guthrie (2015)	1	5	Sequence variable
F28	Natural environment	Curti et al. (2008), Li X J (2008)	1	5	Sequence variable

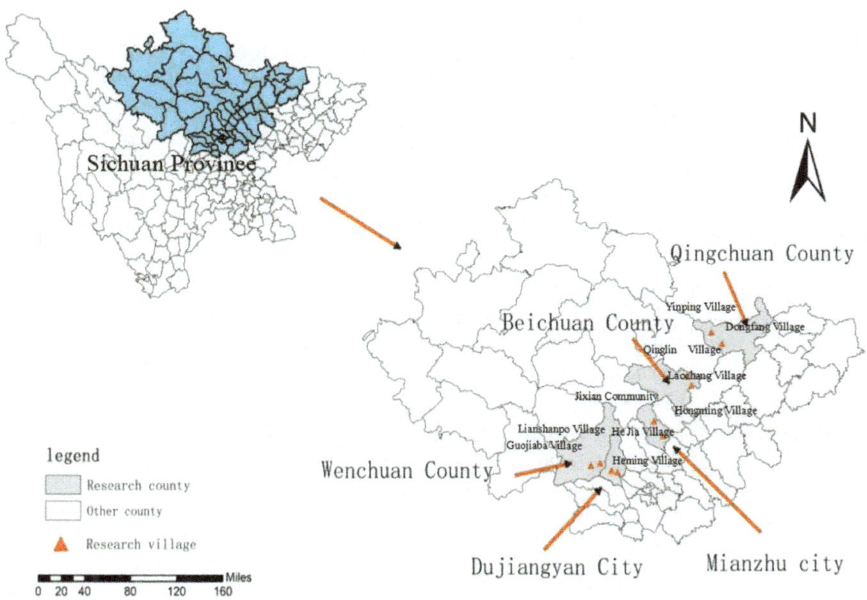

Fig. 11.2 The location of the sample villages

main sources of household income before and after the earthquake is shown in the Fig. 11.3.

Table 11.4 The sample villages and the questionnaires collected information

Investigation site		Disaster degree	Reconstruction of settlement patterns	Questionnaire quantity
Deyang city	Jixian Community, Hanwang Town, Mianzhu City	Severe disaster	Centralized residence	50
	Hongming Village, xinan town, Mianzhu City	Severe disaster	Decentralized residence	49
Dujiangyan city	Heming Village, LiujieTown	Severe disaster	Centralized residence	48
	He Jia Village, Anlong Town	Severe disaster	Decentralized residence	47
Guangyuan City	Dongfang Village, Qingxi Town, Qingchuan County	Severe disaster	Centralized residence	52
	Yinping Village, Qingxi Town, Qingchuan County	Severe disaster	Centralized residence	45
Aba Autonomous Prefecture	Guojiaba Village, Shuimo Town, Wenchuan County	Severe disaster	Centralized residence	49
	Lianshanpo Village, Shuimo Town, Wenchuan County	Severe disaster	Decentralized residence	42
Mianyang city	Laochang Village, Chenjiaba Town, Beichuan Qiang Autonomous County	Severe disaster	Centralized residence	51
	Qinglin Village, Chenjiaba Town, Beichuan Qiang Autonomous County	Severe disaster	Centralized residence	50

11.4 Results and Discussion

11.4.1 Exploratory Factor Analysis

In this study, SPSS23.0 was used to conduct exploratory factor analysis (EFA) on 28 measurement variables of post-disaster reconstruction satisfaction, so as to determine the influence of each factor on the overall satisfaction of post-disaster reconstruction. In order to test the applicability of factor analysis, this study uses kaiser-meye-olkin

Table 11.5 Statistical table of basic information of respondents

Variable	Variable declaration	Percentage (%)	Number of samples	Variable	Variable declaration	Percentage (%)	Number of samples
Gender	Man	45.25	219	Academic degree	Without education	14.49	70
	Woman	54.75	264		Primary school	30.85	149
Age	15-30year old	16.56	80		Junior school	30.85	149
	30–45 year old	21.33	103		Technical secondary school	4.97	24
	45–60 year old	35.20	170		Senior high school	10.14	49
	60–75 year old	22.57	109		Junior college	5.59	27
	Over 75 years old	4.34	21		University and above	3.11	15
Number of residential floors	1st floor	36.44	176	Source of family income before the earthquake	Farming/ Fruit and Vegetable Planting	30.23	147
	2 floors	48.45	234		Poultry/ aquaculture	13.04	63
	3–6 floors	14.70	71		Farmhouse tourism	3.52	17
	Layer 7 and above	0.41	2		Land circulation	1.24	6
Annual household income after earthquake	Under ten thousand	22.57	109		Working outside	36.65	177
	Ten thousand to fifty thousand	62.32	301		Others	15.32	73
	Fifty thousand to one hundred thousand	12.84	62	Sources of household income after the earthquake	Farming/ Fruit and Vegetable Planting	15.32	74
	Over one hundred thousand	2.27	11		Poultry/ aquaculture	3.73	18

(continued)

Table 11.5 (continued)

Variable	Variable declaration	Percentage (%)	Number of samples	Variable	Variable declaration	Percentage (%)	Number of samples
Changes of annual household income after resettlement	Some increase	54.24	262		Farmhouse tourism	3.73	18
	No change	34.37	166		Land circulation	3.52	17
	Some decrease	9.32	45		Working outside	50.31	243
	Significant decrease	2.07	10		Others	23.39	113

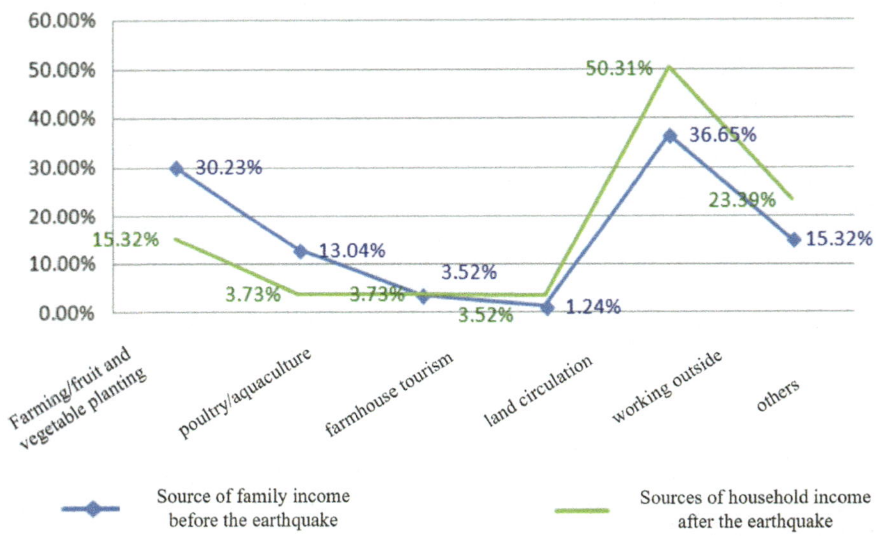

Fig. 11.3 Trend chart of major sources of income for households before and after the earthquak

(KMO) test and Bartlett test to explore the applicability of factor analysis in 28 satisfaction measurement variables. The test results show that KMO value is 0.882 and *P* value is 0.000. The results show that there is a high correlation among 28 satisfaction measurement variables, which indicates that this data is suitable for exploratory factor analysis (EFA) method. The factor analysis results of 28 satisfaction indexes are shown in Table 11.6, in which the variables with factor load lower than 0.4 are considered as insignificant variables, so F15 is deleted from 28 satisfaction measurement variables. Exploratory factor analysis finally determined six common factors.

Table 11.6 Molecular results of exploratory factors

Common factor	Variable		Load					
X1 The village Committee acts as	F1	Subsidy guarantee	0.444					
	F2	Information channel	0.594					
	F3	Education and publicity of disaster prevention and mitigation	0.829					
	F4	Reconstruction policy propaganda	0.835					
	F5	Reconstruction decision-making participation	0.711					
X2 Housing construction quality	F6	Type of layout of an apartment		0.641				
	F7	House safety		0.693				
	F8	Quality of building materials		0.865				
	F9	Technology of constructors		0.832				
	F10	Reconstruction duration		0.538				
X3 Public services	F11	Drinking water quality			0.798			
	F12	Water convenience			0.871			
	F13	Power supply demand			0.845			
	F14	Communication requirements			0.742			
–	F15	Planting space around housing			–			
X4 Policy of benefiting farmers	F16	Cultivated land distance				0.629		
	F17	Land reclamation				0.758		
	F18	Agricultural training				0.779		
	F19	Agricultural diversification				0.67		

(continued)

Table 11.6 (continued)

Common factor	Variable		Load					
	F20	Non-agricultural industry introduction				0.507		
X5 Cultural environment	F21	Talent education					0.527	
	F22	Policies and systems					0.731	
	F23	Cultural tradition					0.736	
X6 Hardware environment	F24	Earthquake shelter						0.512
	F25	Road planning						0.846
	F26	Road quality						0.799
	F27	Sanitary environment/ Village appearance						0.456
	F28	Natural environment						0.469

11.4.2 Multiple Collinearity Analysis

The multiple collinearity problem may lead to the low level of significance of various spatial variables. Therefore, it is necessary to investigate the multiple collinearity of these independent variables (Ding et al., 2017; Yang, Chau, et al., 2020a; Yang, Chu, et al., 2020b; Yang et al., 2022; Zhao et al., 2020). In order to test multiple collinearity, this chapter mainly uses the variable expansion coefficient (VIF). With the increase of VIF value, a specific explanatory variable is more likely to be expressed by the linear function model of other explanatory variables, and there may be multiple collinearity problems in the model. The maximum VIF value of the explanatory variable in this study is 2.094, which shows that there is no multiple collinearity problem. The results of multivariate multiple collinearity test are shown in Table 11.7.

11.4.3 Ordered Logistic Regression

In this study, ordered Logistic regression is used to analyze the influence of the above factors on the satisfaction of rural residents in post-disaster reconstruction. Two models were fitted in this study, Model 1 contains all of the above variables, and model 2 was obtained by re-fitting after deleting insignificant variables in model 1. The results of model 1 and 2 are shown in Table 11.7. The -2log-likelihood values are 1360.778 and 1178.072 respectively, and both of the Sig. values are 0.000, which

Table 11.7 Ordered Logistic regression results of satisfaction

Variable	Model 1		Model 2		Collinearity test	
	B	*P*-value	B	*P*-value	Tolerance	VIF
Gender	– 0.208	0.237	–	–	–	–
Age	– 0.004	0.573	–	–	0.562	1.778
Education level	– 0.004	0.959	–	–	0.554	1.804
Is the current place of residence the birthplace?	– 0.255	0.195	–	–	0.952	1.050
Participate in the reconstruction decision-making process?	– 0.035	0.846	–	–	0.887	1.128
Have you received education on disaster prevention and mitigation?	– 0.081	0.658	–	–	0.870	1.150
Have you experienced secondary disasters after the earthquake?	– 0.360	0.106	– 0.411[a]	0.053	0.887	1.127
Understanding of post-disaster reconstruction management regulations	0.138	0.244	–	–	0.518	1.930
Understanding of seismic fortification level of buildings	– 0.137	0.272	–	–	0.478	2.094
Understanding of post-disaster reconstruction methods	0.212[a]	0.051	0.223[b]	0.012	0.623	1.604
Is the rebuilt house completed on time?	– 0.341	0.124	–	–	0.937	1.068
Post-disaster reconstruction mode of housing is overall construction	0.018	0.935	–	–	0.854	1.171
Annual income of families after resettlement	0.024	0.316	–	–	0.946	1.057
Family income stability after resettlement	– 0.049	0.783	–	–	0.942	1.062

(continued)

Table 11.7 (continued)

Variable	Model 1		Model 2		Collinearity test	
	B	*P*-value	B	*P*-value	Tolerance	VIF
Changes of annual household income before and after resettlement	− 0.226[a]	0.052	− 0.217[a]	0.057	0.937	1.067
X1 The village Committee acts as	0.548[c]	0.000	0.525[c]	0.000	0.979	1.021
X2 Housing construction quality	0.447[c]	0.000	0.434[c]	0.000	0.973	1.028
X3 Public services	0.365[c]	0.000	0.372[c]	0.000	0.955	1.047
X4 Preferential agricultural policy	0.393[c]	0.000	0.393[c]	0.000	0.950	1.052
X5 Cultural environment	0.595[c]	0.000	0.594[c]	0.000	0.950	1.053
X6 Hardware environment	0.649[c]	0.000	0.637[c]	0.000	0.982	1.018

Note [a], [b], [c] represent significant levels of 10%, 5% and 1% respectively

means that model 2 fit the data better. Finally, model 2 was selected to interpretate and analyze. Meanwhile, the Cox and Snell and Nagelkerke R2 values are 0.315 and 0.335 respectively, which means the model fit the data well and has statistical significance.

11.4.3.1 The Influence of Demographic Variables on Post-disaster Reconstruction Satisfaction

Villagers' understanding of post-disaster reconstruction methods (B = 0.223, P = 0.012) had a positive and significant influence on the overall satisfaction level at the significant level of 5%, that is, the more fully rural residents' understanding of reconstruction methods of residential areas, the higher their satisfaction level. Whether the residents have experienced the secondary earthquake disaster (B = − 0.411, P = 0.053) has a negative correlation with the overall satisfaction at the 10% significant level, which indicates that the residents who have not experienced the earthquake disaster in the current reconstruction residential area have high satisfaction with the reconstruction residential area. Indirectly explain the importance of reconstruction of residential areas in avoiding secondary earthquake disasters.

In addition, the change of annual household income before and after the post-disaster reconstruction (B = − 0.217, P = 0.057) has a negative correlation with the overall satisfaction at a significant level of 10%. This option in the questionnaire of this study is designed as (significant increase = 1, some increase = 2, no change = 3, some decrease = 4, significant decrease = 5), that is, the annual household income after resettlement is higher. This is consistent with the research results of Shi et al. (2018) when they studied urban–rural migration and resettlement and found that the income increase has a positive correlation with residents' life satisfaction (Shi et al., 2018).

11.4.3.2 The Influence of Six Common Factors on Satisfaction

The more satisfied the villagers in the post-disaster reconstruction area are with the village committee's actions (X1, B = 0.525, $P = 0.000$) in the post-earthquake resettlement process, the higher their satisfaction with the post-disaster reconstruction of residential areas. The implementation of government policies is directly related to the style and ability of village cadres, which shows that improving the village cadres' sense of responsibility and working ability is an important content of improving rural residents' satisfaction with post-disaster reconstruction of residential areas. Huang et al. believes that village cadres should improve their own skills, use information means to improve the efficiency of rural community governance, and achieve the goal of rural governance informatization. The more stable the cadres network is, the higher the rural residents' evaluation of village cadres will be.

The quality of housing construction (X2, B = 0.434, $P = 0.000$) significantly affects the overall satisfaction of villagers in post-disaster reconstruction. The better the quality of housing construction in post-disaster reconstruction areas, the higher the overall satisfaction of villagers in post-disaster reconstruction. It shows that the strong earthquake resistance and comfort of the house itself are one of the important factors that determine the satisfaction of rural residents to the post-disaster reconstruction of residential areas. Xiao et al. (2014) has found that in the built environment, the higher the completion degree of residential reconstruction is, the shorter the construction period is, and the disaster-affected people are more satisfied with the post-disaster reconstruction. Therefore, the quality and efficiency of housing construction in the post-disaster reconstruction area play an important role in improving villagers' satisfaction with the post-disaster reconstruction.

The public service in the post-disaster reconstruction area (X3, B = 0.372, $P = 0.000$) has a significant positive correlation with the villagers' overall satisfaction with post-disaster reconstruction, which shows that rural residents pay more attention to the public service level of resettlement sites. The higher the public service level, the higher the satisfaction of rural residents with the post-disaster reconstruction settlements (Wang, 2019). In the study of rural medical and health services, Wang et al. found that rural residents with better self-rated health status were more satisfied with public health services, and the higher the satisfaction of rural residents with medical and health services, the higher their life satisfaction. Therefore, in the process of post-disaster reconstruction, not only the construction work should be done well, but also the role of public services should not be ignored.

The preferential agricultural policies (X4, B = 0.393, $P = 0.000$) have a significant positive impact on the overall satisfaction of villagers in the post-disaster reconstruction area, indicating that the higher the implementation of the preferential agricultural policies, the higher the overall satisfaction of villagers in the post-disaster reconstruction settlements. This is consistent with the view Tian et al. (2010) that the implementation intensity of agricultural benefit policies and the two exemption and one subsidy policies have a great positive impact on farmers' life satisfaction.

The higher the satisfaction of residents in the construction of human environment (X5, B = 0.594, $P = 0.000$) in the post-earthquake reconstruction area, the higher the overall satisfaction of villagers in the post-earthquake reconstruction residential area. This shows that improving the satisfaction of rural residents from the perspective of rural policy environment, rural talent environment and rural cultural environment will be conducive to improving the satisfaction of rural residents to post-disaster reconstruction settlements. Yė believes that there are many problems in the construction of rural cultural environment, which reduce the life happiness of rural residents to varying degrees, and proposes that corresponding policies should be formulated for different problems to improve the life satisfaction of rural residents, which is consistent with the conclusion of this study (X, 2015).

The higher the villagers' recognition in the construction of hardware environment (X6, B = 0.637, $P = 0.000$), the higher the villagers' overall satisfaction with the post-disaster reconstruction, and the greatest influence of hardware facilities construction in the post-disaster reconstruction area. This shows that the infrastructure construction of post-disaster reconstruction settlements is the most important content that affects rural residents' satisfaction with post-disaster reconstruction settlements. Fast evacuation routes and earthquake shelters are the basic guarantee to improve the level of seismic resilience in the new era. Giving full play to the reliability and timeliness of evacuation routes and earthquake shelters during disasters will reduce casualties. Therefore, the rationality of road planning, the safety and accessibility of earthquake shelters have a significant impact on the satisfaction of rural residents in rebuilding settlements after disasters, which is consistent with the conclusions of the study of Ma et al. (2021).

11.5 Conclusions

Rural residents' overall satisfaction with post-disaster reconstruction of residential areas is influenced by many factors. Based on the literature research, combined with the current situation of ten post-disaster reconstruction settlements in Sichuan Province, this study summarized the factors that affect the overall satisfaction of rural residents with post-disaster reconstruction settlements. Based on this design of research questionnaire, 483 valid questionnaires were collected by face-to-face questionnaire survey in villages and households. The research discovers the following insights:

(1) The more the villagers in earthquake-stricken areas know about post-disaster reconstruction methods, the higher their overall satisfaction with post-disaster reconstruction. Therefore, attention should be paid to improving rural residents' awareness of reconstruction methods of residential areas and strengthening publicity and education of post-disaster reconstruction methods.
(2) After the earthquake disaster reconstruction, if the annual household income increases more than that before the reconstruction, the villagers in the disaster area are more satisfied with the overall reconstruction. Therefore, we should pay attention to the employment problem after the reconstruction and increase the income of rural residents after the disaster.
(3) Six public factors, such as the village committee acts as, housing construction quality, public service, policy of benefiting farmers, cultural environment and hardware environment, all significantly positively affect the overall satisfaction of residents in post-earthquake reconstruction. This study is of great significance to enrich the theory of residents' satisfaction and the practice of post-earthquake reconstruction.

Through the post-earthquake reconstruction of Wenchuan residential satisfaction survey, this research finds out the influence of rural residents for post-earthquake reconstruction overall satisfaction of the key factors of residential area, reveals the relationship between human activities and residents of the ecological environment, it also reveals the unreasonable planning in the construction of post disaster reconstruction settlements from the perspective of rural residents, complements the theory development of post-disaster reconstruction settlement. More scientifically this research presents the construction status and problems of post-disaster reconstruction settlements. This has certain scientific and practical significance for maintaining the spatial stability of post-disaster reconstruction settlements.

11.6 Limitations

However this study does have certain limitations as elaborated below.

(1) Variable collection is limited. Although this study combines the current situation of post disaster reconstruction settlements and puts forward all variable factors based on the research of existing scholars, the content cannot fully cover all aspects of rural residents' satisfaction with post disaster reconstruction settlements. Therefore, there are limitations in the collection of variables in this study.
(2) Data values are limited. When the respondents evaluate the satisfaction of various indicators, they hesitate to convey vague language information, so the data value has certain limitations.

(3) The scope of the investigation also have limitations. This study selects rural residents from ten sample villages in Sichuan Province as the research object. The research adopts the form of field entry, which is difficult to avoid contingency and regional distribution limitations.

References

Alparslan, E., Ince, F., Erkan, B., Aydoner, C., Ozen, H., Doenertas, A., Ergintav, S., Yağsan, F.S., Zateroğulları, A., Eroğlu, I. and Değer, M., & Ozkan, M. (2008). A GIS model for settlement suitability regarding disaster mitigation, a case study in Bolu Turkey. *Engineering Geology, 96*(3–4), 126–140. https://doi.org/10.1016/j.enggeo.2007.10.006

Ansal, A., Akinci, A., Cultrera, G., Erdik, M., Pessina, V., Tonuk, G., & Ameri, G. (2009). Loss estimation in Istanbul based on deterministic earthquake scenarios of the Marmara Sea region (Turkey). *Soil Dynamics and Earthquake Engineering, 29*(4), 699–709. https://doi.org/10.1016/j.soildyn.2008.07.006

Ao, Y., Zhang, Y., Wang, Y., Chen, Y., & Yang, L. (2020). Influences of rural built environment on travel mode choice of rural residents: The case of rural Sichuan [Article]. *Journal of Transport Geography, 85*, 102708, Article 102708. https://doi.org/10.1016/j.jtrangeo.2020.102708

Ao, Y., Zhang, H., Yang, L., Wang, Y., Martek, I., & Wang, G. (2021). Impacts of earthquake knowledge and risk perception on earthquake preparedness of rural residents [Article]. *Natural Hazards, 107*(2), 1287–1310. https://doi.org/10.1007/s11069-021-04632-w

Aulia, D. N., & Ismail, A. M. (2013). Residential satisfaction of middle income population: Medan city. *Procedia-Social and Behavioral Sciences.*

Bryant, E. (1991). *Natural hazards*. Cambridge University Press.

Cao. (2016). How does neighborhood design affect life satisfaction? Evidence from Twin Cities. *Travel Behaviour and Society.*

Cassidy. (2007). Impacts of prefabricated temporary housing after disasters: 1999 earthquakes in Turkey. *Habitat International.*

Center, C. G. A. M. (2008). Preliminary assessment of Wenchuan earthquake disaster. *Leadership Decision Information, 22*, 26–30.

Cinicioglu, S. F., Bozbey, I., Oztoprak, S., & Kelesoglu, A. K. (2007). An integrated earthquake damage assessment methodology and its application for two districts in Istanbul, Turkey. *Engineering Geology, 94*(3–4), 145–165. https://doi.org/10.1016/j.enggeo.2007.06.007

Curti, E., Podestà, S., & Resemini, S. (2008). *The post-earthquake reconstruction process of monumental masonry buildings: Suggestions from the molise event (Italy).*

Davidson, C. H., Johnson, C., Lizarralde, G., Dikmen, N., & Sliwinski, A. (2007). *Truths and myths about community participation in post-disaster housing projects.*

Ding, C., Wang, D., Liu, C., Zhang, Y., & Yang, J. (2017). Exploring the influence of built environment on travel mode choice considering the mediating effects of car ownership and travel distance. *Transportation Research Part a-Policy and Practice, 100*, 65–80. https://doi.org/10.1016/j.tra.2017.04.008

Ergonul, S. (2005). A probabilistic approach for earthquake loss estimation. *Structural Safety, 27*(4), 309–321. https://doi.org/10.1016/j.strusafe.2005.01.001

F, Y. (2014). Study on the satisfaction of rural residents' concentrated residence in post-disaster reconstruction: A case study of Chongzhou City, Sichuan Province. *Rural Economy and Science and Technology, 25*(03), 145–148.

Inneke, K. H., Vidya, N. F., & Nur, V. Y. (2013). Application of remote sensing and geographic information system for settlement land use classification planning in bantul based on earthquake disaster mitigation (Case Study: Bantul Earthquake, May 27th 2006).

Jansen, S. J. T. (2014). The impact of the have-want discrepancy on residential satisfaction. *Journal of Environmental Psychology, 40*, 26–38. https://doi.org/10.1016/j.jenvp.2014.04.006

Kun, Z. (2013). Reconstruction of transportation infrastructure service post-disaster recovery and reconstruction. *Ya'an Daily* (07), 34–38.

Li, X.-M., & Tian, S.-Z. (2005). Shaking geoseismic science [M]. *Tsinghua University Publishing House, 35*(12), 1495–1501. https://doi.org/10.13249/j.cnki.sgs.2015.12.001

Li X J, S. H. N. (2008). Research on the development of agricultural areas from the perspective of farmers. *Human Geography* (01), 1–6.

Ma, C., Wang, P., & Zhang, K. (2021). Study on the evaluation and selection of the capacity of the shock absorber and evacuation passage in small mountain cities—Taking Dongchuan city of Kunming as an example. *Journal of Earthquake Engineering, 43*(05), 1112–1122. https://kns.cnki.net/kcms/detail/62.1208.P.20210916.0917.002.html

MacAskill, K., & Guthrie, P. (2015). A hierarchy of measures for infrastructure resilience - learning from post-disaster reconstruction in Christchurch, New Zealand. *Civil Engineering and Environmental Systems, 32*(1–2), 130–142. https://doi.org/10.1080/10286608.2015.1022728

Morimoto. (2012). A preliminary proposal for urban and transportation planning in response to the Great East Japan Earthquake.

Peng, Y., Shen, L., Tan, C., Tan, D., & Wang, H. (2013). Critical determinant factors (CDFs) for developing concentrated rural settlement in post-disaster reconstruction: A China study. *Natural Hazards, 66*(2), 355–373. https://doi.org/10.1007/s11069-012-0488-7

Peng, Y., Zhu, X., Zhang, F., Huang, L., Xue, J., & Xu, Y. (2018). Farmers' risk perception of concentrated rural settlement development after the 5.12 Sichuan Earthquake. *Habitat International, 71*, 169–176. https://doi.org/10.1016/j.habitatint.2017.11.008

Shi, L., Chen, J.-S., Jiang, H.-T., & Wang, H.-Y. (2018). Influencing factors of residents' life satisfaction in rural-urban resettlement communities. *Acta Agriculturae Jiangxi, 30*(04), 141–146. https://doi.org/10.19386/j.cnki.jxnyxb.2018.04.27

Shi, M., Cao, Q., Ran, B., & Wei, L. (2021). A conceptual framework integrating "building back better" and post-earthquake needs for recovery and reconstruction. *Sustainability, 13*(10), Article 5608. https://doi.org/10.3390/su13105608

Song, Y., Li, Z., Zhang, X., & Zhang, M. (2019). The long-term indirect impact of natural disasters on economic growth: A synthetic control method based on county level data in Wenchuan earthquake area. *China Population, Resources and Environment., 29*(09), 117–126.

Speare, A. (1974). Residential satisfaction as an intervening variable in residential mobility. *Demography, 11*(2), 173–188. https://doi.org/10.2307/2060556

Steinberg, F. (2007). Housing reconstruction and rehabilitation in Aceh and Nias.

Sun, J., & Chen, T. (2016). Empirical analysis on the farmers' resettlement satisfaction in the M Town of Jiangsu Province in the process of urban-rural integration [江苏省m镇农户城乡一体化动迁安置满意度实证分析]. *Acta Agriculturae Shanghai, 32*(2), 117–121, Article 1000–3924 (2016)32:2<117:Jssmzn>2.0.Tx;2–1. <Go to ISI>://CSCD:5698968

Tian Y., & Zhao X.-F. (2010). Analysis and evaluation of influencing factors of farmers' satisfaction in the construction of new countryside-Based on survey data of Hubei Province. *Finance and Trade Research, 21*(06), 39–47. https://doi.org/10.19337/j.cnki.34-1093/f.2010.06.006

Wang Z, L. J. M., Liao, B. H., Ren, J. C., Huang, J., & Wang, R. H. (2012). A survey of residents' satisfaction with post-earthquake reconstruction of community health services in Mianzhu city and analysis of influencing factors. *Chinese Journal of Evidence-Based Medicine, 12*(06), 647–650.

Wang, B. (2019). Analysis of influencing factors of rural residents' satisfaction with medical and health services. *Medicine and Society, 32*(07), 28–31. https://doi.org/10.13723/j.yxysh.2019.07.007

Wen, J. (2001). From existential rationality to social rational choice: A sociological analysis of the motivation of contemporary Chinese farmers to go out for employment. *Sociological Research* (06), 19–30. https://doi.org/10.19934/j.cnki.shxyj.2001.06.002

Wu, M., & Wu, G. (2020). An analysis of rural households' earthquake-resistant construction behavior: Evidence from pingliang and Yuxi, China. *International Journal of Environmental Research and Public Health, 17*(23), Article 9079. https://doi.org/10.3390/ijerph17239079

X, Y. H. (2015). Path analysis of building rural harmonious society. *fujian Agriculture* (07), 1.

Xiao, F., Zhai, G., Wan, B., Yuan, C. C., & Li, W. C. (2014). Reconstruction Assessment on the Perspective of Aborigines--A Case Study of Beichuan County and Yingxiu [J].

Xue, Y. (2009). The series of large earthquakes in Longmenshan Seismic Belt—Discussion on the prediction of Wenchuan M8.0 earthquake in Sichuan. *Journal of Institute of Disaster Prevention and Technology, 11*(01), 133–135.

Y, G., & J, F. (2019). Research on the influencing factors of farmers' satisfaction with rural basic public services-based on logistic-ISM model. *Technology and Industry, 19*(07), 100–107.

Yang Z.-Y., & Jie, P.-H. (2017). Evaluation on residents' satisfaction of human settlement environment in small towns reconstructed after disaster: A case study of Shuimo Town in Wenchuan County. *Hubei Agricultural Sciences.], 56*(21), 4165–4168. https://doi.org/10.14088/j.cnki.issn0439-8114.2017.21.041

Yang, L., Chu, X., Gou, Z., Yang, H., Lu, Y., & Huang, W. (2020). Accessibility and proximity effects of bus rapid transit on housing prices: Heterogeneity across price quantiles and space [Article]. *Journal of Transport Geography, 88*, 102850, Article 102850. https://doi.org/10.1016/j.jtrangeo.2020.102850

Yang, L., Chau, K. W., Szeto, W. Y., Cui, X., & Wang, X. (2020). Accessibility to transit, by transit, and property prices: Spatially varying relationships [Article]. *Transportation Research Part D: Transport and Environment, 85*, 102387, Article 102387. https://doi.org/10.1016/j.trd.2020.102387

Yang, L., Liu, J., Liang, Y., Lu, Y., & Yang, H. (2021). Spatially varying effects of street greenery on walking time of older adults [Article]. *ISPRS International Journal of Geo-Information, 10*(9), 596, Article 596. https://doi.org/10.3390/ijgi10090596

Yang, L., Liang, Y., He, B., Lu, Y., & Gou, Z. (2022). COVID-19 effects on property markets: The pandemic decreases the implicit price of metro accessibility. *Tunnelling and Underground Space Technology, 125*, 104528.

Yang, L., Wang, X., Sun, G., & Li, Y. (2020c). Modeling the perception of walking environmental quality in a traffic-free tourist destination [Article]. *Journal of Travel and Tourism Marketing, 37*(5), 608–623. https://doi.org/10.1080/10548408.2019.1598534

Z, H. Y. (2008). Dialogue with Professor Edward Blakely: Reconstruction of New Orleans in America and reconstruction of Wenchuan earthquake in China. *International Urban Planning.*

Zhao, R., Zhan, L., Yao, M., & Yang, L. (2020). A geographically weighted regression model augmented by Geodetector analysis and principal component analysis for the spatial distribution of PM2.5 [Article]. *Sustainable Cities and Society, 56*, 102106, Article 102106. https://doi.org/10.1016/j.scs.2020.102106

Chapter 12
Synthesis and Outlook: Contributions, Challenges, and Future Research Avenues

Abstract This book conducted a comprehensive exploration of human responses to disasters, spanning the stages of disaster management from pre-event preparation to post-catastrophic recovery. It addressed a notable gap in existing literature by offering a holistic understanding of individuals' reactions to disasters, filling a crucial void in prior studies that often focused on isolated aspects of human behavior. Encompassing diverse disaster cases, including earthquakes and the COVID-19 pandemic, the research provided robust findings applicable across various contexts. Its recommendations extended to multiple audiences, offering insights for both managerial levels and individuals, with a special emphasis on pandemic response. The book contributed to a nuanced understanding of human behavior during disasters, touching on topics such as disaster preparedness, risk perception, livelihood capital, and the impact of the COVID-19 pandemic on evacuation behavior and mental health. While acknowledging limitations, including data collection challenges and ethical considerations, the study emphasized the potential for future research to refine methodologies, apply comprehensive human response frameworks, focus on vulnerable populations, and enhance disaster information dissemination.

Keywords Disaster response · Human behavior · Risk perception · Evacuation decision-making · Emergency preparedness

12.1 Research Contributions

This book has undertaken a comprehensive examination of human responses to disasters, encompassing various stages of disaster management before, during, and after catastrophic events. The distinctive characteristics and contributions of this research can be outlined as follows:

1. Addressing a knowledge gap: The chapters within this book fill a significant void in the existing literature on human responses to disasters. Prior studies tended to focus on isolated aspects of human behaviour during disasters, whereas

Y. Ao and H. Bahmani, *Navigating Complexity: Understanding Human Responses to Multifaceted Disasters*, https://doi.org/10.1007/978-981-99-8207-3_12

this book takes a valuable step towards bridging this gap by offering a holistic understanding of individuals' responses to disasters.

2. Diverse disaster cases: This book encompasses a wide range of disaster types, from earthquakes to the COVID-19 pandemic, making its findings more robust and applicable across different contexts. This diversity enhances the reliability of the results for generalization and practical application.
3. Relevance to multiple audiences: Each chapter's recommendations and policy implications extend their applicability to both managerial levels and individuals and households. The book provides valuable insights for enhancing coping mechanisms and response efficiency when confronted with disasters.
4. Insights for pandemic response: Five chapters of this book delve into the impacts of COVID-19 pandemic restrictions on society. Given the global increase in pandemics, this collection offers valuable insights into the short-term and long-term effects of lockdowns and mobility restrictions. Policymakers can draw upon the book's findings to enhance disaster management protocols and mobility regulations during pandemics, taking into account the genuine needs of the public.
5. Comprehensive insights on human behavior in disasters: The research presented in this book offers valuable insights into the complex dynamics of human behavior during disasters. Its contributions span a range of critical topics, from disaster preparedness and risk perception to the impact of livelihood capital and the COVID-19 pandemic on evacuation behavior and mental health. These findings hold the potential to inform and improve disaster management strategies, public health preparedness, and community resilience efforts, benefitting emergency management agencies, public health authorities, community leaders, educational institutions, researchers, policymakers, NGOs, and the general public alike.

12.2 Research Limitations

Due to the high cost of data collection, resource and time limitations, and mobility restrictions (in COVID-19 case studies), this study has the following limitations:

1. Data collection challenges: Researching disaster-affected areas presents logistical hurdles. Accessing affected regions can be hindered by damaged infrastructure, security concerns, and limited resources. In this monograph, the chapters taking COVID-19 as the case study face additional challenges due to the lockdown mobility limitations obstructing access to the affected population. Researchers may struggle to collect data efficiently, leading to smaller sample sizes that might not accurately represent the diverse population affected by the disaster. This limitation can affect the reliability and generalizability of the findings.
2. Ethical constraints: Ethical considerations become particularly complex in disaster research. Obtaining informed consent from individuals in emergencies can be challenging, especially when they are distressed or facing urgent threats.

Balancing the need for research with ethical standards is crucial but can lead to potential limitations in the scope and depth of data collected.

3. Generalizability and contextual variability: Research findings from one disaster or region may not be directly applicable to others. This book's cases are located in China, with unique cultural norms and socio-economic conditions, which could vary significantly from other locations. These contextual differences can limit the extent to which research findings can be generalized beyond the specific disaster context in which they were obtained.
4. Limited pre-event data: Collecting data on individuals' behaviors and characteristics before a disaster occurs is essential for understanding changes during and after a disaster. However, obtaining baseline data can be challenging, as researchers may not have anticipated the disaster. A lack of pre-event data can limit the ability to make robust comparisons and draw definitive conclusions about the impact of the disaster.

These limitations highlight the complex and often unpredictable nature of disaster research. Researchers must navigate these challenges to ensure their studies' validity and ethical conduct while acknowledging the potential constraints that may affect the interpretation of their findings. Addressing these limitations requires creative methodological approaches, collaboration with local communities and organizations, and acknowledging the potential constraints while interpreting research findings.

12.3 Future Research Directions

As we reflect on the limitations of the current research and its implications, it becomes evident that there are several promising avenues for future research in the realm of disaster studies and human behavior, as follows:

1. Future research endeavors can significantly benefit from the refinement of data collection and analysis techniques. Embracing cutting-edge methods such as GIS-based data mapping, video recordings, Virtual Reality simulations, Artificial Intelligence algorithms, drone technology, remote sensing tools, and sophisticated simulations can greatly enhance the precision and depth of our understanding of human responses to disasters. By leveraging these innovative approaches, researchers can capture and interpret data more effectively, shedding new light on the intricate dynamics of disaster response.
2. A promising avenue for future research is the application of the comprehensive human response framework proposed in this study. This framework offers a systematic and holistic approach to studying how groups of individuals react to disasters across various phases of the response process. By employing this framework, researchers can gain practical, real-time insights into human behavior throughout the entire disaster management cycle, offering valuable data for improving preparedness and response strategies.

3. Vulnerable populations, such as the elderly and children, require special attention in disaster research. Given the global aging trend and the need to ensure the safety and well-being of diverse age groups during disasters, future studies should focus on analyzing the unique responses and needs of these vulnerable demographics. This research could lead to tailored training programs and preparedness initiatives that address their specific challenges, ultimately enhancing their resilience in the face of disasters.
4. Disaster information dissemination is a critical component of effective disaster response. Future research should concentrate on designing and evaluating disaster information systems tailored to specific disaster types and building occupants. Understanding the effectiveness of various communication channels and evacuation notices, especially in relation to different age groups, is a promising research direction. This investigation can inform the development of more efficient and targeted information dissemination strategies, ensuring that vital instructions reach those in need during emergencies.

This book represents a significant step in understanding human responses to disasters. However, it's crucial to acknowledge that our research is still in its preliminary stages. The complexities of disaster response offer a vast landscape for future exploration. As we face evolving socio-economic conditions, technology advancements, and new disaster scenarios, there is a clear need for continued research in this field.

The manufacturer's authorised representative in the EU is Springer Nature Customer Service Centre GmbH, Europaplatz 3, 69115 Heidelberg, Germany. If you have any concerns regarding our products, please contact ProductSafety@springernature.com

Printed and bound by CPI Group (UK) Ltd, Croydon, CR0 4YY

30/09/2026

02225322-0010